Portable Food
Proceedings of the Oxford Symposium on Food and Cookery 2022

Portable Food

Proceedings of the Oxford Symposium on Food and Cookery 2022

Edited by Mark McWilliams

Prospect Books
2023

First published in Great Britain in 2023 by Prospect Books, 26 Parke Road, London SW13 9NG.

© 2023 as a collection Prospect Books.
© 2023 in individual articles rests with the authors.

The image on the title page shows a Mexican street vendor and his bicycle, with a basket holding *tacos de canasta* and a jar of salsa (photo taken in Ciudad Universitaria, 2022).

The authors assert their moral right to be identified as authors in accordance with the Copyright, Designs & Patents Act 1988. No part of this publication may be reproduced, stored in a retrieval system or transmitted in any form or by any means, electronic, mechanical, photocopying, recording or otherwise, without the prior permission of the copyright holders.

ISBN 978-1-909-248-809

Design and typesetting in Gill Sans and Adobe Garamond by Catheryn Kilgarriff and Brendan King.

Printed and bound in Great Britain.

Contents

Foreword
Mark McWilliams — 9

Dietary Advice for Travellers from the Mid-Sixteenth Century to Early Seventeenth Century with Two Test Cases: Michel de Montaigne and Fynes Moryson
Ken Albala — 11

We Are Open, Says the City
Tamar Babuadze — 19

Urban Nomads and Vagabond Food: Eating Away from the Table in Nineteenth-Century France
Janet Beizer — 29

When the Lord Eats Away from the Table: Outdoor Food in Indian Gaudiya Vaishnavism
Tanushree Bhowmik — 43

Cunard and Collins: Portable Food and Evolving Notions of Customer Service at Sea, 1840-1880
Andrea Broomfield — 55

From 'Peanut Weddings' to 'Beef Stands': The Socio-Culinary History of Chicago's 'Italian Beef'
Anthony F. Buccini — 67

Achar: An Unexpected Stowaway on the Ships of the Dutch East India Company
Kathleen Burke — 77

Food for Travel, Travel for Food: The Story of *Ekiben*
Voltaire Cang — 88

The Raw and the Disordered: Portable Feasts, Blood, and Raw Meat Salads in Contemporary Political Movements in Thailand
Nattha Chuenwattana and Chonlatorn Wongrussame — 99

Savour the Flavour: A Portable History of Instant Coffee
Julia Fine — 107

Mobile Gastronomies: A Case Study of Food on Wheels Culture in Hyderabad, South India
Kashyapi Ghosh and V. Vamshi Krishna Reddy — 117

Portable Food

An Exploration of Food Mobility through the Cultural Practice of Koseli in Contemporary Nepal
Binti Gurung — 127

Fair Food to Walk With
Peter Hertzmann — 134

Dispatches from the Chicken Sandwich Wars
Michael Johnson — 145

Istanbul's Portable Food Practices: The Case of *Dolma*
Pırıl Kadırgan — 156

Provisioning for Adventures: The Meanings of Paddington Bear's Marmalade Sandwiches
Laura Kitchings — 163

Serving the Global Nation: Menus from the Royal Dutch Airlines
Charlotte Kleyn and Aimée Plukker — 172

'No kind of food? Just rubbish?': Food Options and Family Meals of Visitors to Canadian Prisons
Else Marie Knudsen — 183

Eating in Public in Antiquity: Honorable or Obscene?
Joshua Lovinger — 194

Food for the Final Journey: Feeding the Soul in Hindu Death Rituals
Priya Mani — 206

Picnicking in Iran: From Cemeteries to Ski Slopes, Caves, Traffic Circles, and Beaches
Nader Mehravari — 217

Food on the Move: Commensality in an Indian Train Compartment
Shirin Mehrotra — 231

An Army Marches on Its Stomach: Portable Foods in War and Peace
Johanna Mendelson Forman — 240

Going to the *Gahambar*: Tradition and Invention in Parsi Community Feasts
Meher Mirza — 251

The Perils and Promises of Portability: Sweets and Snacks in Modernizing Japan
Tatsuya Mitsuda — 261

Portable Food

A Rational Approach: Portable and Practical Eating 'Beyond the Seas'
    Jacqui Newling — 272

Hoosh, Dogs, and Seal Meat: The Role of Food in the Race to the South Pole
    Diana Noyce — 284

The '*tacos de canasta*' (Basket Tacos) in the University: An Example of the Close Relationship between Informality and Mobility
    Ayari G. Pasquier Merino and M. Fernanda Estrada González — 296

From the Steppe to Space: The Portable Power of Qurut
    Simi Rezai-Ghassemi — 306

Cooking Summer by the Beach: Rulebreaking, Modernity, and Wellness Policies in Three 1960s Egyptian Cookbooks
    Salma Serry — 311

Pemmican: An Ideal Trail Food
    C. Thomas Shay — 321

'The Clever Dining Car Conductor': Creating a Luxury Dining Experience on the Move in Britain, 1879-1948
    Chloe Shields — 331

Connected Food: Preserving Traditional Food Practices via Portable Foods
    Sevgi Mutlu Sirakova — 337

Portable Poetic Commensality: Reflections on the Use of Food Language in Ancient Israel's Pilgrimage Songs (Psalms 120–134)
    Michelle A. Stinson — 349

Travelling Silver for Those Not to the Manor Born: Old Sheffield Plate and Electro-plated Silver in Travel Equipage and Cutlery, from 1730 to the Belle Epoque
    Carolyn Tillie — 357

Food for Walking: On the Camino Real de Tierra Adentro
    Jaime Iram Vargas Barrientos — 368

Contributors — 378

Foreword

One of the highlights of each year's Oxford Food Symposium is a wonderfully messy exercise in pure democracy: the closing session features lively deliberation to choose the topic that will govern our discussions at our meeting three years later. Thus, at the close of a lovely weekend in 2019, symposiasts selected Portable Food – Food Away from the Table as the 2022 topic. No one could have known, then, that our ideas about the table, indeed our very conception of shared space, would undergo such terrific assault before we gathered again to consider the topic.

But at least we could gather. Finally able to reconvene in person at St Catherine's College, we were also determined to continue the virtual symposium which had delivered such unexpected benefits: wildly increased accessibility, especially for symposiasts across the Global South, more time for discussions of individual papers, and even an odd kind of intimacy, meeting not in the hierarchical halls of academia but rather in our own dining rooms and kitchens. Though committed to meeting both in person and online, we were also driven to avoid the now well-known difficulties of the hybrid: we would be at St Catz or on our screens, never both at once.

When we chose the theme of Portable Food back in 2019, I was excited by promised explorations of the foods of travel and of travelling foods. Fascinated by the meals possible on rolling trains or zooming jets, I looked forward to analyses of the history and culture of dining in mobile spaces. And indeed we had spectacular papers on such topics, whether negotiating competing claims of luxury and safety on nineteenth-century trans-Atlantic steamships, creating luxury – and celebrating imperialism – in twentieth-century Dutch airplanes, attempting to maintain class barriers in nineteenth-century British railway dining cars, or breaking down differences in twenty-first-century Indian train compartments. And there were amazing studies of the foods of exploration and nomadic life, from pemmican and penguin meat to achar and *qurut*, not to mention the industrial ration that fuelled settler colonialism.[1]

But other topics had renewed urgency after the pandemic experience of closed restaurants and – far more devastating – closed borders. And yet building walls hardly stops conflicts. In 2022, the number of refugees and displaced people topped 100 million – a previously unimaginable figure – as Ukrainians fleeing the Russian invasion swelled floods from Afghanistan, Colombia, Ethiopia, Syria, Venezuela, Yemen, and many other countries.[2] Those fleeing conflicts and oppressive regimes join millions more who have simply refused to accept losing the 'lottery of birth'. Over the past fifty years, the number of people 'living in a country other than that of their birth' tripled, and that trend has only accelerated over the past few years despite covid-driven travel bans and politically motivated exclusions.[3]

While many of us celebrated restaurants' attempts to repurpose public spaces on sidewalks and parking lots, countless others were engaged in reclaiming foods central to their identities, foods banned by repressive governments, prohibited by oppressive bureaucracies, or left behind in the all-too-human quest for a better life. Symposiasts dedicated their time to these crucial conversations too, hearing about foods from minority cultures used in political protests in Thailand and families of imprisoned loved ones struggling to recreate family meals in Canada. Many papers focused on people clinging to their culinary traditions across miles and generations, whether bringing carloads of traditionally preserved foods from Bulgaria and Turkey to Germany, or reinventing Italian traditions on the streets of Chicago. And there was a throughline of sharing food outside, whether at the impromptu tables of unregistered food vendors in the tech centres of India, on university campuses supplied by bicycle basketfuls of tacos in Mexico, or on the roadsides – and apparently any other open space – in Iran.[4]

The remarkable resilience we have all witnessed, at least to some extent, over the last several tumultuous years seems central to the Oxford Food Symposium's mission: Change the Conversation, Expand the Table, and Improve the Plate. While many politicians seem hell-bent on ignoring pandemic lessons, we have learned, and we know that making things better will take an awful lot of us working tirelessly – and working together.

Here I would like to thank some of those who made this extraordinary journey possible: Elisabeth Luard, Cathy Kaufman, Ursula Heinzelmann, David Matchett, Gamze İneceli, Naomi Daguid, and Carolyn Steel. I'd also like to recognize those who helped this volume see print: Catheryn Kilgarriff, Brendan King, Jake Tilson, Caroline Johnston, Claire Goins, and of course all the authors. These papers, after all, are why we gather.

Mark McWilliams
Editor, Oxford Symposium on Food and Cookery

Notes

1. Lizzie Collingham, 'Portable Power: How the Industrial Ration Projected European Dominance', *Oxford Food Symposium*, 9 July 2022 <https://www.youtube.com/watch?v=OrvofieivlE> [accessed 25 May 2023].
2. United Nations High Commissioner for Refugees, 'Global Trends', *United Nations*, June 2022 <https://www.unhcr.org/global-trends> [accessed 25 May 2023].
3. International Organization for Migration, 'World Migration Report 2022', *United Nations*, 2023 <https://worldmigrationreport.iom.int/wmr-2022-interactive/> [accessed 25 May 2023].
4. For a wonderful overview of such traditions, see Claudia Roden, 'Picnic Traditions', *Oxford Food Symposium*, 10 July 2022 <https://www.youtube.com/watch?v=Rm32mgM1ohc> [accessed 25 May 2023].

Dietary Advice for Travellers from the Mid-Sixteenth Century to Early Seventeenth Century with Two Test Cases: Michel de Montaigne and Fynes Moryson

Ken Albala

Introduction

In the first two centuries of printing, dietary tracts were among the most popular forms of self-help literature. Between 1450 and 1650, over a hundred distinct works were targeted to medical professionals as well as the general public in Latin and in vernacular languages. These texts include explicit advice concerning what one should eat while on the road. Whether this is an indication of an intensification of travel is impossible to say, but at least it is clear that authors believed their audience was particularly eager for such information. In one case, there is even an entire book devoted to the topic of food while travelling – Guglielmo Grataroli's 1556 *Regimen omnium iter agentium*. The details of physicians' advice offer tantalizing clues about how readers may have interpreted recommendations and put them into practice. Ultimately the question I would like to try to answer is whether the prescriptive recommendations in the dietary literature was ever followed by actual travellers?

Dietary Advice and Travelling

Dietary advice was rarely geared toward people who habitually travelled in groups, such as courtly entourages and military operations. That such people brought their own chefs and equipment is apparent from literary and pictorial sources. Bartolomeo Scappi's *Opera* of 1570 illustrates not only equipment that would be needed for outdoor cooking on route or campaign, but even a man turning an outdoor spit and waving to the viewer.

More often the dietary literature addresses individuals who are travelling under atypical circumstances and are clearly not used to doing so regularly. In fact, this was a leitmotif of nutritional advice of the time: the danger of suddenly changing one's air and climate, and the food that one should eat to counteract such alteration. That is, suddenly encountering a hot climate, a traveller should seek out foods that were considered cold, so as to temper and balance the body's humours. Contrariwise, a sudden cold spell could be countered by hot spices and warming foods, some of which the traveller could bring along in a kit of remedies.

Added to this was the effect of simply suddenly changing one's habits and customary diet, the time of meals, the quantity of food, and so forth – all variables one might be forced to deal with while travelling. Incidentally, this also became a major concern for colonial enterprises. Could people actually survive in strange climates and on unusual foods? The answer usually arrived at was no, which is precisely why they imported food from home or did their best to grow familiar plants and rear animals they knew well, like cows and pigs.

We also have to imagine these were travellers dealing with physical exertion, either walking or on horseback, with constant jostling, the digestive process hindered by stress, as well as the emotional toll taken by merely being in strange places, looking for food and perhaps not finding something suitable to one's palate. From the advice it is easy to glean that the projected audience were wealthy enough to be travelling for business or perhaps leisure and were probably more often than not on official ambassadorial missions or scholars in their typical perambulations. They might also be exiles forced to move for political or religious reasons, as was the case with Grataroli, an Italian Protestant who escaped to Basel.

There were also believed to be various accidental challenges to maintaining health, principal among which was encountering foul odours. If inhaled, these might cause the entire digestive process to derail, corrupting the humours and ultimately leading to illness. Rather than any kind of effected delicacy, holding a pomander up to the nose or a sachet of warming herbs was a prophylactic. They functioned precisely as they would in times of pestilence. Guinterius Andernacus recommended bringing pepper, galangal, angelica, saffron, or best of all theriac to smell: '*Tutissimum esset peregrinari, & mutare loca, quia vero plerique, obrerum agendarum necessitatem, & officia civilia*' ('It would be safest for travellers and change of place, for the most part, due to the need for business and official duties').[1]

Fresh water and the difficulty of transporting was also a perpetual concern. Most physicians recommended wine instead, not only because it is less prone to corruption but also because it acts as a preserver of the body and it is an easily digested nutrient, perfect for a body under stress. Wine is not only an analogue of blood and most easily converted into it, but provides the quickest form of nutrition for those on the road and especially early in the morning. By the same logic, *aqua vitae* provides an even quicker and practically incorruptible form of nourishment.

In lieu of these, a concentrated essence of food, a rich stock, offers the same benefits, but because it cannot be transported, was a less viable option. Eggs, however, contain all the elements of generation and growth, are easily assimilated and agglutinated in the body, to use their terminology, and thus are an ideal food for the weary and depleted traveller.

The more pressing danger for most authors, though, was having to deal with strange and unfamiliar foods, being forced to quickly alter one's dietary habits and our accustomed ways of eating. Custom becomes second nature, and the sudden and violent change of the elements that make up the body tends to impair health. Standard theory posited that even pernicious customs must be changed slowly and gradually. The traveller must not do as the Romans when in Rome and worst of all is affecting foreign customs long term. Thomas Cogan warned of a traveller who returned home and 'as it were despising the olde order of England, would not begin his meale with potage, but instead of cheese would eat potage last'.[2] Where this particular gentleman picked up the custom isn't stated, but the idea is

Dietary Advice for Travellers

that, if at all possible, it is important to stick to the particular foods with which your body is familiar, and also to the accustomed times and order of the meal. If forced to deal with these inconveniences temporarily, it is best to return to old habits once home.

The greater danger was of course those moving permanently, or even colonists going to completely foreign places. William Vaughan's *Directions for Health* of 1600 addresses this problem directly, which makes sense since Vaughan was an investor in the Newfoundland colonies. He suggested that, if you have to venture into a climate very different to your own, it is best to only drink boiled water (sound advice!) or take a 'wholesome diet-drinke' for travellers made of barley water with liquorice, violet seed, parsley seed, rose, hyssop and sage – and most importantly figs and raisins.[3]

Strangely, while most physicians recommend eating very lightly while travelling and then only foods that are easily digested, Vaughan says the opposite. Because of the great amount of exercise they get, their stomach heat is intensified and powerfully strong – comparable to that of a labourer – so that they can digest iron. And as Rabelais says, their stomachs are 'as hollow as Saint Bennet's boote', implying that they can eat as much as they like too.[4]

Another curious comment in Vaughan is that, while he is generally much against the use of tobacco, he does contend that it can be very useful for travellers for slaking hunger and thirst, or if you happen to find yourself straggling in a desert. Otherwise it is 'an outlandish weede, it spends the braine and spoiles the seede'.[5]

The work that provides the most information pertinent to eating on the road is Grataroli's *De regimine iter agentium*, printed in Basel in 1561. The author had considerable personal experience in the topic, having fled Italy, moving to Strasbourg and eventually Basel. Like most authors he includes standard advice to alter customs gradually, paying particular attention to climate changes, such as one would encounter crossing the Alps.[6]

He specifically advises that travellers bring foods with which they are familiar, especially if they expect them not to be found *en route*. Excess must be avoided, as should odd foods and strange customs: '*Ego his scribo qui sanitatis curam habent, non his quorum Deus venter est, & fruges aut vinum consumere natis.*' ('I write this for those who care about their health, not those whose belly is a God, and who eat fruits or wine with snow.')[7] This was indeed a fashion of the day, presumably in places one would have access to snow like Switzerland.

My impression is that Grataroli would really not have been fun to travel with, telling people not to try the local food, but to bring your own if possible. Just as bad, you should eat food that's nourishing, of good substance, as Avicenna would say, but in small quantity. This is to avoid indigestion from being bumped about constantly.[8]

Of course the difficulty of bringing your own food presented its own logistical problems. Fresh food or anything too heavy would be impossible, so Grataroli recommends what are essentially super concentrated portable snacks – the sort that one might find today, ironically in a roadside gas station. He suggests *globuli* and *pastilli* (candies), marzipan, as well as *placentae* and *nebulis* (flat cakes and cookies). One pastry made of almonds and marshmallow root sounds delicious. *Biscocti* – or twice cooked biscuits – are ideal, and he doesn't mean ships biscuit either, but proper sweetened and spiced, though quasi medicinal, cookies: '*non diu est inventam esse hanc artem*' ('not long ago these arts were invented'), and they're best dipped in wine.[9] You can even dip these in water, and they'll make it taste of wine because the original biscuits were made with wine and oil.

If by chance you can't afford such luxuries, there are little pills made of suet and violet oil. Most importantly you should not press on in a deserted place or if the local food seems suspect, which is why these portable snacks are so important.

Grataroli's worst fear, as an Italian on the other side of the Alps, was what happens if you can't find wine? And you are forced to drink beer? He suggests you carefully examine it first. If it's bitter, it won't be nourishing at all; if it's sour, it will increase cold phlegmatic humours. And you should try to find out if it was made with old or new grain, because 'who can trust the inn keeper?'[10] Even if you do drink it, don't overindulge, because beer courses through the body slowly, and will leave you drunk much longer than wine, which is more volatile and evaporates much more quickly.

Thirst is of course the greatest danger in general while travelling, so to avoid it, you should not talk too much, nor eat salty foods which are excessively hot and drying. Rather, cold and moistening foods are best, such as oranges and lemons.

Grataroli was also concerned about getting food poisoning, how to cure sore feet, how to prevent frost bite, how to rid yourself of lice. Specifically, all this advice reflected his own personal experience of going over the Alps, from Northern Italy, through Switzerland and into Burgundy. (Coincidentally I've walked through many of the same places and have gotten sick from drinking the local water.)

The Case of Michel de Montaigne

Now to return to my central question: did anyone follow this kind of advice? The logical place to look would be a travel journal, and the ideal person to serve as a test case is Michel de Montaigne, partly because he was well versed in medicine, but also because he spoke so frankly about his own experiences and was very interested in food. Montaigne, a native of Bordeaux, travelled to various spas through Europe in an attempt to cure his kidney stones.

Montaigne, I think, also will reveal an honest interaction with the medical advice rather than just blindly doing whatever he had read, mostly because of his sceptical attitude in general, but even more specifically for his own distrust of medicine and preference for his own personal experience. He mentions the number of gentlemen who, though still young and sound, have made themselves into prisoners through the folly of their physicians: 'Let us leave the making of dietaries to almanac makers and physicians.' He quipped, 'If your doctor does not think it good for you to sleep, to take wine, or some particular meat, do not worry; I will find you another who will disagree with him.'[11]

More importantly, Montaigne's *Travel Journal* contains numerous anecdotes about the foods he ate, many of which contradict the prevailing medical advice. Concerning the advice not to change customs rapidly, Montaigne flaunted this dictum immediately. Upon arriving in Basel he drank his wine undiluted (against French practice), he sat at a meal three or four hours and ate slowly, contrary to his own personal habit of wolfing down his food and biting his fingers. In fact he considered these foreign customs to be healthier: '*Les moindre repas sont de trois ou quatre heures pour la longueur de ces services; et á la verité ils mangent aussi beaucoup moins hâtivement que nous et plus sainement*' (Lesser meals are three or four hours for the length of service, and the truth is they also eat much less quickly than us and more healthily).[12] It is not that Montaigne is careless or unconcerned with health, quite the opposite, he just doesn't believe quickly changing his custom could be harmful.

Elsewhere his secretary mentioned:

> M. de Montaigne, in order to thoroughly understand the diversity of manner and customs, allowed himself, in every place he visited, to be waited upon after the particular fashion of that place, however troublesome it might appear at the time, or however different from what he had been accustomed to.[13]

This approach generally caused him no inconvenience, though he was confused by the Swiss custom of using only a small napkin and never opening it up, and everyone eating with their own knives and not putting their fingers in their food.

Apart from these awkward moments, Montaigne regretted that he hadn't brought a cook along with him to learn how to prepare all the delicious foods he found there, a great variety of soups and sauces that he liked so much. Moreover, conforming to local custom with none of the apprehension of sourpusses like Grataroli, as his secretary commented, here 'he so entirely conformed himself to their customs and manners as to drink his wine without water'.[14] That is something no cautious traveller would have dared to do, but Montaigne, in the interest of research and absorbing local culture, dared it. It's not that he was unfamiliar with medical opinion: he chose to ignore it.

Of course this cavalier attitude waned once he got sick in his travels. After a bout of colic, his meals became decidedly simpler: he more often ate poached eggs and increasingly lost his taste for escargot. Even the truffles in Northern Italy he could do without – they're not bad he claimed, though that might be a matter of French prejudice.[15] As he made his way into Italy he could not curb his enthusiasm for the oranges, and lemons and olives. Those of course would have been good for travellers, and, while there's no indication that his opinion was driven by health concerns, he did grow increasingly cautious as the trip progresses. For example, once the good wine he brought from Germany ran out, he was forced to drink inferior new wine in Italy and notes that it was bad for his colic. He was also served wine with sage in it, which he contended was pretty good, once you got used to it.[16]

Eventually however Montaigne grew quite cautious: 'he frequently avoided taking meals, omitting sometimes supper, sometimes dinner'.[17] This might have been just a natural reaction to getting sick and is really not evidence that he had internalized any nutritional advice, but it does suggest that he was thinking twice about his indulgence and gourmandizing up to this point. His randomly eating anything put in front of him was one of the chief targets of the medical literature: eating a great variety of meats at one sitting was a chief target of physicians' warnings.

However, it seems that the moment Montaigne was feeling better when he got to Rome, he went back to his serious consideration of the local produce. He found the pike inferior, the mutton very bad, but the oil was excellent and left none of that unpleasant burning sensation in the back of the throat. Nonetheless, he seemed to want his readers to think he hadn't thrown all caution to the winds. One day he ate only three slices of toast buttered and sugared and nothing to drink.[18]

Montaigne's sceptical attitude toward the whole value of dietary advice is revealed in a brief episode. He met a Cremonese merchant who consulted twenty different physicians, each of whom had prescribed a different regimen. Each doctor was also told what the previous one had said, whereupon in a rage each doctor called the previous one a homicidal maniac.[19] Montaigne appeared ambivalent though: on the one hand he had a serious and understandable distrust of physicians, but on the other he also knew through his own experience that eating whatever he liked without any regulation made him sick. That is, he often followed dietary advice in spite of his basic mistrust.

Not that this always worked. One case in point: midsummer while in Florence, Montaigne was suffering from terrible thirst, and he decided to eat nothing but fruit and a sugared salad. The coldness of these foods should have counteracted the awful heat. He tried the recommended remedy, to no avail: 'Notwithstanding this temperate diet, I continued very unwell.'[20]

The remainder of his journey was spent in various baths suffering unspeakable agony in passing kidney stones, but he was still tempted by local foods and unusual fashions. Again, this was in Florence, where '[t]hey have a custom here of cooking their wine by putting snow in the glass'. This was a practice universally condemned by dieticians, as Montaigne almost certainly knew. He remarked that 'I myself put [in] very little, for I was far from well'.[21] That is to say, Montaigne remained sceptical of dietaries but had nonetheless internalized and accepted many of their ideas and repeatedly tried to follow physicians' advice, albeit unsuccessfully.

The Case of Fynes Moryson

In comparison to Montaigne, the Englishman Fynes Moryson is particularly apt, as recorded in his *Itinerary*. He visited many of the same places, about a decade apart: Montaigne was travelling in the early 1580s, Moryson in 1591.

Although he gives some very detailed descriptions of what he ate *en route*, Moryson's concern was less gastronomic than economic. He gives prices for practically every meal, but comments less frequently on their wholesomeness. He does nonetheless have certain ingrained prejudices many of which were clearly derived from the dietary literature. While in Germany he was amazed at the amount of beer the locals drank and comments freely with standard jokes such as *'plures crapula quam ensis'* ('more die by gluttony than by the sword').[22] Moryson finds himself continually complaining about drunkenness, and the specific context of this comment was a table in Dresden from which people have to be carried away in a drunken stupor.

Dietary Advice for Travellers

That Moryson was familiar with dietary rules for travelling is revealed in a later work on 'Travelling in General' where he recounts some basic rules. For example he dissuades old men from travelling because their 'custome is growne to another nature, shalle never be able to endure frequent changes of diet and aire, which young men cannot beare without prejudice to their health except it be by little and little (as it were) insensible degrees'.[23] He also goes on at length against picking up foreign customs, especially Italian ones and then bringing them home: 'So many home bred Angles, returne from Italy no better than Courtly Divells.'[24] This was the exact opinion of this countryman Thomas Cogan, the author of *The Haven of Health*.

But the larger question also concerns Moryson: should one indulge in the local fare, or play it safe and remain abstemious? He says, 'Some perswade a traveler to use himself first to hardness, as abstaining from wine, fasting, eating grosse meates.'[25] Ultimately he believes that suffering these hardships first only weakens the body – it is much easier to steer a middle course, eat well but not in excess, and imitate those 'who cherish their body while they may'.[26] That is, do what you can given the constraints of being in a strange place.

But what of the strange food? While on the road from Hesse to Brunswick he regretted having tried the local specialty of sauerkraut. He tells readers, 'your diet shall be for the most part of cole worts, which was so strange to me, and so hard of digestion, as it greatly troubled me, and wrought upon my body like physicke'.[27] I assume that means it had a purgative effect.

Moryson was also careful to have wine with his breakfast while on the road, to fortify his body – something Montaigne did not follow.[28] He was also careful to choose foods that according to standard dietary theory were considered light and easily digested, foods like lamb, hens, pigeon, rabbit, and he consciously avoided all the salted meats he found in Germany 'which being (upon ill disposition of my body) once displeasing and unwholesome for me'.[29]

Trying to keep an open mind and trying most everything he encountered, Moryson turns out to be a cautious but still adventurous eater. For example, while in Padua he bought turkeys, mutton, veal, pork, hens, little birds, eels, shrimp, pike, cockles, and scallops. He especially loved the local bread and cheese – a *piacentine*, that is Placentia from Piacenza, which was at the time more popular than Parmigiano. He even tried goats' testicles 'which we cast away, being a very good meate fried'.[30]

In general Moryson was impressed by the frugality of the Italians, what he calls their 'sparing diet' which he thought was good for health. On one occasion in Naples, the *Vetturines* or porters were supposed to supply food for the road and apparently didn't bring enough. And Moryson simply laughed it off: 'here our supper was so short, as we judged our vetturines good Phisitians, who persuade light suppers'.[31] He was obviously being sarcastic, and he wasn't about to stand for this again. On the way to Capua he 'refused to be dieted at their vetturines pleasure, and chose rather to feast themselves as they list'.[32]

These comments suggest that Moryson was familiar with the dietary advice of the day, but was willing to ignore it when his gastronomic interests took precedence. Not that the food was always good, mind you. At a country inn he was served 'a pudding as big as a man's legge'.[33]

Conclusion

Like today, travellers have an idea of how they should eat informed by nutritional theory of the day. Negotiating that while hoping to sample local delicacies, and at the same time avoiding getting sick, is a careful balancing act. But these two examples do provide evidence that people did know of dietary advice, even if they did not always follow it.

Notes

1. Johannes Guinterius Andernacus, *De victus et medicinae ratione e cum alio, tum pestilentiae tempore observanda* (Strassburg: Wendelin Richel, 1542), fol. B8vo.
2. Thomas Cogan, *The Haven of Health* (London: Thomas Orwin, 1589), qq. 4.
3. William Vaughan, *Directions for Health* (London: T.S. for Roger Jackson, 1600, 5th edition 1617, p. 39.
4. Vaughan, p. 86. This reference comes from a scene in *Gargantua and Pantagruel* wherein the giant invites Friar John to come sit and eat with him. The friar says though he's already supped he'll eat anyway, because his stomach is 'as hollow as a butt of malvasie, or St. Benedictus' [St Benet] boot and always open like a lawyer's pouch'. Rabelais, *Gargantua and Pantagruel* (London: H.G. Bohn, 1854), book 1, p. 226.
5. Vaughan, pp. 139, 146.
6. Grataroli, *De regimine iter agentium* (Basel: Nicholas Brylinger, 1561), p. 8.
7. Grataroli, pp. 11–12.
8. Grataroli, p. 15.
9. Grataroli, p. 24.
10. Grataroli, p. 22.
11. Michel de Montaigne, *On Experience* in *Works of Montaigne*, trans. by William Hazlitt (NY: Derby and Jackson, 1859), IV, pp. 368, 391.
12. Montaigne, *Journal de voyage en Italie* (Paris: Garnier Freres, 1955), p. 19.
13. Montaigne, *Works of Montaigne*, IV, p. 210.
14. Montaigne, *Works*, IV, p. 223.
15. Montaigne, *Journal*, p. 64.
16. Montaigne, *Works*, IV, p. 267.
17. Montaigne, *Works*, IV, p. 298.
18. Montaigne, *Works*, IV, pp. 325, 381.
19. Montaigne, *Works*, IV, p. 93.
20. Montaigne, *Works*, IV, p. 409.
21. Montaigne, *Works*, IV, p. 410.
22. Moryson, *An Itinerary* (London: John Beale, 1617), Part I, p. 10 <https://quod.lib.umich.edu/cgi/t/text/text-idx?c=eebo;idno=A07834.0001.001> [accessed 15 April 2023].
23. Moryson, Part III, p. 2.
24. Moryson, Part III, p. 5.
25. Moryson, Part III, p19.
26. Moryson, Part III, p. 20.
27. Moryson, Part I, p. 36.
28. Moryson, Part I, p. 39.
29. Moryson, Part III, p. 83.
30. Moryson, Part III, p. 115.
31. Moryson, Part III, p. 105.
32. Moryson, Part III, p. 106.
33. Moryson, Part I, p. 203.

We Are Open, Says the City

Tamar Babuadze

Introduction

Tbilisi: at the crossroads of cultures, the East's gateway to the West, the West's gateway to the East, the capital and cultural hub of the Caucasus – these assessments from different eras sound cliché, almost like tourism branding or text for promotional videos. And yet I also use these phrases when talking with tourists, when there is so little time but so much to tell, when my goal is to introduce the city's trademark features and character. At such moments, I feel myself transforming into a tourist attraction, relishing – as though for the first time – salad with sweet and juicy tomatoes, purple basil, *jonjoli* (pickled Colchis bladdernut), and Kakheti oil; or slicing into hot *mchadi* cornbread fresh out of the oven, putting a piece of salty August cheese inside, and – while the cheese is melting – savouring chilled golden *Khikhvi*, a wine with sweet and sour flavours fermented in a Georgian *qvevri* earthenware vessel. At such moments, I am an attraction myself because I am the one setting an example for my guests, attempting to help them get a taste for Tbilisi, to help them feel what you cannot possibly convey about your hometown over just one evening. Others have struggled with such overflowing descriptions too, as Grigol Robakidze – a Georgian author who emigrated to Germany in the 1920s – reveals in his novel, *The Snake's Skin*:

> In Tbilisi, no one goes home *per se*. Every home here is *your* home.
>
> Georgians, Armenians, Tatars, Ossetians, [Caucasian] highlanders, Jews, Poles, Estonians, Lithuanians, Russian officers, colonels, self-governing *zemstvo* hussars... Khakis, greatcoats, all kinds of epaulettes, and a legion of sisters of mercy, and a crowd of officers' wives, blue-blooded landowners and demi-monde... cigarette smoke, coffee, and cognac... Kakheti wine... roaring and commotion, feasting and having a high old time... the latest romance ballad... someone's tenor... and someone's baritone... frenzy and bacchanalia...
>
> the city is rocking. Tbilisi is inebriated.
>
> The city-boat is rocking. Tbilisi is delirious.[1]

Robakidze's work is fictional, of course, but the archival press of that time – I am referring to the period around Sovietization in 1921 – features both news stories and notes from foreign visitors testifying to the truthfulness of the fact that, before modernity and industrialization, Tbilisi was an open city where life 'happened' on its balconies, in its courtyards, through its public gardens, and along the banks of its River Mtkvari. Consequently, *al fresco* dining did not necessarily mean carrying food outdoors. Instead, it involved improvised, impromptu endeavours, such as setting up a makeshift *makali* barbecue and skewering *mtsvadi* shish-kabob right in a public garden, spreading a rug on the ground, loading it with *tone,* similar to tandoor-baked bread, and salty aged cheese – complimented with chilled Eastern Georgian wine. This was done wherever you could find a place, naturally preferring a shady nook or a cosy grove by the river over dust-covered, unpaved city streets.

I can only imagine how many things you could see with your own eyes in Tbilisi of that time – though this does not imply that the city did not have its own secrets, its own mysterious life; of course, it did, and to the hilt at that – observing bread-baking in a *tone* tandoor, leafy greens drying in the sun, Turkish coffee brewed by localized Persians in hot sand, wine in a clay jug chilling in the river, wheat ground in the numerous mills on the Mtkvari. Each of these ingredients or foods is easy to carry, portable by nature, simplifying people-to-people interaction on the streets, especially in a warm-climate city like Tbilisi. Better still, before Sovietization Tbilisi offered a unique case to observe a mix of Near Eastern influences – Turkish and Persian, deep-rooted due to years-long conquests – and innovations from the Russian Empire and Europe – because the European business community did take an interest in this outpost of the Russian Empire, among others. From my observations, this melting pot, both as process and result, is inherently embedded in Tbilisi's character, explaining why, for me, Tbilisi means constant movement itself – though, admittedly, the same likely goes to many other cities with millennial histories, abundance of influences, and self-determination efforts continuing to this day. Consequently, with portable foods as a signpost, we might as well describe how food – and cultural symbols with it – travels to Tbilisi from other countries, how it transforms here to shape Tbilisi's locality in the here and now.

Sir Oliver Wardrop, a British diplomat and the United Kingdom's first Chief Commissioner of Transcaucasia from 1919 to 1921, describes this peculiar openness of Tbilisi:

> Here we see small open-fronted Oriental shops in which dark Persians ply their trades, making arms, saddlery, jewellery, selling carpets, and doing a hundred other things all before the eyes of men and in the open air. There is a strange confusion of tongues and dresses; a smart little grammar-school girl rubs shoulders with a veiled Mussulman woman, and occasionally you see the uniform of a Russian officer elbowing his way through a crowd of Lesghians, Armenians, Georgians, Persians; through the midst of all this confusion runs the tram-car. We are not beyond all the influences of civilization.[2]

We Are Open, Says the City

Useful Historical Context

The mid-eighteenth century saw the beginning of a long process when the Georgian kingdoms and the Russian empire drew closer, seen by the kings of Kartli-Kakheti and Imereti as the only deliverance from Muslim khanates. Raiding Leki highlanders, also known as Dagestanis, were cutting a wide swath through Kartli and Kakheti. In 1783, the Treaty of Georgievsk was signed with Russia. Nonetheless, in 1795, Iran's ruler Agha Mohammad Khan Qajar claimed victory, with the Russian Empire repeatedly going back on its word and Persian troops wreaking havoc on Tbilisi and burning it down, slaughtering many of its citizens and enslaving others. As a result, the king's palace was razed to the ground and destroyed beyond repair, 'together with its wings, auxiliary facilities, and marble fountains adorned with lion's head decorations, also its printing house, seminary […]'.[3]

Although it was clear from the outset that relations with the Russian Empire would not be on an equal footing, mulling over this truth was a luxury the rulers of the Georgian kingdoms could no longer afford. In 1801, the Kingdom of Kartli and Kakheti was annexed by the Russian Empire, marking the beginning of a long process of complex and ambivalent relations. Clearly, closer relations with Christian Russia gradually eclipsed Tbilisi's unmistakable Asian hues and flavours. Sticking with the culinary dimension, new trade emerged in this period, along with the first boulangeries opening.

Tbilisi back then remained a major commercial hub with a variety of bazaars carrying Persian, European, and Russian goods. Camel caravans, also those with horses and oxen, delivered fish, rice, honey, olive oil, olives, sugar, and honeydew and watermelons. Numerous caravanserais operated here, with a handful of them surviving to this day, albeit architecturally rather than functionally. There are also *dukani* taverns and *kulbaki* trade workshops.

In the 1850s, the Dariali trade route opened – a direct route to Russia – enabling further commercial ties with Russia. Things grew even more diverse in Tbilisi. While the names of several bazaars continued to reflect Asian affiliations and one was a dedicated Armenian marketplace, soon a 'Soldier's Bazaar' became part of the city, referring to the Russian Empire's troops by the word *soldier*. A Tbilisi history noted that '[o]ne Petersburg-based journalist likened this marketplace to the Petersburg bazaar, pointing out: "The Petersburg bazaar is a drop in the ocean, while Tbilisi's bazaar is an ocean in the drop"'.[4]

In 1897, Tbilisi-based artist and author Karapet Grigoryants described one aspect of the bazaar's overall picture: 'Simply browsing through or glancing at wagons, you are sure to be invited to a thousand of places, and everyone, just to spite the others, will fill your glass of wine to the brim. Some dine right there by their wagons. Bread and boiled fish are sold and served hot nearby.'[5]

At the same time, buffets, boulangeries, confectionery stores, cafes, and restaurants were becoming just as popular as Tbilisi's so-called Asian-style eateries, such as coffeeshops, *khashi* soup taverns, kebob and *kharsho* soup diners, barrelhouses, wine shops, and Muslim *ashpashkhana* food vendors. Every newly opened hotel boasted a European-style restaurant, and their menus were advertised as Georgian and European/Asian. In an attempt to take advantage of Tbilisi garden culture, indoor restaurants often featured outdoor garden areas. Evenings were enriched by string quartets or choirs. According to one author, Asian-style eateries that could not afford orchestras kept canaries in cages to entertain patrons.[6]

Feasting Artisans

Tbilisi is also a city of artisans. Its marketplaces abound with locally manufactured goods, such as tableware, kitchenware, implements, and various devices. And, needless to say, there is also food. Artisans run their own trade unions called *amkari*, and member admission ceremonies often turn into major celebrations in Tbilisi, naturally ending with festive meals. Such gatherings at tables honouring new trade union members come across as ephemeral and yet recurring settings, with food doled around and taken outside, consumed together with wine, and guests enjoying themselves and then parting ways to get back to work. Everyone in town is familiar with this new member 'baptism' ritual. But the menu never changes. At least, this is how this celebration is described by Ioseb Grishashvili, Tbilisi's very own poet and a unique researcher of the city's history:

> The feasters and drummers arrive in the garden where they receive instructions from the trade union's seasoned and reliable member. On one side, *shilaplavi* pilaf is cooked in an enormous cauldron. On the other side there is a festive table covered with a blue tablecloth, with grape, fig, or cabbage leaves served instead of plates. Clay bowls are lined up to replace glasses. Sacred candles are shining brightly around the table. Customarily, these candles must burn to the end.
>
> [… T]he feast carries on for two or three days. The festive reception on the first day is believed to bring about special blessings. The second day is reserved just for seniors. The third day is known as 'crumbs from the table', alluding to leftover food.
>
> The celebration's grand finale is as follows. Around sunset, the participants gather around, holding wreaths of flowers, lit candles, torches, and *zurna* pipes in their hands, leaving the garden and heading back to the workshop from where they proceeded earlier.
>
> And now it is time to enjoy a festive meal, to sing, dance, and have fun, in front of the *dukani* establishment. If the first toast during the feast in the garden was raised to commemorate a deceased master craftsman, a custom accompanied by flipping a bowl of wine over a loaf of bread, now the participants, sufficiently inebriated, can no longer maintain the orderliness of the first reception, explaining why the dinner ends with a ruckus dragging out for quite some time.[7]

Shilaplavi pilaf mentioned in the quote above is an imported dish betraying Asian influences. Nowadays, it is made with lamb or beef, or even veal. Recipes of this and other types of pilaf are offered by Georgian feminist journalist Barbare Jorjadze in her recipe book published in 1875. Her *shilaplavi* recipe uses lamb, with finely chopped onion added to boiling rice. Next, as the rice thickens, the ingredients are mixed well, with the addition of only black pepper and cumin.[8] Today, *shilaplavi* is an inseparable dish of memorial receptions, having even developed into a butt of a joke of sorts: *shilaplavi* is considered the last dish served at such dinners, prompting the participants to raise the final toast in memory of the departed.⁹

Tbilisi's Urban Carnival

Prior to Sovietization, Tbilisi also boasted its own carnivals that, again, ended with lavish meals outdoors. One such festival is *Keenoba*, an attempt to ridicule the Muslim ruler, or rather the collective character thereof, according to many researchers – a sort of carnival-style revenge on the conqueror, in retaliation for the destruction brought by him.

Georgian ethnographer Julieta Rukhadze describes a typical *Keenoba* festival in Tbilisi:

> On Clean Monday, the first day of Great Lent, designated participants proceed to dress the Keeni and his retinue in the Dabakhana Gorge (near the Tbilisi Botanic Garden). The Keeni in a colorful Persian costume wears a conical hat, with his face covered in soot. Some Keeni characters wear a sheepskin coat inside out. The Keeni's retinue is dressed in colorful clothing, wearing sheepskin coats inside out or garish capes, also with their faces covered in soot. Besides the Keeni, the city also has a king. The king and queen are dressed in royal attire with tree-branch crowns on their heads. The city is divided into two opposing camps, with the king's retinue and supporters on one side, and the Keeni and his party on the other.¹⁰

Around this time, every typical Tbilisi musical instrument was put to use. Taverns and diners were closed, with feasting and partying in full swing across the board. Until noon, the Keeni ruled over the city. Then the city revolted, and staged clashes ensued, as Ioseb Grishashvili writes in *The Literary Bohemianism of Old Tbilisi*:

> The grand finale of the *Keenoba* festival is as follows: At noon, the Keeni is led toward the river and dumped into the water, in a safe spot preventing him from drowning, of course [...]. Next, the citizens chip in contributions, and the participants of *Keenoba* buy enough food to engage in a week-long celebration and endless feasting in Ortachala, by the Mtkvari River, the gardens of the Vera neighborhood, Didube, and other locations.¹¹

The menus of these lavish meals are not specified, though what is always a must at festive tables includes *shoti* bread from the *tone* tandoori, fresh green herbs – with tarragon adding a special flavour – and boiled meat.

These details, though not related directly to the notion of portable food, are nonetheless essential for getting a sense of the Tbilisi-style carnival. After the establishment of Russia's tsarist rule, the empire's local governors tried to pass *Keenoba* off as a demonstration of tsarism's might. The idea was not only rejected by Tbilisi's residents, but they also introduced and ridiculed new characters, such as 'the drunk captain', general, colonel, and others. Needless to say, this new version *Keenoba* was unacceptable to tsarist authorities: *Keenoba* was banned in 1894. Yet even though the festival was gradually lost to obscurity,

Tbilisi residents still kept it alive in a previously banned form known as *Keenobis Kudebi* (*Keenoba Tails*), holding the festival outside the city, in secret, and enjoying festive meals outdoors: 'The feasting lasted until sunset, after which the feasters headed back to the city, playing *zurna* pipes and dancing *kintauri*.'[12] It follows that Tbilisi-style portable food was sometimes incorporated into these illegal social functions.

Islands Disappeared, so Did the Raft-feasting

As we focus on Old Tbilisi's gardens, before proceeding down to the riverbank, let me offer another small excerpt about public gardens and feasts in the public areas, again from Sir Wardrop's book:

> The view from the churchyard is a splendid one; the whole city, with its wonderful diversity of form and colour, lies at your feet; on the right you can see far along the Kakhetian road, and on the left the great highway to Vladikavkaz follows the winding course of the Kura [*Mtkvari*]. In the evening we often climbed to the top of a bare crag not far from the church, carrying with us a large earthenware flagon of wine, a roast leg of mutton, fruit, cucumbers, and other delicacies, and spreading out our cloaks on the ground lay there making merry, singing and telling tales until long after midnight; the lights of the town below us seemed like a reflection of the bright stars above us, and the music and laughter of many a jovial group came up the hillside to mingle with our own.[13]

In 1869, *Droeba* newspaper article likened the Ortachala Gardens to their 'European' counterparts:

> Nothing compares to this poetic chaos and confusion, these vine-covered pergola, the sweet aroma of trees laden with walnuts, the array of eye-catching and mouthwatering fruits, and the mellifluous sounds of barrel organs. Singing accompanied by pipes and dayerehs, chilled mulberries and cherries [...]. These are superior beyond comparison to those gardens with uniformly trimmed trees and alleys with a packed brick pavement. You cannot possibly find a rug or a cushion there to sprawl and catch your breath for a second.[14]

Nowadays, the romanticism of Ortachala is gone: in the 1950s, the Ortachala Hydropower Plant was built there, followed by brand new riverbank highways, requiring one of the Mtkvari's tributaries to be drained, so that the island eventually vanished.[15] Thus the grand Soviet engineering project also changed the memory of the city's main river. The taverns along the Mtkvari and the river's grain mills disappeared, and so did the traditions of using rafts to ship produce into the city and enjoy festive meals while gliding down the river.

As for raft-feasting on the Mtkvari, the way this makeshift dining venue was set up is best illustrated by Grigol Orbeliani, a nineteenth-century Georgian Romantic poet, in one of his personal letters:

> We young people had a gathering recently [...] and, one evening, went down to Ortachala, enjoying a meal on the Mtkvari riverbank, accompanied by *zurna* [pipes], *chianuri* [bowed musical instruments], and *tari* [lutes], and a skillful singer. The air was fresh, and the night was moonlit. Out of the blue, a raft carrying a group of people appeared and stopped ahead of us in the middle of the Mtkvari. Their leader shouted, 'Light the sacred candles!' And the raft's perimeter lit up with brightly shining sacred candles. And then he shouted, 'Set up the festive table!' And a festive meal was served. And a plate with four glasses was brought to the leader, and he skillfully downed all four. Then he refilled the four glasses which he handed to his fellow feasters one by one. This marked the beginning of a vibrant dinner, with us enjoying ourselves in our spot, and them reciprocating in the middle of the river. It turned out that the raft's leader – admiral as we nicknamed him later – was a butcher, one Kundzua's brother. As the partying picked up, the admiral ordered: 'Serve shish-kebobs now!' Immediately, one of the feasters plunged into the river, performing a headstand, then re-emerged, swimming and propelling his body by means of strokes with his massive arms, and finally reaching the shore. Bare-chested, he approached a nearby tavern, fetching sizzling skewered shish-kebobs, and swimming back to the raft. Now these shish-kebobs, together with four glasses, were handed to the feasters, one at a time. The singing of our musicians excited and galvanized the feasters, prompting them to request: 'Do it again, please!' My companions delighted in watching them enjoy their meal with dignity. Around 1 a.m., we left, and so did they, slowly rafting toward the upper bridge.[16]

In this letter, one of the participants swims across the river to deliver shish-kebob to his mates on another raft. The meat for shish-kebobs back then could have come from a variety of animals, such as a gazelle, for example, judging from a poem by Grigol Orbeliani:

> My Iarali,
> I bless the time when we,
> Seated amidst a green valley,
> Enjoyed our meal lavish.
> [...] With sizzling gazelle cubes
> On skewers rotating over the fire.
> [...] And bowls with wine from Kakheti
> Elegantly placed in our hands.
> And crumbly cheese, and fish,
> Adorning our festive table along with leafy greens.
> [...] Beautiful skies,
> Our native skies,
> Above us ravished with delight;

> ....
> You, a seasoned aged man,
> Tell me, an untrained lad,
> Stories of old,
> How lionhearted Georgian heroes
> Lived their lives and
> Put up a valiant fight.[17]

Thus, the Tbilisi examples of portable foods enjoyed in makeshift venues were enriched with scenes depicting feasts with sizzling skewered shish-kebobs on rafts in the middle of a river. This imagery persists until 1924, when Soviet authorities issue a new ordinance banning raft-feasting on the Mtkvari.

Before leaving the river, we note that Grigol Orbeliani's poem above was later countered in a feuilleton by the poet Akakai Tsereteli, who criticized this approach of weaving patriotic sentiments into festive meals:

> Who says I am not a patriot? Wrong! [...] How can I watch sizzling shish-kebobs without watering my mouth with patriotic saliva? How can I imbibe Kakheti wine without commemorating our ancestors! I'm halfway through feasting when the apparitions of my country's enemies materialize before my eyes. And that's when I snap. Enraged, I toss my hat into the ceiling and spill red wine over the tablecloth.[18]

Tsereteli's celebration of patriotism includes a menu that perfectly embodies the portable, movable feasting and dining of Old Tbilisi: hot *tone* bread, fresh fish, tarragon and other green herbs, cheese aged in a sheepskin sack, and chilled Kakheti wine.

Feast in the Bathhouse

The bathhouse is another place in Tbilisi where people share meals in a temporary space. In the nineteenth century – though we could go back much further – lavish meals were common in bathhouses. Grishashvili notes that '[b]athhouse feasters, these Asian gastronomes, had a peculiar kind of taste perception. Dried Georgian bread and Armenian cheese rinsed in sulfur water boasted these specific flavors that old artisans known as *karachogeli* reminisce about to this day. Watermelons chilled in the tub were also commonplace'.[19]

Tbilisi diversified even further when external influences on the country expanded in the period after the advent of the Russian Empire. The expansion of Russian policy and soft power was inevitable. Written sources, including documents from public archives, from that time tell of preparations for yet another major event, this time to greet Emperor Nicholas I of Russia in 1837. The Prince of Samegrelo commissioned a silver platter with bread and salt, famous Russian (Slavic) greeting symbols, to welcome the honorary guest. This ritual was repeated as the Emperor entered Tbilisi, when a special delegation met him in a settlement near the capital, kneeling in an open field and, again, holding a silver platter with bread and salt in their hands. Naturally, now it is impossible to ascertain direct ties, but it remains a fact that the phrase 'bread and salt' (*Pur-Marili*) is still used in everyday Georgian speech as a synonym of feasting, hospitality, and dining – to some extent even becoming a synonym for the Georgian sacral symbol of 'bread and wine' (*Pur-Ghvino*).

Conclusion

The goal of this essay is to remind us that portable food does not vanish without a trace. Each 'serving' of this kind transforms local flavours and leaves its stamp on them.

Ancient cities with experiences similar to that of Tbilisi not only have a genetic memory of collectively shaping their mixed vernacular language in their streets and homes, but also mould their flavours in the same manner. Both these processes of creation, for their part, are driven by the law of constant change. Sometimes, especially in ancient eras of conquests, the reason behind change is outside their control, enforced by conquerors slowly or rapidly but always surely leaving lasting marks on conquered countries' cultural identities, something consequently taking root in the very behavioural patterns of the speech and way of life of the locals. Finding evidence of this process in cuisine and everyday culture can be like catnip to researchers.

This prism allows for easily identifying links between a new vernacular language and flavours resulting from the mix of portable foods, both imported and indigenous. In *The Literary Bohemianism of Old Tbilisi*, Ioseb Grishashvili admits doting over these 'rich, quintessentially Tbilisi-style words', attempting to keep this 'linguistic rosary' together by turning to his own work, *The Urban Dictionary of Tbilisi*, whenever unable to find Russian equivalents. Significantly, after Georgia's incorporation into the Russian Empire, Russian lexical borrowings usually failed to take root in the Georgian language, so Grishashvili once again turns to urban Tbilisi figures of speech for help: '*Charozi* for dessert, *ortomeli* for a wineglass, and *laylaghi* for jasmine'.[20]

The words *dessert, wineglass,* and *jasmine* reflect new circumstances and new dining specifics. It follows that, if before Tbilisi residents savoured *charozi* – something sweet, such as *gozinaki*, a dish using honey and walnuts that, since the Soviet era and probably even earlier, has been inseparable from a traditional New Year's Eve meal – now the advent of the term *dessert* has shifted the meaning and context of sweets. This term has ushered in European, most likely French, and Russian meanings, edging out the Eastern (Asian) influence found in local sweets. This process of mutual replacement of external influences is best illustrated by contemporary names of tableware and kitchenware utensils listed by Grishashvili as parts of a dowry.[21] When I first read his list of twenty-six items, I recognized only eight words. Some of the remaining utensils may have gone out of use, which would account for their names vanishing as well. But others are nonetheless utilized to this day, though their names have changed.

When we say *Tbilisi flavour*, we paradoxically describe not one unchanging flavour attached to one particular locality, but rather a flavour constructed through the recognition that food is constantly portable: otherwise urban tastes would have never existed in the shape and to the extent as we know it. Tbilisi's food and the urban taste of this city are constantly inventing and reinventing themselves by blending layers of culinary influences, old and new, that have ended up here for various reasons (negative, less negative, more positive…), and that locals in Tbilisi's neighbourhoods, along with daily interaction between them (sometimes distant, sometimes antagonistic, then more intense…), have transformed to become quintessentially Tbilisian tastes. So, on a regular food tour that I used to guide in Tbilisi, I had this 'mission impossible' to show people all this colourful essence and all the diversity of these senses that the term *Tbilisi flavours* holds. Sometimes I had a feeling I could manage conveying this broad, bodily, and soulful knowledge to my guests by just biting *tonis puri* (bread baked in a *tone* tandoor oven). Or maybe it was all my imagination: imagination that was revived by the simple and banal effect of wine. We already know that 'Tbilisi is inebriated. […] The city-boat is rocking. Tbilisi is delirious.'

Notes

1. Grigol Robakidze, *The Snake's Skin* (Tbilisi: Merani Publishing, 1989), pp. 38–45.
2. Sir Oliver Wardrop, *The Kingdom of Georgia. Notes of Travel in a Land of Women, Wine and Song* (London, St. Dunstan's House, 1888), p. 26.
3. D. Gvritishvili and Sh. Meskhia, *Tbilisi History* (Tbilisi: State Publishing House of Children and Young Adult Literature, 1952), p. 145.
4. Gvritishvili and Meskhia, p. 201.
5. Karapet Grigoriants, *Rare Stories of Old Tbilisi* (Tbilisi: Museum of Literature Publishing, 2015), p. 22.
6. Gvritishvili and Meskhia, p. 176.
7. Ioseb Grishashvili, *The Literary Bohemianism of Old Tbilisi* (Batumi: Soviet Adjara Publishing, 1986), p. 38.
8. Barbare Jorjadze, *Complete Kitchen* (Tbilisi: Shorapani Publishing, 1914), p. 43.
9. *The Ethnographic Dictionary of Georgian Material Culture* (Tbilisi: Georgia National Museum and Meridiani Publishing, 2011), p. 610.
10. Julieta Rukhadze, *Georgian Folk Rituals: Berikaoba-Keenoba* (Tbilisi: Soviet Georgia Publishing, 1966), p. 48.
11. Grishashvili, p. 20.
12. Grishashvili, p. 24.
13. Wardrop, p. 25.
14. 'A Promenade in Tbilisi', *Droeba*, 19 June 1869, p. 3.
15. T. Kvirkvelia, *Old-Tbilisian Neighborhoods* (Tbilisi: Soviet Georgia Publishing, 1985), p. 102.
16. Grigol Orbeliani, *Personal Correspondence with Taso Oglobdzhio* (Tbilisi: State Publishing House, 1936), VIII: 1831-1884, p. 79
17. Grigol Orbeliani, 'My Iarali', *Poetry Collection* <https://poetry.ge/poets/grigol-orbeliani/poems/1337.iaralis.htm > [accessed 15 March 2023].
18. Akaki Tsereteli, 'Patriot's Confession', *Poetry Collection* <https://poetry.ge/poets/akaki-tsereteli/prose/c-patriotis-agsareba-motxroba> [accessed 15 March 2023].
19. Grishashvili, p. 58.
20. Grishashvili, p. 10.
21. Grishashvili, p. 46.

Urban Nomads and Vagabond Food: Eating Away from the Table in Nineteenth-Century France

Janet Beizer

Visitors to twenty-first century Paris who have wandered down the rue des Rosiers in the Marais with an eggplant-laden falafel pita in hand, grazed on a lemon-sugar crêpe while shopping the Raspail Marché Bio on a Sunday morning, or licked a scoop or two of fruit-infused Berthillon ice cream while crossing the Île Saint-Louis might be surprised to hear that eating on foot was a forbidden behaviour in France two centuries ago, a borderline activity reserved, in the popular imaginary, for children, animals, and vagrants. I'm interested in those adult humans who nonetheless ate on the streets in urban nineteenth-century France, the nature of their consumption, and how they were regarded and, especially, imagined by French bourgeois society.

As I pursue representations of French urban alimentary nomadism, I want to zero in on the conjunction of comestibility and mobility. Both the notion of street food and its reception by the French should perhaps be placed in the context of broader sociocultural ideas about vagabonds, vagrancy, and street culture. Conversely, definitions and assumptions about vagrancy may be influenced by traditions of the table and perceived violations of such traditions. The French high table is, we know, highly formalized, and its various components – from individual ingredients to their orchestration as cuisine, to the order of the meal, to the number and attire of guests, to the setting of the table – are intricately regulated, and have been for at least two hundred years. The UNESCO recognition of the French Gastronomic Meal as an Intangible Cultural Heritage in 2010 sanctified these culinary rites, but did not address their margins. However, the official table – the scene of gastronomy acknowledged by UNESCO – has long been shadowed by a subculture of food consumed away from the table by eaters who either do not have a seat there or chose to reject the one that falls to them. It is this category of roving food and itinerant eaters in nineteenth-century Paris – the time and place of French cuisine's codification in the hexagon and in the world, and so also the demarcation of all that lies outside the accepted codes of the table – that is my subject. I borrow the phrase 'vagabond food' from art historian Frédérique Desbuisson's '*nourriture vagabonde*', but I enlarge her concept to include wandering eaters as well as their moveable objects of consumption.[1] My exploratory corpus is split between Manet's 1862 painting, *La Chanteuse des rues* (*The Street Singer*), and Zola's novel of a decade later, *Le Ventre de Paris*

(*The Belly of Paris*); these two examples allow me to work synecdochically to elaborate on the larger visual and literary scenes of representation.

A word is in order first to take us both toward and away from current concepts of 'street food' and their network of connotations. The analytic framework used today to discuss what the United Nation's Food and Agricultural Organization defines as '"ready-to-eat" foods [...] prepared and/or sold by vendors [...] especially in the streets' is nominally relevant to my nineteenth-century focus in two ways: first in that it considers 'the streets' less in terms of physical roads and more in the sense of public space, and second in that it understands 'ready-to-eat' in the sense of requiring no additional culinary effort on the buyer's part.[2] Both these recent specifications describe nineteenth-century alimentary street vending as well. But contemporary approaches – rightly, for our era – emphasize increasing global mobility of populations and their diets, or, in other words, the influences of migrations and multiplying ethnicities on the fare being sold; they look at the evolving variety of this fare for its role in shaping and reflecting national, cultural, and socio-economic identities within discrete and diverse geo-social spheres.[3] My own focus on mobile food, however, homes in on the microlevel of the phenomenon: it follows representations of isolated ambulatory eaters rather than tracing the foodways of groups on the move.

Singing for One's Supper: Manet's *Street Singer*

Manet's *Chanteuse des rues* or *Street Singer*, sometimes called *La Femme aux cerises* or *Woman with Cherries*, is a painting whose charged title almost matches its audacious content and fraught facture. Street singers in nineteenth-century Paris were tantamount to prostitutes in their mythologized presentation in the popular imaginary; they walked the streets just like streetwalkers did, but heightened the offense by having mouths wide open as they did so. Like *saltimbanques* (acrobats) and other street performers, *chanteuses des rues* were associated with gypsies, bohemians, and other socially marginalized figures. In the period when Manet conceived and executed this work, in early 1860s Second Empire Paris, Louis Napoleon's government had cracked down on street singers in a generalized attempt to purge the streets of circulating performers, singers that might inspire political unrest, and vagabonds. Manet's father Auguste was a civil judge whose office included among its duties prosecuting street entertainers; he'd hoped his son would follow his path in the legal profession, but Édouard preferred to follow the footsteps of the vagrant population under his father's surveillance. One element of this judicial scrutiny was requiring performers to be registered (like prostitutes) and to carry a medallion that included identifying information.[4]

The nickname – *Woman with Cherries* – popularly given to Manet's painting highlights the other outrageous element of its subject, the cherries the woman carries and draws to her mouth. This street performer (clearly marked as such by the guitar she holds) is not caught in the act of singing, but in – or almost in – another oral activity.[5] With her right hand, she holds to her face two cherries – of the common *bigarreau* variety, as recognized by countless period commentators – from the cluster supported by her left arm, blocking her mouth from view but simultaneously emphasizing it by the fruits, whose destination is clear. Not quite eating and not quite not eating, she, along with the unofficial title of the painting pointing to the cherries, draws even more attention to consumption by arresting the act in a materialized moment of eternal unfolding. Linda Nochlin, among others, has signalled the in-between-ness of the woman's physical bodily position, 'captured [...] caught, as though by the clicking of the lens of a camera, between the to and fro of the barroom door. [...] Time is simply suspended here'.[6] Also hanging in the balance is her gesture of conveying the *bigarreaux* to her mouth and their incipient ingestion.

When we turn to the critical record, we find the hinged scandal of the painting magnified by the repetition, over history, of a series of comments on the walker and the eater. Whether these remarks are proffered as reproaches to the artist or as disdain for his subject, or offered in the guise of neutral observations, a pattern of echoes emerges that rather oddly does not evolve significantly over time. As I review the resonances, I therefore avoid a fixed diachronic progression, and freely juxtapose critical voices from different eras.

Manet's much maligned *Street Singer* has been accused of factural flaws ranging from gracelessness to crudity, to misguided artistic efforts, to lighting and perspective, to ill-conceived intents. It is often difficult when reading the criticism to know when latter-day critics are voicing their own opinions and when they are simply rehearsing past judgments. All the history of this painting seems to hover and re-emerge in each new appraisal. In sum, the artist has been cited for the way he has his singer hide her mouth, and hence her expression, by the cherries; for her inept grasp on these fruits, a gesture that to many appears graceless, kinetically illogical (not at all conducive to eating the cherries); for the lumpy clumsiness of the fingers of her left hand and the awkward grasp of her instrument in that multi-tasking hand that also helps to balance the wrapper of cherries in the crook of her arm while it holds the '*tirettes*' or skirt lifters that raise her dress up from the ground. The literature refers to the crude highlights of the painting, sometimes with specific notes (Carol Armstrong mentions the 'white highlights upon the bunches of loose cherries' as evidence of illusionism); it points out perspectival angles that, as Michael Fried observes, '[don't] quite work', making for an 'effect of [...] "wrongness"'.[7]

Critics have vociferously drawn attention (whether in their own names or in acknowledgement of broader social condemnation) to certain improbabilities, inconsistencies, contradictions, and improprieties of the painting. Vivien Perutz speaks to the 'implausibly balanced', if succulent, cherries on the woman's arm. Thérèse Dolan recapitulates a tradition of speculation on why a street singer, usually dressed in rags, in this case 'resembles a fashion plate'.[8] Charles Moffett echoes the conundrum around fashion and class, adding gender to the socio-economic misalignment with style as he remarks on the ambiguity of a woman wearing the sort of hat, a *toque*, commonly associated with men.⁹ He finds the cherries in particular to be a puzzlement, as do the many observers who conjecture on why they are there in the picture, offering explanations that range from the practical (perhaps the singer was paid in fruit for her cabaret performance; perhaps she is starving) to the symbolic (they connote properties ranging from sensuality to licentiousness, eroticism, immodesty, naiveté, eternity, and vagrancy).¹⁰ Moffett along with others, however, concludes that they are in the end simply 'inexplicable'. For Frédérique Desbuissons as well, they do not make sense within any conventional representational framework.¹¹ The *bigarreaux* become a condensation of all that frustrates viewers who want to understand the painting and find it ultimately indecipherable. Adolphe Tabarant attributes the unusual depiction of '*cette fille sortant d'un cabaret en mangeant des cerises*' ('this girl leaving a cabaret while eating cherries') to modern life, which of course is a convenient (if evasive) catch-all explanation for the new and inexplicable; Collins links the woman's expressionless 'empty stare' (a platitude in Manet commentary that cuts across the wide swathe of representations of Victorine Meurent, Manet's model here as elsewhere) to her 'unreadability', while Nancy Locke speaks of a lack of narrative sense and of meaning.¹²

These various allusions to assorted elements of the painting's illegibility are crowned by one of the earliest critical reactions to the work at the time when it was first exhibited at the Martinet Gallery in 1863. Way back then, Paul Mantz summed up – in advance – a century and a half's future questioning of the *Street Singer*'s meaning by the snide detailing of '*une singularité qui nous trouble profondément, les sourcils renoncent à leur position horizontale pour venir se placer verticalement le long du nez, comme deux virgules d'ombre [...] l'effet est blafard, dur, sinistre*' ('a peculiarity that troubles us profoundly, her eyebrows renouncing their horizontal position to come alongside her nose vertically, like two shadowy commas [...] the effect is ghastly, hard, sinister').¹³ If I linger on Mantz's comment, it is not because I agree with his critique of Manet's irregular figuration of the singer's brow, but rather because his comma metaphor is prophetic of a constellation of faults that have been regularly ascribed to the painting and that may best be described, in Mantz's terms, as its defective punctuation. Punctuation, we know, is what the OED defines as 'the practice of inserting points or marks in writing or printing, in order to aid the sense'. These marks are crucial to the disambiguation of meaning. The particular sense-making character to which Mantz refers, the comma, or, to turn back to French, *la virgule*, '*marqu[e] une courte pause*': its function is to indicate a short pause that helps to parse language – in this case, visual syntax.¹⁴ But those *deux virgules d'ombre* – the twinned shadowy vertical commas that Mantz perceives in the place of the singer's eyebrows – seem to indicate an extended pause: fallen commas, or a pause that has become a freeze, a suspension of meaning rather than an aid to understanding.

Might we re-punctuate this painting? Could we reimagine these fallen brows as parentheses instead of vertical commas? The *Robert* dictionary defines parenthesis as the insertion '*Dans le corps d'une phrase d'un élément qui [...] interrompt la construction syntaxique*' ('an element in the body of a sentence that fractures its syntax). We need change very little to define parenthesis analogously within the visual syntax of *The Street Singer*: there is an element in this body that interrupts its syntactical construction. If a parenthesis is signalled typographically by two rounded brackets that enclose an element defined as accessory or digressive, this enclosure is paradoxically also defined as central: a frame or emphasis for what is set off. Merriam Webster calls parenthesis 'an amplifying [… inserted] word, phrase, or sentence'.

It is the singer's eyes that are amplified here, not only in and of themselves, but for the way their colour and form make them rhyme with the two cherries held just beneath, which echo the wrapper of cherries below and right, which ring faintly again in the glint of the coral-red earring poised on the visible ear. The cherries, the eyes – turned red-brown by association with the fruits – and the pale reddish jewels are the points that draw the beholder's eye in a painting whose hues are otherwise subdued. Framed by the parenthetical markers that dip down the side of the nose, the eyes and the resonant cherries are marked as both central and accessory. They are the traits of the face and the painting that commentators find unfathomable or disruptive of meaning – yet no one can resist pondering them. It's a platitude of Manet criticism to note – and to bemoan or applaud – the way his model Victorine Meurent withholds her gaze, makes herself vacuous, distant, unreadable, and she is the model for *The Street Singer*, as she was for his *Olympia* and many other paintings. I'll note in passing that in Meurent's recently rediscovered self-portrait, her eyes are very different: the MFA's blurb on the painting highlights the artist's 'sidelong gaze of self-assessment in a mirror'. Far from withholding, her look there is canny and incisive as she holds herself in scrutiny.

The incomprehensible heart of the widely circulated critical literature on Manet's *Street Singer* is that myriad disparate features of its craft and conception have to do with all that is called awkward, crude, rude, wrong, implausible, threatening, and sinister in the painting, to echo a handful of recurrent terms. All these adjectives, while ostensibly applied to technical and aesthetic aspects of Manet's work, can and should be understood on a moral register as well. Where he is overtly charged with flouting artistic convention, Manet is often also (often surreptitiously) being accused of social and even ethical transgression; the aesthetic and moral affronts cannot easily be separated. When commentators take umbrage at the unconventional obstruction of the singer's mouth by the paired cherries, objecting explicitly to a flagrant violation of codes of facial expressivity in painting, we might add, as Locke reminds us, that this was 'an era when a proper women would never lift a fruit to her lips in public'.[15] When mention is made of the 'vulgarity' of the *bigarreaux* and of their offensive unreadability, we might remember that cherries were inexpensive fruits, perceived therefore as common and associated with the poor.[16] When the violet-tinged shadow cast by the singer's skirt serves as springboard to imaginings about her dirty underwear and heavy boots – neither visible – we need to consider not only prevalent assumptions about women and gender, but also discourses about vagrancy in nineteenth-century France, and how they pair with discourses about eating.[17]

Dolan's observation about *The Street Singer* that 'the rude gesture of eating cherries on the go rather than dining at a well-appointed table suggest[s] the uncouth manners of a street type' harks back to conventions of socialization that disciplined the streets and the classes in such a way as to rigorously separate alimentary and ambulatory activities.[18] An etiquette manual of the period spells out the rules in no uncertain terms: '*Il y a des gens mal élevés qui se permettent de manger dans les rues et dans les promenades; c'est un signe de gourmandise du plus mauvais ton. On ne tolère cela que dans les jeunes enfants qui n'ont pas encore l'âge de la raison*' ('There are badly raised people who dare to eat in the streets and while walking; it's a sign of the worst kind of gourmandise. This can only be tolerated in young children who haven't yet reached the age of reason').[19] Desbuissons expands on the indecorous act of street eating, targeting not only the singer's exit from the cabaret but also the cherries she consumes in their rawness. They are ready to eat, as is, straight from the tree, and fall clearly on the side of nature rather than culture; they are an emblem of the simple and the popular, the sign of an 'anti-cuisine' that requires no preparation. Such a diet, she points out, is the very negation of the complicated and expensive culinary procedures from which imperial France drew pride.[20] The fruits, on the contrary – like the young woman – are understood to be outside the space of sociability, and place her similarly against culture, locating her in the sphere of nature.

In the Street

Scholars of portable food in France contrast the snack (*le casse-croûte*) from the meal (*le repas*) partly on the basis of its site; if the meal is situated around a home, a table, a timetable, the snack takes place usually outside, without pretence and without formality. However, this doesn't mean that it follows no rules. Julia Csergo states in no uncertain terms that '*on ne casse-croûte pas dans la rue*' ('one does not snack in the street'), and she adds: '*De même, on ne casse-croûte pas en marchant*' ('one does not snack while walking').[21] While portable food, snack food, is often associated with the working-class poor, it remains subject to social rules which, for Csergo, have to do with community and sociability, and is even loosely ritualized in that eating together means imposing a pause on the unfolding of a day. Although the *casse-croûte* is usually taken outdoors, whether in the countryside, as at a picnic, or in an urban setting, Csergo and others are adamant about distinguishing it from '*le nomadisme alimentaire*', alimentary nomadism, which they associate with solitary eating that not only is not marked off from the rest of the day, but also presents as an intimate biological gesture that becomes obscenely public when viewed but not shared.[22]

We are brought back, then, to that object of massive bourgeois censure, the figure of the roving urban eater illustrated by Manet (and, we shall see, by Zola). For Alberto Capatti, the prejudice stems not only from the fact that this food is consumed outdoors and outside of normative meal hours, but also, that it is '*à manger avec les mains et salissante*' ('to be eaten with the hands [and is therefore] soiling'). For Alfred Fierro, eating in the street is historically '*l'apanage des misérables*' ('the prerogative of indigents')

– which does not mean that it is an accepted prerogative.[23] We read consistently that eating in the streets can only be tolerated as a behaviour of children and animals, and that even those who are obliged to buy food in the streets normally take it elsewhere to eat, that is, to their home. Those who are seen consuming in the streets, then, are assumed not to have another place to eat their food: they are understood not to have a home. Secondarily, but equally significantly, those who are perceived to be street feeders are deemed to be indigent, miserable, rootless, barbaric, savage, not worthy of representation – and certainly not representation in the *beaux arts*.[24] And it is here that taboos against public eating are conjoined with phobias about vagrancy. Those who indulge their physical need for food on public streets must not have a domestic life and the property that goes with it – a roof, a table, a family – they must be vagrants. The conjunction is well illustrated by Arlette Farge's report from a century earlier that indigents arrested by the police were frequently found to be carrying a wooden spoon that they used to eat soup on occasion.[25] We might say that such individuals had premeditated intent to eat in the streets.

If the terms of vagrancy and the vagabond had already emerged centuries earlier, in thirteenth-century France (and England), it is in the late nineteenth century, in conjunction with industrializing society, that the fears that Dominique Kalifa groups under the broad rubric of '*urbaphobie*' crystallized around vagrancy as a reaction against a notion of poverty that was being reconceived as an individualized rather than as a collective phenomenon.[26] By 1870, vagrancy – vagabondage – was deemed a serious social problem, and in 1885 a series of anti-vagrancy laws were passed. As Ian Hacking has argued in his work on the 'mad traveler', that pathologized voyager was situated between the legal category of the vagabond and the bourgeois activity of the tourist: the criminal pole and the fantasy pole defined the broad parameters of the problem.[27] The literature on vagrancy in France is immense; I allude to its vastness not only to explain my lack of mastery, but to signal the importance of the classification in French criminal and legal systems, which is to say, in the larger ideological infrastructure.

Urban Foragers: Zola's Street Feeders in *Le Ventre de Paris*

Writing in 1901, Zola's friend, journalist Antonin Proust, mockingly recalls that one of the administrators at the École des Beaux Arts had recently claimed that Manet drew the subject for his *Street Singer* from Zola's *L'Assommoir* (*The Drinking Den*) – clearly an absurd pretension, as that novel was written fifteen years after Manet's painting.[28] Less preposterous would have been a claim that Zola was influenced by Manet, whose work he zealously championed, though the sharper resonance would be between Manet's cherry-eating *Street Singer* and Zola's 1872 novel, *The Belly of Paris*, which features, in a secondary but visually dramatic role, the nubile young fruit seller La Sarriette whose cherries are called 'red kisses' and who adorns herself with these fruits, draping a pair over each ear like the vermillion earrings Manet's *Street Singer* wears to rhyme with the two *bigarreaux* held up to her face.[29] Yet the intersections of Manet's street painting and

Zola's street novel are more far-reaching than a handful of cherries.

The peripatetic eaters of Zola's novel advance its narrative course along with its alimentary processing. The novel's well-known catalogues and tableaux present massive accretions of food that are interspersed with scenes of ambulatory eating that put product into process. A series of wanderings serves the dual purpose of steering the reader through the commercial spaces of the *quartier* while also quickening the prose.

Just a handful of intermittent feeders walk the produce-laden, fish and cheese-reeking, lard-greased stage of Zola's Halles.[30] Given their paucity, it is worth considering who they are, why they have been assigned their consuming role, and how this role is shaped.[31] Pride of place goes to the omnipresent Mlle Saget, who criss-crosses the novel and the neighbourhood of Les Halles, ogling the market displays with a cavernous basket in hand in which she collects the butt ends of each merchant's wares, the dribs and drabs that have fallen off by day's end and can be begged or cheaply acquired. When this mode of provisioning falls short, she resorts to the merchants who vend the passed-down repasts of the well-to-do along with the specious lure of an equally transmissible social status. That the material goods she collects are simultaneously crystallizations of word scraps, rumours, *faits-divers* and whispers of calumny – what today might be called 'sound bites' – is one of those hermeneutic 'gifts' with which Zola plies his reader, and which often feel like a barrage of packaged metaphors calling out for unwrapping.

Marjolin and Cadine, the itinerant urchin duo of Les Halles, join the cast of visible eaters in this novel. Marjolin, explicitly designated as unofficial tour guide – '*le Quasimodo de mes* Halles' – in Zola's plans for the novel, serves also to relieve a congestion that might otherwise weigh the narrative down.[32] Although Zola slyly pretends ignorance about the child's naming, *Marjolin* meets the ear like a masculinized form of *marjolaine*, marjoram, the herb used medicinally as an antidote for indigestion, flatulence, and all manner of digestive disorders.[33] The two cabbage-patch children are discovered separately in their infancy, Marjolin literally under a cabbage leaf (625); they become community wards who are fed and lodged by a series of vendors.[34] Eventually raised together by the vegetable seller, *la mère* Chantemesse, in exchange for their labour, the foster siblings roam the marketplace freely, bedding down shamelessly, and feasting openly on pilfered morsels. This last activity warrants a closer look: the alimentary license seamlessly clinches the sexual and ambulatory abandon the nubile adolescents accord themselves, confirming their animal-like ways and assuring their popular status in the eyes of any reader privy to the bodywork of nineteenth-century social discourses. Together with the young charcuterie apprentice, Léon, who comes bearing stolen offerings of cornichons and goose fat, saucisson slivers and ham ends, Marjolin and Cadine seclude themselves in pavilion cellars or attic bedrooms for cobbled-together repasts whose other found objects are cajoled, nipped from the stalls, or nudged out of strategically jostled baskets.[35] The improvised spreads obey only the law of accessibility; whatever goods fall into the nimble fingers of the young thieves land in a heap on their plates in a mockery of culinary harmony and order.[36] In the novel's commanding judgment filtered through the bohemian persona of artist Claude Lantier, such breaches of decorum are morally reproachable but aesthetically laudable: '*Il éprouvait, malgré lui, comme une admiration pour ces bêtes sensuelles, chipeuses et gloutonnes, lâchées dans la jouissance de tout ce qui traînait, ramassant les miettes tombées de la desserte d'un géant*' (784 – 'He felt, in spite of himself, something like admiration for these sensual beasts, who were light-fingered and gluttonous, abandoned to the pleasures of whatever lay around, picking up the crumbs that had fallen from the remains of a giant's feast').[37]

These miscellaneous crumbs cleared from a giant's table are not conceptually different from the leftovers Mlle Saget negotiates on the cheap, begs from shop owners, or trades in kind – a scrap of gossip for a smudge of pâté or a sliver of ham, for example – from the earliest pages of the novel on (634; 668). Together they announce the more conventional institution of cycled-down dinner remnants sold in the later scenes by vendors of scraps ostensibly fallen from high tables. Well before he presents the official merchants of leftovers, then, Zola has introduced the concept of piecemeal eating and exposed it as a generalized mode of consumption among the working-class poor.

Muche and Pauline, the young children of the warring fishmonger and *charcutière*, respectively, extend the file of flagrant eaters in the novel when they evade their bilateral maternal interdictions on associating with the enemy and join forces. Their illicit forays through muddy streets and their collecting of confectionary dregs are the juvenile equivalent of Marjolin and Cadine's nomadic marketplace circuits, and yield bargain booty of similarly haphazard form, notably, paper cones filled with the sugary rubble of candy displays.[38] Further degraded by Pauline's sticky fingers shuttling between her salivating mouth and her bulging pockets, the sweet crumbs disintegrate into confectionary dust and chocolatey stains.

Most (and chronologically first) exposed of all consumers in the novel, however, are the bottom feeders: the anonymous soup guzzlers introduced by Claude to the newly-returned Florent.[39] Early in his tour of the renovated marketplace, Claude presents a soup merchant surrounded by her famished dawn customers. Descending the tiers of the lowest levels of Parisian society from the tidied 'upper crust' to the murky depths – because even the underworld of misery is hierarchized – the narrator, from Claude's perspective, describes the huddled clientele hunched furtively over their soup cups on the sidewalk, dribbling and drooling their foul-smelling cabbage soup.[40] The soup eaters – not only anonymous and faceless, but also synecdochically reduced to a bodily need ('*toutes les faims matinales*', 'all these morning hungers') – divert their eyes instinctively. We may wonder if it is the glance of passers-by in the street or that of the virtual narrator that they shirk while they gulp down their rank meal 'with animal distrust'. We are perhaps meant to understand the qualifier as an indication of a wariness induced by what we would today call their 'food insecurity', which causes them to wolf down a meal in hand, however humble. But the animal modifier tinges the soup consumers integrally as well as their distrust, suggesting that the impulse to look away be assigned to primitive instinct. The averted gaze might be ambiguated as a sign of human embarrassment and humiliation as much as caginess. Florent, in turn, looks away from the soup eaters with an equally vaguely defined discomfort: is he pained to see others eat in the face of his own ravenous hunger, or ashamed to be watching a scene of voracious feeding with which he so readily identifies? The equivocation may be Zola's way of having his wretched soup seen and eaten too: he pairs a show of sympathy for hunger and misery with the unwavering voyeuristic gaze of social privilege.[41]

The particular chronological locus of this soup stand scene within Claude's introductory tour of Les Halles makes it foundational. Positioned at the nadir of alimentary practices, these ambulant soup eaters are the summum of all the Paris underworld consumers Zola parades through his novel: by virtue of the form of their food as well as their mode of ingesting it, Zola's soup guzzlers are extreme prototypes of the other active consumers in this novel. On foot, fingers balancing spoon and bowl, they epitomize the trail of itinerant eaters we encounter throughout, crouched on straw or chicken feathers in cellars, cramped together on attic bedroom floors, making the round of vendors in the streets, and eating out of hand. They exemplify all those who eat on the run because they do not have a place at the table, that emblem of civilization in the bourgeois era that, Carine Goutaland reminds us, is not merely a convenient alimentary prop but a marker of social class modelled on the separation of biological species: '*la table matérialise la distance entre humanité et animalité dans l'acte alimentaire*' ('the table materializes the distance between humanity and animality through the act of eating').[42]

As Goutaland along with Roland Barthes, Geneviève Sicotte, and others have argued, meals and the people ingesting them constituted one of the last great areas of repression for literature and the other arts in the nineteenth century, one of the last frontiers of corporeal representation remaining to be crossed.[43] It was a challenge that Zola and other naturalists gladly took up. But I believe it would be an oversimplification to declare victory to Zola for '*cette volonté d'exposer le corps sans tabou*' ('this will to expose the body with no taboos'), as Goutaland frames the novelist's feeding forays, or even to qualify his project as one aimed at breaking all taboos.[44] For while Zola may display the eating body in *Le Ventre de Paris*, and may gesture toward its digestion and (even more prominently) its indigestion, he does not zero in on the *bourgeois* eating body, which remains largely untouched and unviewed in this novel.[45] The eaters who are exposed – we might say publicized or even, anachronistically, 'outed' – displayed in the act of eating in *Le Ventre* – are defined by their low or marginal social status: they are children, animals, and the working-class poor, those lower-class humans infantilized or likened to beasts because of their poverty and déclassé rank.[46] And it is crucial to emphasize as well that their style of feeding does not fit under the rubric of a *meal*, if we understand this term to describe an act that unfolds in time, implies regular occurrence, and entails a round measure of food.[47] Their consumption is represented instead as scavenging: that is, as a discontinuous, sporadic activity targeting part objects, and to which no measure of duration, frequency, or wholeness can be assigned. Zola's feeders are his doubly-determined bit players: the vital secondary actors to whom he relegates the function of eating and its styling in dribs and drabs. Just as what defines the food we see ingested in *Le Ventre* is its fallen status (it is second-hand, residuary, and mixed), what characterizes the active eaters is their lowly social grade.

Like Manet, Zola crosses accepted aesthetic conventions by representing street people in the act of street eating, transgressing social and aesthetic taboos, and in doing so, places his subjects outside the sanctioned social sphere. While it might be reductive to come to conclusions on the basis of observations rooted in one painting and one novel, I want to speculate briefly about these two works and the nature of their sociocultural breaches. If Zola violates representational strictures by brazenly laying out the street and its inhabitants in their daily processes of circulation and consumption, he does so in a way I will cautiously suggest is less subversive than Manet's. Where Zola dares to shock bourgeois mores by making the intimate and the uncouth public and visible, he does so in the limited sense of charging such behaviours only to those coded as social inferiors, while Manet complicates his representation by confusing the social clues we look for to identify individuals of various social classes. The critical reaction to Manet's *Street Singer* across the years – that is, the accusations of unintelligibility, meaninglessness, indecipherability, and implausibility, correspond, I suggest, to Manet's ambiguation of preconceived class cues to create an unreadability that is ideological at its base. This comparison along with the related comments needs to be explored and amplified, and serves as a point of departure for further discussion.

Notes

1. Frédérique Desbuissons, '*Yeux ouverts et bouche affamée: le paradigme culinaire de l'art moderne (1850-1880)*', *Sociétés et Représentations*, 34.2 (2012), 49–70 (p. 63).
2. Food and Agriculture Organization of the United Nations. 'Street Foods', *Food for the Cities* <https://www.fao.org/fcit/food-processing/street-foods/en/> [accessed 25 April 2023].
3. See the articles anthologized in Martina Kaller, Markus Mayer, and John Kear (eds.), *Delicious Migration: Street Food in a Globalized World* (Rockville MD: Global South Press, 2017), especially Kaller, Kear, and Mayer, 'Street Food: A Mobile Concept' (pp. 17–46).
4. Thérèse Dolan suggests that this medallion may be in the pouch hanging at the singer's waist that her guitar-grasping hand casually points to. See 'Manet's *The Street Singer* and the Poets' in *Word and Image*, 34.2 (June 2018), 88–110 (p. 93).
5. The guitar was not accepted at the Paris Conservatory until 1969. See Frédéric Frank, 'Gauguin and Manet: Tradition Reinvented' in *In Concert! Musical Instruments in Art 1860-1910*, ed. by Frédéric Frank and Belinda Thomson (Giverny: Hazan, 2017), p. 31.
6. Linda Nochlin, *Realism* (New York: Penguin Books, 1971), p. 160.
7. Carol Armstrong, 'To Paint, to Point, to Pose: Manet's *Le Déjeuner sur l'herbe*, in *Manet's Le Déjeuner sur l'herbe*, ed. by Paul Hayes Tucker (Cambridge: Cambridge University Press, 1998), p. 112; Michael Fried, *Manet's Modernism, or The Face of Painting in the 1860s* (Chicago: University of Chicago Press, 1996), p. 292.
8. Vivien Perutz, *Édouard Manet* (Lewisburg, PA: Bucknell University Press, 1993), p. 88; Dolan, p. 91.
9. Charles S. Moffett, '*The Street Singer*' in Françoise Cachin, Charles S. Moffett, and Michel Melot, *Manet 1832–1883* (Catalogue: Galeries Nationales du Grand Palais, Paris/The Metropolitan Museum of Art, New York), (New York: Harry N. Abrams and The Metropolitan Museum of Art, 1983), p. 109.

MFA curator Katie Hanson's research, however, shows that the *toque* became fashionable for women at this time (personal conversation, 21 April 2022).

10. See Margaret Mary Armbrust Seibert, 'A Biography of Victorine-Louise Meurent and her Role in the Art of Edouard Manet' (unpublished doctoral dissertation, Ohio State University, 1986), I, p. 98; Desbuissons, p. 9; Moffett, p. 109.
11. Desbuissons, pp. 58–65
12. See Adolphe Tabarant, *Manet et ses oeuvres* (Paris: Gallimard, 1947), p. 48; Bradford R. Collins, 'Manet's Rue Mosnier Decked with Flags and the Flaneur Concept', *The Burlington Magazine*, 117.872 (November 1975), p. 713; Nancy Locke, *Manet and the Family Romance* (Princeton: Princeton University Press, 2001), pp. 70–71.
13. Paul Mantz, '*Exposition du boulevard des Italiens*', *Gazette des Beaux-Arts,* 14.1 (January 1863), p. 383.
14. Le Robert, *Dictionnaire Historique de la langue française*.
15. Locke, p. 71.
16. Antonin Proust, 'The Art of Édouard Manet', *The International Studio*, 12.48 (February 1901), p. 233; Zola speaks about cherries in his *Ventre de Paris*, listing them among cheap fruits, '*les fruits à bas prix*' (in *Les Rougon-Macquart*, ed. by Armand Lanoux and Henri Mitterand (Paris: Gallimard, 1960), I, p. 823. Subsequent citations from Zola will be taken from this edition; translations are mine.
17. It is true that the edge of her petticoat is visible, and it is worth noting that when the painting was cleaned in 2018 by the Conservation staff of the MFA, layers of dirt adhering to the resin were removed. Were they already present in 1881 when the review was written, less than twenty years after Manet painted *The Street Singer*? Or was it simply assumed that a *chanteuse des rues* would have soiled underwear by virtue of her sex, class, and loose morals? For the gift of viewing the painting up close in the conservation space of the MFA, I owe thanks to Conservator Rhona Macbeth, European Paintings Curator Katie Hanson, and Coordinator Cara Wolahan. The author of the article reviewing '*L'Exposition Manet*' of 1881 in *Le Radical*, A.F., suggests that the street singer inspires the viewer with pity and apprehension, '*car, sous cette robe à tirettes, le linge est fatigué, terne, la pauvre fille! Elle doit porter de grosses bottines […] des bottines d'homme. Ces femmes, enfin, sont ce qu'elles sont*' ('for, under that robe with skirt lifters, the underwear is worn and drab, the poor girl! She must be wearing heavy boots […] men's boots').
18. Dolan, 'Manet's *The Street Singer*', p. 102.
19. Jules Clément, *Traité de la politesse et du savoir-vivre* (Paris: Bernardin-Béchet, 1879), p. 179.
20. Desbuissons, pp. 60–65.
21. Julia Csergo, '*Avant-propos*', in *Le Dossier: Casse-croûte: Aliment portatif, repas indéfinissable*, ed. by Julia Csergo (Paris: Éditions Autrement, 2001), p. 14.
22. Csergo, '*Avant-propos*', p. 14; André Rauch, '*Le Cercle des affamés*', in *Le Dossier: Casse-croûte*, ed. by Csergo, p. 38
23. Alberto Capatti, '*Le Pizza: Quand le Casse-croûte des misérables passe à table*', in *Le Dossier: Casse-croûte*, ed. by Csergo, p. 57; Alfred Fierro, '*Manger dans la rue à Paris sous l'Ancien Régime*' in *Le Dossier: Casse-croûte*, ed. by Csergo, p. 132.
24. Former MOMA curator William Rubin notes that those figures believed not to belong in serious painting in the 1840s and 50s (like ragpickers and street singers) began to appear in photographs which influenced minor painters (*A Curator's Quest: Building the Collection of Painting and Sculpture of the MOMA 1967-1988* (New York: Overlook Duckworth, 2011), p. 519).
25. Arlette Farge, *Vivre dans la rue à Paris au dix-huitième siècle* (Paris: Gallimard, 1979), p. 24.
26. Dominique Kalifa, *Les Bas-fonds, histoire d'un imaginaire* (Paris: Éditions du Seuil, 2013), pp. 26, 77.
27. Ian Hacking, *Mad Travelers: Reflections on the Reality of Transient Mental Illnesses* (Cambridge, MA: Harvard University Press, 2002).
28. Proust, p. 233.
29. Zola, *Le Ventre* p. 822. For an extended analysis of those who walk and eat the streets in Zola's Paris, see chapter 4 in Janet Beizer, *The Harlequin Eaters: From Food Scraps to Modernism in Nineteenth-Century*

30. *Paris* (Minneapolis: University of Minnesota Press, 2023).
31. I emphasize that these eaters walk Zola's planks, for it is crucial to note the correlation of consumption and ambulation in this novel. Zola, like Manet, and later Huysmans, highlights the itinerant nature of working-class consumption. Desbuissons, writing about Manet, dubs the food represented in such paintings as *La Chanteuse des rues*, where the street singer exits a café munching on cherries, *la nourriture vagabonde*. Both 'vagabond food' and its corollary, which we might call vagabond or nomadic eating, are useful concepts to hold onto as we scan the modalities of nourishment taken in away from the table by the urban poor: see Desbuissons, pp. 63–65.
32. While it is not very complicated to run through the small troupe of visible feeders in *Le Ventre*, it is a bit thornier to make sense of the casting, which does not align with the consecrated division of the novel's characters into *les gras* and *les maigres* ('the fat and the lean ones'). The most extended treatment of this opposition, enunciated by Claude Lantier (Zola, *Le Ventre* pp. 804-806) as a parable that reduces all humankind (and specifically the main players of this novel) into two types, is in Marie Scarpa, *Le Carnaval des Halles: Une ethnocritique du* Ventre de Paris *de Zola* (Paris: CNRS, 2000). See, too, Scarpa, '*Retour ethnique sur les modalités du ventre dans* Le Ventre de Paris' in *La Cuisine de l'œuvre au XIXe siècle: regards d'artistes et d'écrivains*. ed. by Éléonore Reverzy and Bertrand Marquer (Strasbourg: University Press of Strasbourg, 2013), pp. 203–15, and Geneviève Sicotte, *Le Festin lu: Le Repas chez Flaubert, Zola et Huysmans* (Paris: Liber, 1999). Though one might logically expect to find the eater's part attributed to the more corpulent personae, beginning with the fleshy embodiment of the titular belly, the *charcutière* Lisa (née Macquart) Quenu, this is not at all correct, but I dispute the case that Zola's consumers align neatly with the lean and hungry, as has been argued. See Scarpa, *Carnaval*, p. 247; Scarpa, '*Retour ethnique*', pp. 208-09; Carine Goutaland, *De régals en dégoûts: Le naturalisme à table* (Paris: Classiques Garnier, 2017), pp 67-68. For Éléonore Reverzy, the *gras/maigre* binary is paired with sex and gender roles, with the forces of devouring women ready to set upon Florent (*La Chair de l'idée: Poétique de l'allégorie dans Les Rougon-Macquart* (Geneva: Droz, 2007), p. 120). I find this opposition too schematic.
33. See Émile Zola, *Ébauche*, in *La Fabrique des Rougon-Macquart: Édition des dossiers préparatoires*, ed. by Colette Becker (Paris: Honoré Champion, 2003-2017), folio 62/16; folio 94/48.
34. '*Une belle fille rousse, qui vendaient des plantes officinales, l'avait appelé Marjolin, sans qu'on sut pourquoi*' (p. 762 – 'a pretty redhead, who sold herbal remedies, named him Marjolin, *though no one knew why*'; my emphasis).
35. The fact that Florent rides into Paris among the cabbages as he is reborn there on his return from Guiana assimilates him to Marjolin, suggesting that he, too, is a marginalized being, a metaphorical foundling.
36. '*Le fromage blanc … était un cadeau. Un friteur … avait vendu à crédit les deux sous de pommes de terre frites. Le reste, les poires, les noix, les crevettes, les radis, était volé aux quatre coins des Halles*' (p. 783 – 'The *fromage blanc* was a handout. A fry cook had sold him on credit two sous' worth of fried potatoes. The rest, the pears, the nuts, the shrimp, the radishes, were stolen from the four corners of les Halles').
37. '*Ils allongeaient la main en passant le long des étalages, chipant un pruneau, une poignée de cerises, un bout de morue … ramassant tout ce qui tombait*' (p. 783–84 – 'They reached out when traversing the stalls, nicking a plum, a handful of cherries, a bit of cod … collecting whatever fell').
38. Priscilla Parkhurst Ferguson calls Cadine and Marjolin's food forays '*flâneries gourmandes*, substitutes for consumption [... for] a perpetually frustrated gourmandise' ('The Sensualization of Flânerie', in *Dix-Neuf*, 16.2 (July 2012), 211–23 (pp. 220–21). It is true that the pair do not acquire all the delicacies they covet, but they are among the few characters seized in the act of indulging their gourmandise and enjoying its fruits.
39. '*De minces cornets de papier, où les épiciers mettent les débris de leur étalage, les dragées cassées, les marrons glacés tombés en morceaux, les fonds suspects des bocaux de bonbons*' (p. 816 – 'Thin paper cones, in which grocers put the debris of their stalls, broken *dragées*, glazed chestnuts crushed into pieces, the questionable bottom layer of candy jars').

39   For a succinct but evocative history of the place of soup in culinary culture, with an emphasis on its historically low status, see Claude Thouvenot, 'La Soupe dans l'histoire', in *Cultures, Nourriture, Internationale de l'imaginaire, Nouvelle Série* – No. 7 (Paris: Babel/Maison des Cultures du Monde, 1997), pp. 153–61.

40   '*Il y avait là des marchandes très propres, des maraîchers en blouse, des porteurs sales, le paletot gras des charges de nourriture qui avaient traîné sur les épaules, de pauvres diables déguenillés, toutes les faims matinales des Halles, mangeant, se brûlant, écartant un peu le menton pour ne pas se tacher de la bavure des cuillers. [...] Mais cette diablesse de soupe aux choux avait une odeur terrible. Florent tournait la tête, gêné par ces tasses pleines, que les consommateurs vidaient sans mot dire, avec un regard de côté d'animaux méfiants*' (pp. 624–25 – 'There were very clean vendors, vegetable sellers in their work blouses, dirty porters whose coats were greasy with traces of the food they had borne on their shoulders, poor raggedy devils, all the morning hungers of les Halles eating, burning themselves (with the hot liquid), sticking out their chins a little so as not to get stained by the drooling spoons. [...] But this cursed cabbage soup had a terrible smell. Florent turned his head, troubled by the full cups that the consumers emptied silently, with a sideways gaze of suspicious animals'). The foul smell is explicitly associated with the soup, but it attaches to the wretches reduced to eating it as well. On nineteenth-century conceptions of the '*foule putride*' or the stink of the people, the idea that the proletariat reeked of their living conditions (including overuse of grease and oil, likely to have been the prime ingredient of this cabbage soup) see Alain Corbin, *Le Miasme et la jonquille: l'odorat et l'imaginaire social XVIII-XIXe siècles* (Paris: Flammarion, 1986 [1982]), pp. 163–88 (p. 174). Cabbage soup holds a special place in the iconography of nineteenth-century French poverty, and its lingering stink pervades the century's literature: it is perceived not only a common vegetable, but also a vegetable for and of the common people. If raw cabbages transfuse the novel's landscape from its opening scene onward, here, not very far in, we find their remains, introduced in the altered state Jean-Pierre Richard has named 'the residual' in his analysis of Joris-Karl Huysmans's *A Vau-l'eau*, in which text he notes '*le dégoût complexe du résiduel, de l'engorgé, du coagulé, de l'aliment anal*' ('*Le Texte et sa cuisine*', *Microlectures* (Paris: Seuil, 1979), p. 147 – 'the complex disgust of the residual, the congested, the coagulated, the anal'). On the connotations of cabbages in art, see Allison Deutsch, *Consuming Painting: Food and the Feminine in Impressionist Paris* (State College: Pennsylvania State University Press, 2021), pp. 113–43.

41   By the time he was writing *Le Ventre*, in 1872-73, Zola was enjoying a modestly comfortable bourgeois existence. His early years in Paris were certainly marked by penury, though he was never reduced to soup stands or other such forms of portable eating. See Henri Mitterand, *Zola* (Paris: Fayard, 1999-2002), I, p. 592 and following.

42   Goutaland, *Régals*, p. 12.

43   Goutaland, *Régals*, pp. 14–22; Roland Barthes, *Sade, Fourier, Loyola* (Baltimore: Johns Hopkins University Press, 1997); Sicotte, *Le Festin*.

44   Goutaland, *Régals*, p. 15.

45   It is worth emphasizing that Zola indicates digestion or indigestion by naming one or the other periodically, usually metaphorically, but he does not show either.

46   Among the many animals represented in the act of eating in the novel are crabs, pigeons, cats, horses, snakes, lice, and mosquitos.

47   The *Petit Robert* (1984) gives as the first definition of *le repas* the following: '*Nourriture, ensemble d'aliments divers, de mets et de boissons pris en une fois à heures réglées*' ('Nourishment, collection of diverse foods, of dishes and beverages taken together at regular hours'). The *OED* gives as first (obsolete) definition, significantly, 'a measure', and, as second, 'any of the occasions of taking food which occur by custom or habit at more or less fixed times of the day, as a breakfast, dinner, supper, etc'.

When the Lord Eats Away from the Table: Outdoor Food in Indian *Gaudiya* Vaishnavism

Tanushree Bhowmik

Indian culinary culture, with its regional and local variations, expresses itself as much in form of what mortals eat as in the form of what the divine is believed to eat. This expression is manifest in the practice of offering food, *bhoga*, to gods and goddesses in Hinduism. While on one hand the offerings reflect the geography and social fabric of the land from where the sect originates, and hence the divine is being worshipped, on the other hand the practice is a projection of what is perceived as perfect and to which the believer aspires.

The Sanskrit saying '*Yatha dehe; Tatha deve*' means 'whatever the mortal body experiences, so does the divine'. It signifies that the mortal body is one with the divine; it is also the basis for anthropomorphizing the divine. Thus, the ritual of offering food to the gods, alongside other daily rituals, transforms the divine into human form and makes the divine more relatable and personal. The idol, *arca*, is god's forms in human shape. The belief is that god descends into the idol, making the divine alive and accessible to devotees. Hence, serving the *arca* in a daily human routine is the primary duty of the devotee.

While most sects of Hinduism follow similar practices, this anthropomorphism is perhaps most developed in the worship of Vishnu ('One who Pervades' or 'The Immanent'), and it attains its fullest form in the worship of one of his incarnations, *avatāras*: Krishna, the pastoral god.

Understanding *Gaudiya* Vaishnavism
Vishnu is one of the principle deities of Hinduism, and Vaishnavism is the worship and acceptance of Vishnu or one of his many incarnations as the supreme manifestation of the divine. The concept of Vishnu has origins in ancient Vedic texts; the basis of Vaishnava practice is the extensive mythology attached to Vishnu, largely that of his incarnations. Of all his incarnations, Krishna and Rama are the most common objects of devotion and daily ritualistic worship.

Krishna entered the pantheon of Hindu gods in the sixth century BCE in the *Chhandogya* Upanishad. In *Indica*, the Greek envoy to the court of a Maurya King, Megasthenes, writes about the worship of Heracles (Krishna) in Mathura, in the present-day state of Uttar Pradesh, still revered as the birthplace of Krishna.

Through a long, complex development, Vaishnavism evolved as a philosophy and as a religious practice. Of the many Vaishnava religious leaders and philosophers, one of the most well-known was the reformer from the region of Bengal, Sri Chaitanya Mahaprabhu (1484-1534), who established the *Gaudiya* Vaishnava sect (*Gaud* is the ancient name of Bengal).¹ To his followers, the *Gaudiya* Vaishnavs, he is a combined incarnation of Krishna and his consort, Radha.²

Mahaprabhu preached Krishna *bhakti* through devotional songs (*kirtans*) that extolled the love of Krishna and Radha. In his teachings, Krishna has infinite powers, *shakti,* and interacts with the mortal world through his divine acts, *leelas*. Krishna is *rasa,* an aesthetic/emotional impression, as well as *rasik,* a connoisseur of *rasa*.³ His devotees are Rasik Vaishnavs, who serve the divine couple in Vrindavan, the *Nitya-Goloka,* the eternal realm of existence.⁴ Vrindavan has seven important temples of *Gaudiya* Vaishnavism established by seven disciples of Sri Chaitanya Mahaprabhu under his instructions.⁵

The offering of *bhoga* or food to Krishna by *Gaudiya* Vaishnavs is considered one of the divine *leelas*. Food is one of the *rasas* enjoyed by the *rasik*. For the *rasik Vaishnavs,* food becomes a means both of expressing devotion and of owning the divine.

Eating Outdoors as One of the Krishna '*Leelas*'

Krishna is a pastoral god, a cow herd from the plains of river Yamuna (Figure 1). He is worshipped in myriad forms, in his different stages of life – as a toddler, as a teenage cow herd, as a lover and consort, as a friend, and as a king and strategist who imparted the knowledge of life and afterlife, documented in the Bhagvad Gita.

Gaudiya Vaishnavs seek divinity and salvation in a teenage, cowherd Krishna who is a lover and a friend. Thus, the pastoral plains of the Yamuna riverbanks and the open gardens and forests of Vrindavan where the teenage Krishna is often depicted play a central role in his *leelas*. The *bhoga* offered to him through the course of the day includes

Figure 1. Krishna with cows near river Yamuna [photograph by the author].

food suitable for consumption outdoors and appropriate for the season. The *bhoga* is prepared and offered according to codes laid down in the *Gaudiya* Vaishnava poetic texts that tell the story of his life in minute detail.

Krishna's *Ashtayam Leela* and Eating Outdoors

Krishna follows the same daily routine as his devotees, which is manifest in what *Gaudiya* Vaishnavs call *nitya leela*, the everlasting and daily divine acts between Krishna and his devotees. Krishna's omnipotence fuses the spiritual realm of *Goloka* into the material realm of Vrindavan. Food forms one of the markers of his *nitya leela*.

In his seminal 1580 text, *Shri Shri Govind Leelamrit*, Shrila Krishnadas Kaviraja Goswami (1496-unknown) elaborated Krishna and Radha's daily *ashtayam leela*, eight acts.[6] He writes, '*Sri Radha-Krsnayoh Asta-kaliya-lila Smarana-mangala Storam*', a sacred song of remembering the acts performed by Sri Radha and Krishna during eight periods of the day. *Ashtayam* comes from the Hindu calendar, where one day is comprised of eight *prahars*. The day is punctuated with eight divine acts, most of which occur outdoors.

This text, and subsequent works based on it, elaborates the daily routine of Krishna and Radha, with sections in each chapter on the *bhoga leela*, including lists of fruits, vegetables, and sweets along with recipes for certain preparations. The food of which Krishna partakes in his acts performed outdoors are described with details of packaging, dinnerware, and serving. These sections written by Krishnadas Kaviraja and other Vaishnav scholars are distilled and translated into the following sub-sections.

Nishant Leela or the Twilight Act from 3:36am to 6:00am

Krishna and Radha have spent the night in the gardens, unknown to their families. Vrindadevi (one who awards residence in Vrindavan) engages the birds, insects, animals, and flora of Vrinadavan to wake them up and send them to their respective homes.

Pratah Leela or the Morning Act from 6:00am to 8:24am

Radha, who has been called over by Krishna's mother, Yashoda, takes a bath, adorns herself with jewellery, and proceeds to Krishna's palace in Nandagram. Here she, along with her *sakhis*, female friends, and Krishna's aunt Rohini, cooks for Krishna and his *sakhas*, male friends. Some foods are consumed by the boys before they leave for the pastures, while others that have been 'cooked well or fried in ghee/clarified butter' (*supakwa*) and would last longer are packed to be consumed later as a forenoon meal for Krishna and his *sakhas*. These include various fried sweets made with coconut, bananas, and milk solids (*pishtaka*); an ear-shaped cake made of rice flour, sugar, and sesame, and fried in ghee (*sashkuli*); a flaky, layered, fried sweet pastry (*feni*); sweet spheres made of flour, sesame, and rice flour (*laddu*); flattened rice and barley fried in ghee; fried sweet stuffed dumplings (*modak*); sweet and savoury fritters made with ground mung beans and ground white lentils (*vada*); some of the *vadas* soaked in twelve syrups of varying degrees of sweetness and sourness made with green mangoes, tamarind, or dates; curd rice; stuffed and deep-fried savoury puffs (*kachori*); chutneys, pickles, and marmalades made with seasonal vegetables and fruits, including gooseberries, green mangoes, lime, berries of Salvadora persica; and ginger candy.

Purvanha Leela or the Forenoon Act from 8:24am till 10:48am

Krishna leaves for the grasslands on the banks of the river Yamuna with his herd of cows. As he prepares to leave, Yashoda tells him that she will send food for him, which he should eat on time and come back home before the sun sets.

As the day progresses, Krishna lets the cows roam freely on the grasslands adjacent to the lake Mansi Ganga and starts playing with his *sakhas*. Just as the boys start feeling tired and hungry after playing and wrestling with each other, Dhanishta, one of Yashoda's assistants, arrives with the food that was cooked in the morning by Radha and a sweet yoghurt drink (*rasala*) to wash it down.

On the cool banks of Mansi Ganga, Krishna sits amidst his *sakhas* like the core of a lotus flower surrounded by petals. Some of the *sakhas* fashion plates and bowls out of flowers, leaves, barks, fruits, and stones. They eat the food brought by Dhanishta while laughing and playing. After finishing his mid-morning meal, Krishna washes his hands and partakes of fragrant betel leaf scented with camphor and betel-nut (*tambula*). He then resumes talking to and playing with his friends.[7]

Madhyanha Leela or the Midnoon Act from 10:36am to 3:36pm

Krishna spends most of this time in company of Radha, her eight *sakhis* and eight assistants (*manjaris*), and his eight *sakhas*. Each of Radha's *sakhis* have their own private gardens (*kunj*), spread around a lake near Radha's abode, Radhakund, which is about 25 km from Vrindavan. They spend this time in the different gardens, engrossed in various sports, walking in the forest, playing in the water of the lakes, dancing while Krishna plays the flute, amorous play, and worshipping the sun god. The gardens are adorned with jasmine, marigold, and cooling vetiver in summer; roses in monsoon; burflower in autumn; and flame-of-the-forest in spring. The air of Yamuna becomes fragrant as it filters through the floral foliage.

Each of the *sakhis* and their gardens have different characteristics.[8] Lalita *sakhi's* Lalitanand Kunj is beautifully overgrown with various kinds of flowers and trees. It changes its size based on the requirements of Radha and Krishna. Lalita and Krishna are very dear to each other, and she brings camphor and *tambula* to him at the end of his meals in the gardens.

Vishakha *sakhi's* Madansukhad Kunj resembles a royal palace, while Chitra *sakhi's* Chittanand Kunj is full of flowers, roaming spotted deer, and swings hanging from branches. She is a skilled gourmet and can sense the taste of food made with honey, milk, and many other ingredients simply by glancing at them. She is proficient at making a variety of drinks with honey and nectar that she prepares for Radha and Krishna during their *leela*. Indulekha *sakhi's* Purnendu Kunj is white – even the fruits are white – and here the divine couple spends full moon nights when she often serves Krishna with nectar-like delicious meals. Rangadevi *sakhi*'s Manohar Kunj is black like the moonless night; Tungvidya *sakhi*, an expert in medicinal herbs, has the red Arunambuj Kunj; and Sudevi *sakhi*'s Harit Kunj is green.

To the south of Radhakund is Champakalata *sakhi*'s golden hued Hem Kunj. Champakalata *sakhi* is expert at collecting fruits, flowers, and roots from the forest. She is the leader of all the *sakhis* appointed as protectors of the trees, creepers, and bushes of Vrindavan. In her *kunj* is a well-appointed kitchen with a huge dining table that caters to the divine couple. She is an expert cook, well versed in all the culinary scriptures that describe the six flavours of gourmet cooking. Her expertise in making candy and sweets earned her the name '*Mistahasta*' (sweet hands). Champakalata *sakhi*, along with Vrindadevi and other *sakhis*, arranges divine feasts for Radha and Krishna in the gardens.

They prepare seasonal food for Krishna. In the winter, they cook *khichdi*, a one-pot savoury porridge made with rice and husked moong beans, with cashews, almonds, walnuts, pistachio, green cardamom, nutmeg, mace, black pepper, cloves, and copious amounts of ghee in winters; in the summer and during the monsoons, they make *kheer*, a sweet porridge made of rice and milk, flavoured with cooling camphor. Spices like saffron, nutmeg, mace, cinnamon, black pepper, and chillies are judiciously used based on the season.

As the *madhyanha* passes, Radha and Krishna, with their retinue of friends, play in the clear waters of Radhakund. Later Vrindadevi leads the couple to a gemstone-encrusted platform in a grove called Padmamandir inside Lalitanand Kunj for their afternoon meal.[9] The table is laid with food served on gold trays and in leaves of flame-of-the-forest, *Shorea robusta*, and the banana plant. Krishna sits on a white floral bed, overlaid with a white mat, with his *sakhas* on both sides. Radha sits facing them and lovingly starts to serve the food that Vrindadevi brings.

Radha first serves water from four kinds of coconuts, red, green, white and yellow. She then takes out the coconut from inside its shells, some of which are tender, some partly mature, and some fully mature. She removes the dark skin and serves the flesh of coconut, white as a conch shell. Next, she serves an assortment of mangoes sorted according to their variety, size, and sweetness. Some mangoes are slightly green; some, fully ripe. Some of the mangoes are made into thick juice, while others are perfect for sucking the sweet flesh out of their skin. She then opens, deseeds, and serves golden, sweet-smelling jackfruit, followed by an array of things like berries of *Salvadora persica*, grapes, pomegranates, fresh and dried dates, palmyra palms, wood apples, java plums, litchi, jujubes, water chestnuts, lemons, mulberries, guavas, pears, sweet limes, star fruit, carandas plums, citrons, monkey fruit, cucumbers, watermelons, yams, jicamas, radishes, lotus stems, foxnuts, bread fruit, almondette kernels, almonds, and other dried seeds. Fruits from all seasons are available throughout the year in Lalitanand Kunj because, by the blessings of Basant or Spring, the king of seasons, all seasons reside at the same time in this garden.

Then come the sweets that Radha prepared at home and brought over. There are floral-shaped sweets made with rock candy and milk fudge; sugar candies in the shape of fruits and vegetables; Krishna's favourite *chandrakanti* and '*gangajali*' laddu made of finely desiccated coconut. The *sakhis* brought milk cream laddus laced with sugar, camphor, and cloves; mango and jackfruit pulp mixed with camphor and rock candy; and crisp fried pancakes called *karpurkeli* and *amritkeli*.

Krishna and his *sakhas* partake of this meal while engaging in laughter and jest. At the end of the meal, the *sakhis* serve camphor-scented cool water, and the *manjaris* serve *tambula*. After Krishna and his friends finish eating, Radha has her meal with her *sakhis*, served by Vrindadevi and the *manjaris*.

The divine couple then retires for an afternoon siesta in each other's arms in Padmamandir. The *manjaris* fan them while chewing on *tambula* and dozing.

Aparanha Leela or Afternoon Act from 3:36pm to 6:00pm

Radha returns to her abode. She does her ablutions, dresses herself in beautiful clothes, and proceeds to make various delicacies to be sent to Krishna's palace for his evening repast and to be carried to the gardens for the night. She makes *amritkeli* with ripe bananas, white lentil flour, desiccated coconut, black pepper, reduced milk, cardamom seeds, and powders of clove, nutmeg, cinnamon, and camphor. These are deep fried in ghee and immersed in sugar syrup. *Karpurkeli*, made with fresh cottage cheese, fragrant rice flour, white lentil flour, curds, black pepper, desiccated coconut, sugar, pulp of a special banana called *amrita-kadali*, and powders of clove, nutmeg, and cardamom are fried in ghee and soaked in reduced milk. She makes fried balls of cottage cheese soaked in curds, milk, ghee, honey, and rock candy called *Amritgranthi*. *Anangvada/Ananggutika* are made with milk cream mixed with sugar, fine rice flour, finely desiccated coconut, and powders of clove, nutmeg, cardamom, and black pepper, then combined with ripe banana and deep fried. *Sidhuvilas*, made with ripe bananas, milk, unrefined sugar, and powders of black pepper, nutmeg and camphor mixed into wheat flour, are fried. She makes laddus with coconut and milk cream, flavoured with cloves, cardamom, camphor, black pepper, and sugar. She prepares *malpua*, a deep-fried pancake made with reduced milk. The *manjaris* pack the food into earthen and gold pots, as appropriate, secure them with cloths, and pack them into wooden chests.

Krishna and his *sakhas* gather their cows and prepare to return home in Nandagram. On the way they engage in mock fights and wrestle with *sakha* Madhumangal, trying to snatch and eat the food that he has hidden inside his shawl.

Sayahna Leela or Evening Act from 6:00pm to 8:24pm

Krishna returns home, bathes, dresses, and partakes of his evening repast that has been sent by Radha. He then checks on the cows in the cowsheds with his father Nand before sneaking out for a quick rendezvous with Radha on the pretext of going to play with his friends.

They meet at Chandrashalika situated within the Kandarpakuheli garden and share a hurried drink of sweet fruit juice from the same glass. Radha offers Krishna a *tambula* made with ripe betel leaf and they return to their respective homes.

Outdoor Food in Indian *Gaudiya* Vaishnavism

Pradosh Leela or Late Evening Act from 8:24pm to 10:48pm
Radha adorns herself in dark clothes, black *bindi*, and jewellery of blue lotuses and sapphires that all blend into the night and quietly sneaks out through the backdoor of her bedroom to meet her lover. The *manjaris* walk behind her with wooden chests on their heads, filled with sweets and *tambula* for the couple's dinner.

Krishna quietly leaves his bed and proceeds to the gardens on the banks of Yamuna to meet Radha.

Ratri Leela or Night Act 10:48pm to 3:36am
Radha and Krishna meet in the gardens of Vrindavan, anxious for each other's company. The amorous couple roam in the forests engrossed in singing, dancing, and playing in the waters of the Yamuna. Tired and satiated, they then move to Arunambuj Kunj for dinner. Vrindadevi and her assistants serve seasonal fruits, cashews, raisins, and dates to the couple, followed by the sweets that were packed in wooden chests by Radha and other *sakhis*. Tired from the day's activities and eager to be with each other, Radha and Krishna repose on a bed of flowers in Hem Kunj and fall asleep in each other's embrace.

Naimityik Leela or Acts during Annual Festivals
Just like their mortal devotees, festivals bring happy changes in the daily routine of Krishna and Radha. The food also changes during festivals that the divine couple celebrate outdoors to enjoy the seasons.

Sometime in mid to early February, when spring starts, mango blossoms, hibiscus, Ceylon ironwood, medlar, asoca, marigold, and flame-of-forest make the air fragrant. This is when *Basant utsav* is celebrated by Krishna and Radha. They gather in the morning to sing and dance with their friends in the mango orchards. It ends late afternoon in a playful fight with flowers. Vrindadevi forages fresh fruits and tubers from the forest, and they have a meal while seated on tender mango leaves and petals that have fallen on the ground. As the monsoons drench Vrindavan in early July, Krishna and Radha celebrate *Nauka-vihar*, their annual boating festival on the Manasi-ganga lake; during the peak rains in August, they celebrate the *Jhulan* (swing) and *Raas Leela* (dance) festivals (Figure 2). The second *Raas Leela* is celebrated during the night of the full moon under the clear winter skies in late November. They have a meal of seasonal fruits, dry sweetmeats, and fruit juices.

Figure 2 (left). Raas Leela [watercolour on paper by author].

Ashtayam Sewa under Gaudiya Vaishavism

The daily ritual and annual festivals of the divine are acknowledged in the material realm by *Gaudiya* Vaishnav *Goswamis* (priests) of the temples with *ashtayam sewa* (eight services).[10] These daily services to Krishna are performed with 'to-the-minute' accuracy, strictly in accordance with the details laid down in *Govind Leelamrit*, and the Lord is regaled with all the necessities and luxuries of life. The offering of food is called *bhoga sewa*.

The *sewa* in the Radha Raman (Krishna's name within the temple) temple of Vrindavan is particularly significant because of its importance within *Gaudiya* Vaishnavism. The temple was established in 1542 CE by Shri Gopalbhatt Goswami, who was chosen by Shri Chaitanya Mahaprabhu as his successor as the next guru of *Gaudiya* Vaishnavism.[11] *Gaudiya* Vaishnavs believe that Gunamanjari, Radha's main assistant, was reborn as Gopal Bhatt. Radha Raman was established under what is believed to be the sacred fig tree within which Krishna had hidden from Radha during *Raas Leela* about 4500 years ago; the temple is still served by the successors of Gopal Bhatt (Figure 3). The importance of the *bhoga sewa* in this temple is evident from the fact that its kitchen fire has been constantly burning for 479 years, after it was lit by Gopal Bhatt, and according to devotees the cooking is overseen by Radha herself (Figure 4).

Radha Raman's idol is taken to a dining chamber next to the *garbhagriha* or *sanctum sanctorum*, where he partakes of the *bhoga* for the *leelas* that he performs indoors. His *garbagriha*, however, is *golaka* and holds within itself all elements of the universe. It can transform to reflect all places, seasons, and times. The different *rasas* are evoked through *sewa* – *bhoga*, music, aesthetics, smell, devotional love – through which the divine *rasik* interacts with mortal *rasiks*.

Shri Gunamanjari Das Goswami (1828-1891) who wrote songs and sang to Radha Raman writes about his *bhoga* during *madhyanha*, 'Shri Radha Ramanji is having his midnoon meal, as he sits drenched in love within a beautiful *kunj*. [...] Vrindadevi and *sakhis* enter the *kunj* with sweet fruits; Gunamanjari serves those with her own hands.'[12] The kind of dishes offered in the *bhoga* is narrated to the deity through these songs every day.

Figure 3 (left). Gopal Bhatt establishing Radha Raman [from plaque in Radha Raman temple].
Figure 4 (right). 479-year-old Radha Raman's kitchen. [Both photographs by the author.]

Outdoor Food in Indian *Gaudiya* Vaishnavism

In Radhavallabh temple of Vrindavan, established in 1585, Krishna is offered *boondi-sev* (sweet droplet-sized deep fried snacks – savoury crispy noodles made of gram flour) while he is about to return home from the pastures.

The *sewa* ensures that Krishna is both exposed to and protected from seasonal changes. The *raga sewa*, or devotional music, follows the code of seasonal ragas in Indian classical music. The hand-painted tapestry, *pichwai*, behind him shows Vrindavan in different seasons (Figure 5). In early summers, the deity sits inside a tableau of fragrant seasonal flowers called *phool bangla*. When the breeze passes through the floral curtains, it creates the illusion of the cool air from Yamuna blowing through the blooming *kunj* (Figure 6). In peak summers he is dressed in a short under-cloth with bare torso; in the winters he is fully clad in warm clothess. Sandalwood is applied on him in summers, essence of roses is applied in monsoons to rise above petrichor, and heavier essences like musk are applied in winters.

The *bhoga sewa* in the summers and monsoons includes watery and cooling fruits; dried melon seeds; sweet sherbets made of seasonal fruits, honey, and yoghurt; and camphor. Radha Raman is served camphor-flavoured *kheer* during the summers, similar to what Radha is believed to have made for him. In the autumns and winters, there is a lot more fried food such as *poori* (deep-fried puffed bread) and *kachoris*; *mohanbhog*, a semolina fudge rich in ghee; pistachio, walnut, dates, and *shikharini* flavoured hung curd; and candy with sesame and milk fudge (Figure 7). Saffron replaces camphor in *kheer*. All seven temples offer *khichdi* in accordance with Champakalata *sakhi*'s *khichdi* for the month between mid-January to mid-February. Sweet *khichdi* laden with saffron and musk is also offered. *Khichdi* is accompanied by pickles, yoghurt, and fritters. Radha Raman is regaled with the verses: 'Radha and her Krishna are sitting inside a warm blanket in the cold morning; they are savouring spicy, hot *khichdi* dripping with ghee.' This *khichdi* has given rise to the saying, '*khichdi, ghee ki nichudi*' – *khichdi* should drip ghee. *Kulia*, sweet thick milk in small earthen pots, and *Tambula* is served with all meals (Figure 8).

Figure 5 (left). Pichwai showing the summertime night sky [photograph by the author].
Figure 6 (right). Phool Bangla [photograph by Avinish Kumar, by permission.]

For the festivals of *Jhulan* and *Raas Leela*, as Krishna is eating outdoors while dancing, dry *bhoga* is offered in all *Gaudiya* Vaishnav temples and in followers' households in earthen or leaf bowls that are easy to carry and dispose of (Figure 9). This *bhoga* is called the '*malsha bhoga*', unique to the *Gaudiya* Vaishnav practice.

It is evident that food as *sewa* in *Gaudiya* Vaishnavism is a key component in the belief system that transforms the closed confines of the temples into the outdoors exposed to time and seasons in the minds of the believer. The aesthetics and music accompanying the service reinforces this transformation.

Conclusion

The reading of the *Gaudiya* Vaishnav texts helps identify the foods that were deemed 'kosher' under the belief system and that were considered appropriate for eating outdoors in the fifteenth century, when the sect was established and its codes of practice laid down. Though the leading scholars were from or had spent considerable time in what is now the states of Bengal and Odisha in the east, they readily absorbed foods from the present-day states of Uttar Pradesh and Rajasthan. This is especially true for the foods they find suitable for outdoor eating. This might be attributed to a few factors. Krishna's early life was in Uttar Pradesh in the Northern Gangetic plains; thus, divinity was attached to the geography. Ancient Bengal had relations with the people from that area through invasions, rule, religion, and trade. By this time, Bengal and Odisha were also exposed to the foods that came in with traders from Rajasthan through Uttar Pradesh. This food may have resonated with *Gaudiya* Vaishnavs because of their vegetarian food habits and affiliation with the land of Krishna's mythology. Many of the easy to store, longer lasting, dried, fried foods that these traders carried with them on their journeys are reflected in the outdoor foods mentioned in the texts. It is almost as if the Vaishnav scholars from the east of sub-continent, who had travelled to Vrindavan and were spreading across the sub-continent, identified the virtues of the

Figure 7 (left). Winter bhoga served in sanctum sanctorum. Figure 8 (right). Kulia Bhoga. [Both photographs by the author.]

Outdoor Food in Indian *Gaudiya* Vaishnavism

Figure 9. Malsha Bhoga [photograph by the author].

foods carried by the travelling traders and adopted them. The food reverse-migrated to Vrindavan with certain influences, changed aesthetics, and a new belief system.

While the food absorbed influences already present in the sub-continent by the mid-fifteenth century, as is evident in the use of many spices, most things that came after are considered foreign, with the exception of fruits like apples. Thus, potato, cabbage, cauliflower, and tomato are still not found in the temple kitchens. However, much of what Krishna eats according to the texts is still eaten in the sub-continent, though the names of many dishes have changed. Deep-fried breads and sweets are carried during travels; *boondi-sev* is considered a snack that is easy to eat; milk fudge and sugar candies are moulded into shapes of flowers, vegetables, and fruits by women in Bengal and exchanged as wedding gifts between the bride's and groom's families; *khichdi* is widely consumed in winters; camphor is used in summers to cool and increase sweets' shelf life; and saffron is added to food in winters. Leaf plates/bowls and earthen pots are widely used during picnics and while serving street food on the go.

When it comes to food codes, it is as if the *Guadiya* Vaishnav scholars epitomized the deity's anthropomorphism by elaborating on the *leelas* and *sewa* of Krishna in their writings; then cast those in stone by establishing the deities within the stone temples. For centuries, the divine has influenced what is considered purest, most aspirational, and aesthetically supreme by the mortal devotees.

Notes

1. Shri Vrindabandas Thakur, *Shri Shri Chaitanya Bhagavat* ([1400s], repr. ed by Anantavasudev Brahmachari, Kolkata: Gaudiya Printing, [n.d.]); Shri Lochan Das Thakur, *Shri Chaitanya Mangal* (Vrindavan: Rasbehari Lal & Sons, 2017).
2. Kavi Karnapura, *Shri Shri Gaur Ganoddesh Dipika* ([1400s], repr. trans. by Narayan Goswami, Vrindavan: Shriman Premanand Brahmachari Sevaratna Publishing, 2001).
3. Shri Haridas Das, *Shri Shri Gaudiya Vaishnav Abhidan* (1957, repr. Kolkata: Sanskrit Book Depot Publishing, 2014).
4. Vrindavan, located in the present-day state of Uttar Pradesh, is one of the main seats of worship in India for Gaudiya Vaishnavs. It has seven major temples established by direct disciples of Chaitanya Mahaprabhu under his directive to revive the holy land of Krishna. In the tradition of Gaudiya Vaishnavism, Goloka is considered the ultimate destination of spiritual endeavour. The word literally means the realm of cows (Sankrit: *Go* – Cows; *Loka* – Realm), a reflection of the pastoralist origin of Krishna.
5. Sukumar Sen, T. Mukhopadhyay, *Shri Krishnadasa Kaviraja's Shri Shri Chaitanyacharanamrita* (Kolkata: Ananda Publishers, 2020).
6. Shrila Krishnadas Kaviraja Goswami, *Shri Shri Govind Leelamrit* (1580, repr. ed. by Kolkata:

Jadunandan Das, 1852).
7 Shrila Krishnadas Kaviraja Goswami, *Shri Shri Govind Leelamrit* (1580, repr. ed. by Giriraj Das, Vrindavan: ShriHarinam Sankrintan Mandal Publishing; 2013).
8 ShrilaRup Goswami, *Radhakrishna Ganodesh Dipika* (1550, repr. trans. and ed. by Narayan Goswami, New Delhi: Gaudiya Vedant Prakashan Publishing, 2003).
9 Shrila Krishnadas Kaviraja Goswami, *Shri Shri Gaur Govind Leelamrit* (fifteenth century, repr. ed. by JaiHari Das, Mathura: [n.p.], 2015).
10 Chacha Shrihit Vrindavandas, *Ashtayam Sewa Samay Prabandha* (Vrindavan: Shrihit Sahitya Prakashan Publishing; 2008).
11 Kavi Karnapura, *Shri Shri Gaur Ganoddesh Dipika* (fifteenth century, repr. trans. by Narayan Goswami, Vrindavan: Shriman Premanand Brahmachari Sevaratna Publishing, 2001).
12 Vaishavacharya Chandan Goswami, *Shri Radha Ramana Geeta* (Vrindavan: Shriradhaman Temple Publishing, 2015), [n.p.].
13 Author's translation of oral Vaishnav songs sung in the temple.

Cunard and Collins: Portable Food and Evolving Notions of Customer Service at Sea, 1840-1880

Andrea Broomfield

The wreck off Newfoundland's Grand Banks of the steamship, *Arctic*, was arguably the most notorious of the nineteenth century. It sank on 27 September 1854, and no women or children survived. The ship's demise and the 1856 disappearance at sea of the *Arctic*'s sister ship, *Pacific*, led to the bankruptcy in 1858 of the United States Mail Steamship Company, or Collins Line, only eight years after it began.[1]

Arctic's shipwreck and its aftermath is remembered because of eighty-six survivors who both found a seat in a lifeboat and actually reached shore. Their fate gets at the heart of one of my concerns: how could survivors drift for days in a lifeboat with no water or ship biscuits, while on-board, passengers' ticket prices included lavish food, wines, and spirits? Commodore Matthew Peary claimed that for $286, ten lifeboats could be equipped with water and 1500 pounds of biscuits along with compasses, oars, yards, sails, and masts.[2] Most ship owners dismissed the proposal, responding that such provisions would cost them 'too much'.[3]

The thirty-one men aboard one *Arctic* lifeboat salvaged a pumpkin and cabbage. Spotted by the Canadian bark, *Huron*, at dusk on the second day after the wreck, the dehydrated, wretched survivors were taken aboard. Forty-five occupants in the port quarter boat and the starboard guard boat had a handful of biscuits that passenger William Gilbert had stuffed in his pocket. Fellow lifeboat occupant, Frederick De Mayer, remembered that 'the gnawings of hunger and the terrors of starvation' compounded their plight.[4] These two boats came ashore off Avalon Peninsula at Cappahayden, Newfoundland.

The horrors increased when Robert Gourlay's empty lifeboat was picked up in mid-November by schooner *Lily Dale* with oars, marked *Arctic*, still inside. The only reasonable explanations were that Gourlay and the other occupants were rescued by a vessel that subsequently sank, or that, adrift without water or food, each man died and was thrown overboard by survivors until the last survivor, in a fit of delirium, committed suicide.[5] Two more empty lifeboats washed ashore at Avalon Peninsula. A reporter from the 21 December 1854 *Newfoundlander* wrote, 'We fear the people who took to her at the time of the loss of the ship must have perished fearfully.'[6]

When it came to Commodore Peary's proposal for provisioned lifeboats, Edward

Knight Collins, desperate to save his company, ordered ones 'able to provide for four hundred persons with water and provisions for several days in ordinary weather, at sea'.[7] Collins's newfound respect for passenger safety came too late, however.

How could shipping companies claim to be unable to provision lifeboats with biscuits and water when they spent lavishly on food? The Collins Line set the culinary bar spectacularly high in an era where food and cooking was carried out without gas, electricity, or refrigeration aside from ice, and when steamships carried only cabin-class passengers. Not only did Collins drive its ships on average three knots faster than Cunard, but the line offered first-class passengers meals that the ticket price of $130 (£30) could not cover. And yet, the deficit that Collins carried was justified by the company because it confidently maintained that its ships' superior dining and luxury would ultimately net it more business than Cunard.

Looking back on the history of passenger steamships, John Gould drew a salient observation about Americans and, by implication, the Collins Line: Americans 'demanded more and more luxurious surroundings and appointments', and they got them 'in every respect' when it came to opulence at sea.[8] However, regarding the ultimate extravagance – the safest possible voyages – American steamships might be blamed for giving passengers the short shrift. Their objective was not to profit by ensuring passenger safety so much as by ensuring passenger deception, helping passengers forget that they were at sea. Fine meals in gorgeous saloons were critical to that deception. Indeed, at the height of the Collins Line's fame, journalists, celebrities, and politicians considered Collins's strategy a success. But at what expense? I move beyond the usual narrative of speed's importance to maritime history and instead consider that question in relation to food and dining.

E.K. Collins's fleet of 'floating palaces' evolved from American sailing packets' reputation for luxury when, after the War of 1812, the United States was determined to establish the supremacy of its North Atlantic merchant marine. Unlike British Post Office Falmouth Packets with unreliable service, American packets promised passengers what today we call 'customer service'. Packet merchants set stunning speed records, departed on time, and captured mail trade and 'cane and glove' passengers 'willing to pay a premium'.[9] Indeed, E.K. Collins's earlier Dramatic Line of sailing packets set high standards for opulence.

Samuel Cunard's British and North American Royal Mail Steam Packet Company, or Cunard Line, threatened American dominance, however, because steam promised faster crossings, and unlike his main British competitor, Isambard Brunel, Cunard secured a British government subsidy to offset the price of running a steamship line to deliver the mails. When Cunard's first paddle ship, *Britannia*, steamed from Liverpool to Boston in 1840, E.K. Collins recognized the threat that steam posed. He vowed to build a steamship line that travelled faster than Cunard's and offered passengers the highest level of comfort and gustatory delight. While Cunard's and Collins's subsequent companies were not the sole genius and property of two men, their personalities, priorities, and upbringings are writ large on their companies' attitudes, including service and dining.

A Near-Monopoly of the North Atlantic: Cunard Line

Samuel Cunard was raised to be cautious. He grew up in Halifax, Nova Scotia, and attended Pictou Academy which emphasized practical arts without the flourishes of gentlemen's educations, such as Latin. After completing his education, Samuel became partner with his father at A. Cunard and Son.[10] To understand Samuel Cunard is to understand Halifax. 'Its political power was appointive,' wrote historian Stephen Fox. Regarding its character, Halifax struck visitors as English, nearer to the mother country regarding customs despite proximity to the United States.[11] A Tory 'tied intimately into the imperial network', Cunard was by nature conservative. He worked within complicated hierarchies of power and leveraged his colonial position to the best of his advantage.[12]

In 1839 when Samuel Cunard, George Burns, and Charles and David MacIver founded their steamship company, Cunard became the company's namesake and public face.[13] He sailed to Britain and secured the British mail contract with few competitors knowing who he was. While I am simplifying a complicated story of governmental and financial manoeuvring, ship historians today concur that Samuel Cunard was famous for a stealth and prudence that influenced the Cunard Line ethos. Regarding the sea, Cunard stood for no nonsense and for strict observance of protocol in the interest of safety and economy. He stressed to Glaswegian shipbuilders Robert Napier and John Wood that he wanted 'a plain and comfortable boat, but not the least unnecessary expense for show. I prefer plain work in the cabin, and it saves a large amount in the cost'.[14]

While other shipping companies kept Standing Orders for the crew's guidance, Cunard's were 'real orders, rigidly enforced'. They stressed that 'the trust of so many lives under the captain's charge is a great responsibility; requiring vigilance night and day'.[15] Cunard officers were some of world's best disciplined seamen and understood safety as the foremost consideration. 'We are certain there are no ships in any navy in the world, be it mercantile or royal, where better order prevails,' managing directors David and Charles MacIver declared.[16] Standing Orders required the crew to 'avoid national observations and discourage them in others; keep yourself always a disinterested party ready to reconcile differences'; be civil to passengers, but *recollect they will value your services on deck looking after their safety more than talking with them in the saloons*'.[17]

While passengers aboard Cunard vessels were arguably the safest, they were also some of the most uncomfortable and indifferently fed. When Charles Dickens published his account of his and Catherine's voyage aboard *Britannia* in 1840, it became the public face of Cunard for decades, one that would play to Collins's advantage when its first ship, *Atlantic*, sailed in April 1850. Dickens described *Britannia*'s all-purpose saloon as 'a long narrow apartment, not unlike a gigantic hearse with windows in the sides; having at the upper end a melancholy stove, at which three or four chilly stewards were warming their hands; while on either side, extending down its whole dreary length, was a long, long, table'. The food was often far from fresh despite the ship's icehouses stuffed with produce, beef, veal, pork, and poultry. Even on the first day, dinner consisted of 'the finest cut of a very yellow boiled leg of mutton with very green capers'. As the ship left coastal waters, the situation worsened with severe winter storms and chaotic conditions ensuing:

> All the stewards have fallen down stairs at various dinner-times, and go about with plasters in various places. The baker is ill, and so is the pastry-cook. A new man, horribly indisposed, has been required to fill the place of the latter officer, and has been propped and jammed up with empty casks in a little house upon deck, and commanded to roll out pie-crust, which he protests.[18]

Dickens's exaggeration aside, such situations aboard Cunarders demoralized passengers; when not seasick, they congregated in the saloon to escape cramped, often flooded staterooms. They craved appetizing meals to distract them from miserable conditions. Bostonian Pelham Warren sailed Cunard's *Caledonia* in 1848. He recorded that spoiled butter and 'wretched' wines made him gag. 'It is difficult,' he continued, 'to tell whether you are eating boiled chicken or roast beef.'[19] Passengers concurred, with seasoned travellers telling Warren that *Caledonia*'s food was the worst that they had encountered at sea.[20]

Cunard resisted improvements to dining and comfort, even though the British Post Office threatened to only renew its contract with Cunard 'on condition that they improve their ships'.[21] Harriet Beecher Stowe and Isabella Bird both sailed separately on Cunard's *Canada* in the 1850s, one of four paddle ships Cunard built in response. 'The coffee was much complained of,' Stowe observed. She noted copious amounts of food, and well-prepared dishes included 'the whole tribe of meats in general'. However, roasted and boiled meats require 'grave conviction and steady perseverance, rather than hope and inspiration', Stowe judged. Dishes requiring imagination and artistry fell short.[22]

When Bird sailed Cunard's *Canada* and *America*, she noted particularly long bills of fare. However, Cunard ships in the 1850s still maintained only one general-purpose saloon. The lack of space for women to congregate without men created anxiety for many. After dinner, men commandeered the saloon for 'comic drinking-songs, and satires on the English' while imbibing 'large bowls of punch'. As such, women had little choice but to return to cramped staterooms or go on deck 'when a refuge from the cold and spray would have been desirable', Bird lamented.[23]

Four rare Cunard Bills of Fare saved by English railroad magnate Henry Pease attest to the low priority that comfort and food took aboard Cunard ships. On the tenth day at sea, 12 July 1856, Pease wrote on his breakfast bill of fare his opinion of the food, also noting what was available and what was not.

That Pease could not have eggs, was limited to three meats when eight were listed (one of which – Cold Meats – was leftovers), and found the mutton chops 'bad', suggests that Cunard placed the lowest priority on breakfast. Passengers' 'Rules and

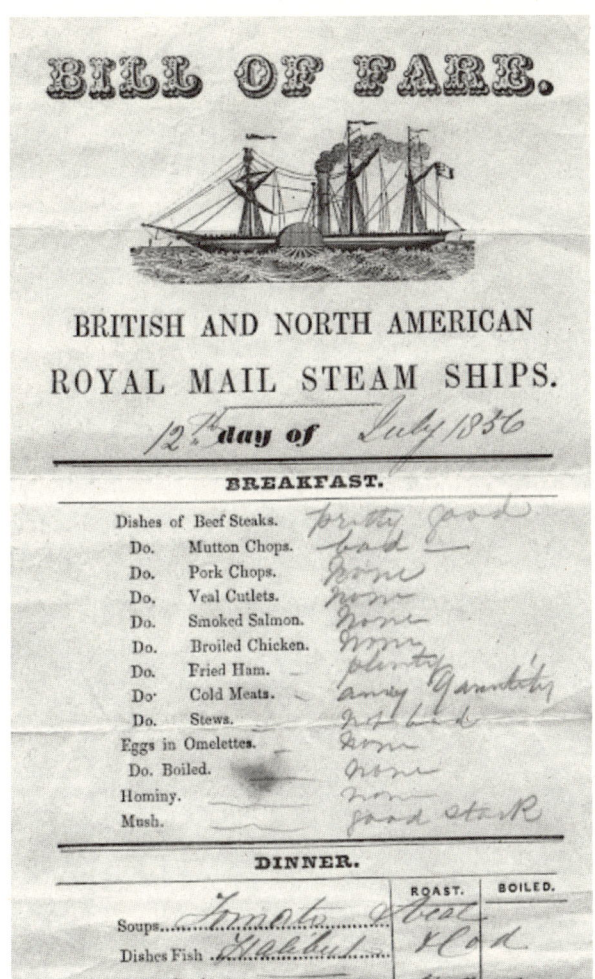

Figure 1. Henry Pease, Breakfast Bill of Fare. Courtesy of Henry Voigt Collection of American Menus.

Regulations' informed them that breakfast began at 'Half-past 8, and Cloths removed by Half-past 9'.[24] The language suggested a Navy frigate, not a passenger ship.

Strict schedules took precedence over creating an amiable environment. From twelve to one o'clock, luncheon was 'to be on the Table'. At half-past three o'clock, passengers dressed for dinner. Even dinner, the 'great event of the day, to be speculated upon before and criticized after', wrote Bird, often disappointed. Inevitably most dishes were roasted or boiled meats.[25] Nor was it leisurely, given that it must be 'on the Table at 4 – the Cloths to be removed the instant it is over'.[26] Again, that military tone. Why such speed? There were no appointments to keep, no theatre performances to attend, just monotonous waves.

In the early 1870s, *Boston Journal of Chemistry* editors wrote that it might seem 'presumptuous' to discuss improvements, but they insisted to Cunard that 'the maximum of safety can be secured, and yet a few more comforts and conveniences supplied', including 'better food'.[27] One editor complained that 'the rattling of knives and plates in the hands of the waiters is incessant from morning to night, and steaming "joints" and dishes of gravy fill the rooms with disgusting odors'. Yes, 'our American ideas of comfort differ essentially from those entertained by the English', the editors allowed, but it is unnecessary 'to subordinate *safety* to any other considerations'.[28]

Dining was an imposition to Cunard's 'Safety First' ethos; as Charles MacIver shruggingly wrote to a complaining passenger, 'going to sea was a hardship'; the company 'did not undertake to make anything else out of it'.[29] Meals interfered with running the ship, and expending money on lavish food and servants took money away from more important priorities. Nonetheless, change was imminent.

In 1879, George Augustus Sala sailed Cunard's *Scythia*. Once settled, he wrote about the marked contrast between this voyage and his previous one aboard Cunard's *Arabia* in 1863. Inevitably, Sala complained about *Arabia*'s food. Dinner was ruined by the 'monotony of the boiled mutton and caper sauce' and reminded Sala of 'the salubrious yet somewhat insipid diet on which, it is stated, Lindley Murray composed his English Grammar' while confined to his room as 'a chronic invalid', able to digest 'nothing more "choleric" in the way of meat than boiled mutton'.[30]

However, by 1879, Sala was delighted to see improvements aboard *Scythia*. Instead of monotonous meals, cabin-class passengers enjoyed fresh meats expertly prepared along with fresh produce, including celery, lettuces, beetroot, and cress; 'Thus, too, was it with the tomatoes, and the rich abundance of fruit provided at dessert.'[31] While the Collins Line was long gone, by 1879 if a transatlantic liner could not offer luxury food, dining, and spacious accommodations, it could not compete. Cunard had learned its lesson. What role might Collins have played in Cunard's transformation?

Collins Line: The American Alternative

Collins Line protocols could not have been more at odds with Cunard's. While Cunard asked its officers to '*recollect that [passengers] will value your services on deck looking after their safety more than talking with them in the saloons*', Collins hired dashing captains, requiring them to be sociable in the saloons.[32] While Cunard worried that stoking nationalism at sea could create unwelcome distractions, Collins played up patriotism to enhance its profit margin at Cunard's expense by claiming American exceptionalism.

Samuel Cunard and his partners deemed E.K. Collins their worst nightmare.[33] Cunard wrote Charles MacIver in May 1847, warning him that the Americans will match Cunard in improvements of speed, and they 'will be alive to every thing;' now the British will 'have National prejudices to contend with' as well.[34] Samuel Cunard understood how nationalism played into the Collins Line's competitive strategy, given how, during lobbying for subsidies, rhetorical flourishes pitted Americans against the British. Delaware Senator James Bayard thundered that Congress had better 'grant a carefully selected American shipping expert a completely free hand to proceed with the absolute conquest of this man Cunard!'[35] In November 1847 Congress awarded Collins $385,000 to help do just that.

Among maritime experts this history is well-known, including the complicated incident involving what Edward Sloan labelled a 'very secret cartel', where it was decided that all-out competition between Collins and Cunard would weaken them both, wasting essential money in rate wars. Hence, both companies fixed minimum rates for cargo and passengers, and also split revenues.[36] This cartel did not mean that the two companies melded into one entity. Outwardly, Collins and Cunard remained distinctive lines, and, to convince Congress and Parliament to continue subsidizing them, they promoted distinctive identities with Collins leveraging food, service, and luxury surroundings that reset wealthy passengers' expectations.

As Fox pointed out, Collins's steamships 'manifestly expressed' E.K. Collins's 'personality'.[37] In 1820s and 30s New York City, young Edward was exposed to high-quality American cuisine. By the time he was a shipping magnate, Collins maintained homes in New Orleans, St. Louis, and Long Island. His firm in Lower Manhattan at 73 South Street was a stone's throw from the fabulous Delmonico's at the intersection of Beaver and William Streets.[38] Delmonico's established the trend among fashionable New Yorkers of dining *à la carte,* and it quickly became the first choice among wealthy merchants.

New Orleans was likewise important to E.K. Collins's culinary exposure. By the early 1800s, the city achieved financial affluence, in part due to a port filled with cotton and food. 'Eating became [for New Orleanians] the key pleasure of both daily life and special occasions,' wrote historian Susan Tucker.[39] Attention given to dining aboard Collins's ships suggests that E.K. Collins had fixed expectations born of eating at establishments like New York's Delmonico's and Alciatore's in New Orleans.

Atlantic, Arctic, Baltic, and *Pacific* exuded American pride and bested Cunard's fleet in all respects – except safety. Along with faster ships, Collins excelled in décor, defined spaces, and passenger comfort. Instead of one saloon, Collins's ships offered dining and grand saloons. Women could retire to a drawing room near the grand saloon.[40]

Recounting his March 1852 voyage aboard the *Arctic*, S.C. Abbott described the dining saloon as 'a large, airy, beautiful room, sixty-two feet long and thirty feet wide, with windows opening upon the ocean as pleasantly as those of any parlor, and where two hundred guests can dine luxuriously'.[41] Most novel was a smoking room housed at the stern that 'communicates with the cabin below, so that after dinners, those passengers so disposed may, without the least exposure to the weather, or annoyance to their neighbors, enjoy the weed of old Virginia in perfection', wrote a *Chambers Edinburgh* journalist.[42]

Collins ensured passengers knew they were aboard an American ship. Saloon walls were 'decorated with paintings of the coats of arms of the various States of the Union'. They 'do exalt and honor our nation', Abbott enthused.[43] Royal Navy Captain Lauchlan MacKinnon's verdict carried significant weight: 'I strongly advise the builders of England to wake up from their lethargy, half composed of prejudice. I tell them again plainly, (however unpleasant to myself), that there are no ocean steamers in England comparable with the *Baltic*.'[44]

Amidst splendour, Collins's passengers enjoyed meals whose variety addressed their physical as well as emotional needs. Thurlow Weed, who sailed the *Baltic* in December 1851, was impressed with 'every conceivable comfort for the sick and every attainable luxury for those who are well'.[45] Collins's meals were also leisurely, allowing passengers pauses between courses to 'take a rest and a smoke', observed Edward Thomson.[46]

Correspondence in the Collins archive suggests how seriously the company took cuisine. William Ezra Bowman, senior partner of Brown Brothers & Company (which

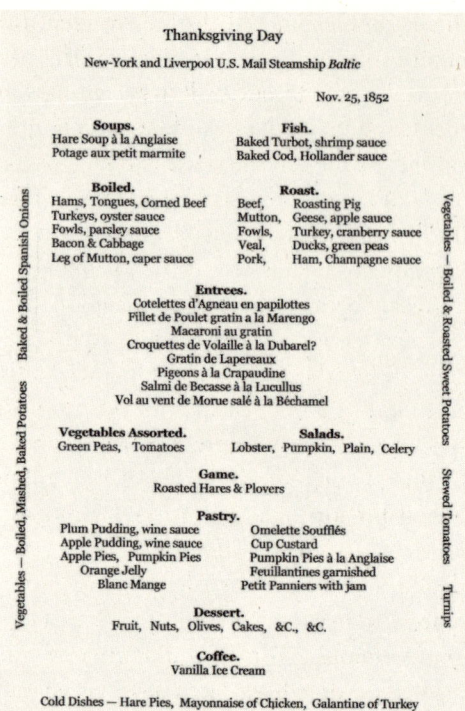

Figures 2 and 3. *The* Baltic *Thanksgiving Bill of Fare and its transcription further illustrate the lavishness of a Collins Line dinner. Courtesy of Henry Voigt Collection of American Menus.*

financed the Collins Line), wrote to William and James Brown about passenger complaints regarding food and service aboard an *Arctic* voyage from Liverpool to New York. 'We hear,' wrote Bowman, 'that on arrival in N.Y. the passengers had a meeting and sent in a protest to Mr Collins.' Bowman continued, 'I am told the fare was no better than to be had at a Common tavern.' In Liverpool and unable to verify the information, Bowman asked James Brown at the New York office to investigate.[47]

James Brown spoke with A. T. Stewart, a real estate magnate who had been on board the *Arctic* and had witnessed the problems. Stewart told Brown that bad weather made some passengers sick 'and of course sour':

> The tea was complained of [...]. The butter was complained of, this is the fault of the providers on the other side [Britain]. Some side dishes, Irish stews, etc. were not cooked well. The principal dishes Roast Beef etc. very good. This [poor food] can be accounted for from the fact that the main cook was sick & the Head Steward had to superintend the cooking, consequently he could not give his personal attention to the passengers as much as would have been desirable.[48]

Brown's prompt response to Bowman indicated genuine concern. Speaking on behalf

Cunard and Collins: Portable Food and Customer Service at Sea

of both Brown brothers, James promised that nothing would stop them from working harder to 'try to make things as comfortable and as sound as we can'.[49]

Captain MacKinnon was impressed enough to print his *Baltic* bill of fare from the night of 9 November 1852, giving us one of a handful of extant Collins artifacts. By means of contrast, Henry Pease's Bill of Fare from an unnamed Cunard Ship from July 1856 is to the right (with the Pastry category deleted):

Collins, *Baltic*
Bill of Fare, 9 November 1852

Soups.
Green turtle soup;
Potage aux choux

Fish.
Cod fish, stuffed and baked;
boiled bass, Hollander sauce.

Boiled.
Hams, tongues, cold corned beef,
turkeys, oyster sauce, fowls,
parsley sauce, leg of mutton,
caper sauce.

Roast.
Beef, veal, mutton, lamb, geese,
ham, champ, sauce, ducks,
pigs, turkeys, fowls.

Entrées
Maccaroni au gratin;
filet de pigeon au Cronstaidt [Cronstadt?];
croquette de poisson a la Richelieu;
salmi de canard sauvage;
poulet's pique sauce tomatoe;
cotellete de veau a la St. Gara;
fricandeau de torteu au petit pois;
d'oyois en cassi;
epegram de agneau sauce truffe.

Cunard [ship unspecified]
Bill of Fare, 7th July 1856

Soups.
Tomato
Gravy

Dishes.
Fish: Haddock and Cod
D[itt]o. Beef, roast
Do. Mutton & Caper sauce, roast
Do. Lamb & Mint, roast
Do. Veal & Bacon, roast
Do. Pork, Apple sauce & Beans
Do. Pigs [crossed out, inserted instead]: Beef a la Mode
Do. Turkeys Sausages & Oyster Sauce, Roast
Do. Geese & Gooseberry. Roast
Do. Ducks, 2 roasted
Do. Fowls. 4 roasted, 2 boiled
Do. Currie Fowls
Do. Stews[ed] Kidneys

Fricasse. Lobsters

Made Dishes.
Lamb Cutletts
Calves' Heads & Brain Sauce
[written in] Hams & Tongues,
[written in] Chicken Pies

Vegetables.
Green corn, green peas.

Salads.
Potato and plain.

Dessert.
Fruit, nuts, olives, cakes, etc. etc.
Coffee, lemonade (frozen).

Vegetables.
Assorted

As the Bills of Fare indicate, the variety and sophistication of Collins's offerings contrast with Cunard's plainer offerings.[50]

Cunard's boring choices (such as gravy soup), few complicated dishes, and last-minute substitutions stressed convenience, while Collins's expressed artistry. Take Calves' Heads and Brain Sauce, a ubiquitous Cunard "Made Dish" offered at least three times when Pease sailed: 4, 7 (above), and 12 July. It was pragmatic, given that brain sauce accompanied the calves' heads from which the brains had been removed. Meanwhile, the cook boiled tongues alongside heads and brains, serving it at dinner, breakfast, and luncheon the following day.

Using the term 'Made Dish' to characterize what Collins called an 'Entrée' was likely Samuel Cunard's personality writ large on the company ethos, his holdout against French fashion in favour of sturdy ships and a shipshape mentality. Perhaps Samuel assumed that British passengers, even the socially elite, would appreciate that same ethos and take pride that Cunard ships shunned Frenchified food, limiting 'Made Dish' selections to as few as two per dinner.

Collins took no issue with French food and hired a French *maître de cuisine* to oversee its Bills of Fare, budgeting $4000 a year per ship to pay each ship's galley staff.[51] As with fashionable restaurateurs, E.K. Collins understood an *entrée* as representative of the chef's sophistication and what elite passengers expected. Compare the sophistication of *Fricandeau de Torteu* and *Epigram d'Agneau Sauce Truffe* to Calves' Heads and Brain Sauce or Hams & Tongues, and we see how Collins leveraged cuisine. Steam enabled ships on a ten-day crossing to replicate the same dining customary on land, and Collins budgeted an unprecedented $15,000, 'to supply passengers and crew for each voyage to England and back', marvelled Captain MacKinnon.[52] By the 1850s, sizable numbers of passengers crossing the Atlantic chose Collins over Cunard 'as the dash and glamour – and occasional hint of rashness – of Collins's ships gained in popularity and appeal', wrote Butler.[53]

Conclusion

British journalists characterized both Collins and Americans as excessive. An 1859 *London Times* editorial concluded that Collins's failure resulted in part from Americans'

Cunard and Collins: Portable Food and Customer Service at Sea

hubris, stupidity, and greed.[54] Collins's cuisine was not mentioned, but editorials pointed to the foolhardiness of lavish expenditure on matters other than safety. Thus, it might appear that the moral of this story can be summed up neatly: speed, luxury, and a distracted crew signal disastrous consequence, while economy, military hierarchy, and vigilance signal reward. However, that moral remains tenuous at best. By the 1860s, numerous European and British competitors had entered the North Atlantic contest. In 1867, Thomas Henry Ismay, with financial backing from Gustav Schwabe, bought the Oceanic Steam Navigation Company, or White Star Line, and hired Edward J. Harland to design a new fleet. White Star invoked the American spirit and boldness that Collins had stood for, appointing Digby Murray, a former first officer on a Collins ship, as White Star commodore, 'thus embody[ing] the links between the two lines', wrote Fox.[55] Collins's and White Star's movement away from maritime traditions that privileged sailors and cargo over wealthy passengers is significant: it represented the direction steamship companies had to take to compete – including Cunard.

The real moral of the Collins Line was that its shipwrecks did not teach officials or passengers the importance of safety so much as it taught them that distraction at sea mattered profoundly, and the better the cuisine and surroundings, the more likely the line would be to profit. Ship biscuits, potable water, and lifeboats? Not much profit in that.

Notes

1. Daniel Allen Butler, *Age of Cunard: A Transatlantic History, 1839-2003* (Annapolis, MD: Lighthouse Press, 2003), pp. 82–83.
2. 'Notes in Regard to Safety', *Hunt's Merchants Magazine*, 32 (1855), 379–80.
3. David Shaw, *Sea Shall Embrace Them: The Tragic Story of the Steamship* Arctic (New York: Free Press, 2022), p. 204.
4. Alexander Crosby Brown, *Women and Children Last: Loss of the Steamship Arctic* (New York: G. P. Putnam, 1961), p. 123.
5. Brown, p. 147.
6. Qtd. in Brown, p. 148.
7. 'Life-Boats Ordered for the Collins Line', *New York Daily Times*, 3 November 1854 <https://www.proquest.com/historical-newspapers/life-boats-ordered-collins-line/docview/95837907/se-2?accountid=2200> [accessed 25 May 2022].
8. French Ensor Chadwick and others, *Ocean Steamships* (London: Murray, 1892), p. 120.
9. Alex W. Roland, Jeffrey Bolster, and Alexander Keyssar, *The Way of the Ship* (Hoboken, NJ: Wiley, 2008), pp. 159–60; William M. Fowler, Jr., *Steam Titans: Cunard, Collins, and the Epic Battle for Commerce on the North Atlantic* (New York: Bloomsbury, 2017), pp. 54, 76.
10. Roland, Bolster, and Keyssar, p. 71.
11. Fowler, p. 81; Stephen Fox, *Ocean Railway: Isambard Kingdom Brunel, Samuel Cunard, and the Revolutionary World of the Great Atlantic Steamships* (London: Harper Perennial, 2004), pp. 46–47.
12. Fowler, p. 80.
13. Fowler, p. 128.
14. Qtd. in Fox, p. 89.
15. Alan Villiers, *Western Ocean: The Story of the North Atlantic* (London, Museum Press, 1957), pp. 220–21; William Shaw Lindsay, *History of Merchant Shipping and Ancient Commerce*, 4 vols (London: Low, Marston, Low, and Searle, 1876) IV, pp. 242–44.
16. 'The Cunard Steamers', *Boston Journal of Chemistry*, 8 (1873), 19.

17 Lindsay, IV, p. 243.
18 Charles Dickens, *American Notes for General Circulation*, 2nd edn, 2 vols (London: Chapman and Hall, 1842), I, pp. 21, 10–11, 42.
19 Qtd. in Fox, p. 208.
20 Fox, p. 208.
21 Haws, pp. 23, 9, 25; T.W.E. Roche, *Samuel Cunard and the North Atlantic* (London: Macdonald, 1971), p. 17.
22 Harriet Beecher Stowe, *Sunny Memories of Foreign Lands,* 2 vols (Boston: Phillips, Sampson, and Co., 1854), I, pp. 8–9, 1, 10.
23 Isabella Lucy Bird, *Englishwoman in America*, 2nd edn (London: John Murray, 1856), pp. 456, 9, 450.
24 'Rules and Regulations', *Official Guide and Album of the Cunard Steamship Company* (1840; revised ed. London: Sutton Sharpe, 1877), pp. 41–43.
25 Bird, p. 450.
26 'Rules and Regulations', pp. 40–43.
27 'The Cunard Steamers', p. 9.
28 'Atlantic Ferry', *Boston Journal of Chemistry*, 8 (1872), p. 55.
29 Qtd. in Francis E. Hyde, *Cunard and the North Atlantic, 1840–1973* (London: Macmillan, 1975), p. 75.
30 George Augustus Sala, *America Revisited*, 6th edn (London: Vizetelly, 1886), pp. 2–3.
31 Sala, pp. 10–11.
32 Lindsay, IV, p. 243; Fowler, p. 37.
33 Fox, p. 116.
34 Qtd in Edward W. Sloan, 'The First (and Very Secret) International Steamship Cartel, 1850–1856', in *Global Markets: Internationalization of the Sea Transport Industries Since 1850*, ed. by David J. Starkey and Gelina Harlaftis (Liverpool: University of Liverpool Press, 2017), pp. 29–52 (p. 32).
35 Qtd. in Butler, p. 73.
36 Sloan, pp. 38–39.
37 Fox, p. 120.
38 Walter Barrett, *Old Merchants of New York City* (New York: Carlton, 1862), p. 141.
39 Susan Tucker, ed. *New Orleans Cuisine* (Jackson, Mississippi: University Press of Mississippi, 2009), p. 6.
40 'Steam Bridge of the Atlantic', *Harper's New Monthly*, 1 (1850), 411–14 (p. 413).
41 S. C. Abbott, 'Ocean Life', *Harper's New Monthly*, 5 (1852), 61–66 (p. 62).
42 'Steam Bridge of the Atlantic', p. 413.
43 Abbott, p. 62.
44 [Lauchan] MacKinnon, 'Growth of Towns in the Far West of America', *Colburn's United Service Magazine*, 295 (1853), 246–55 (p. 253).
45 Thurlow Weed, *Letters from Europe and the West Indies* (Albany, NY: Weed, Parsons & Co, 1866), p. 409.
46 Edward Thomson, *Letters from Europe* (Cincinnati: L. Swormstedt and A. Poe, 1856), p. 30.
47 New York, New York Historical Society, Letter, 9 September 1852, Brown to BBH, Collins Line, Brown Correspondence, 1843-1852.
48 Letter, 9 September 1852, Brown to BBH.
49 New York, New York Historical Society, Letter, 15 September 15, 1851, Brown Brothers & Co. to William E. Bowen, Collins Line, Brown Correspondence, 1843-1852.
50 MacKinnon, p. 254; *Baltic* Thanksgiving Bill of Fare and Cunard Bill of Fare, Henry Voigt Collection of American Menus, Wilmington, DE.
51 Abbott, p. 62; 'Steam-Bridge of the Atlantic', p. 409.
52 MacKinnon, p. 250.
53 Butler, p. 78.
54 'America Is Weeping for Her Steamers', *London Times*, 18 May 1859, p. 8.
55 Fox, p. 241.

From 'Peanut Weddings' to 'Beef Stands': The Socio-Culinary History of Chicago's 'Italian Beef'

Anthony F. Buccini

'Italian beef', or simply 'beef', is a dish made throughout Chicago and its suburbs but hardly found elsewhere in the US. It thus represents a particular specialty of the area and, as it has been celebrated in national popular food media in recent decades, Italian beef has become one of a small set of dishes which for many Chicagoans is associated strongly with local identity and civic pride. For tourists, it is considered one of the city's culinary 'musts'. Nowadays, this preparation is primarily a commercial food, sold in a wide array of informal restaurants and fast-food businesses, including local chains, but it is considered best prepared by small establishments known as 'beef stands', which specialize in this product and are often still Italian-American owned. Domestic preparation is still carried out in some families, especially those of Italian origin and with ties to the Taylor Street neighbourhood of the city, for whom the homemade version is generally reserved for festive occasions.[1]

As for the dish itself, Italian beef is always served as a sandwich, of which the filling is a heap of thinly sliced roasted beef. At first blush then, it appears to be just a take on the roast beef sandwich found throughout the US; consequently food writers all seem to assume that Chicago's specialty arose simply as an Italianized version of the Anglo-American mainstream's dish, and, as such, its 'invention' should be attributable to some individual associated with one of the original commercial producers. In this paper, I show this line of thinking to be wrong, for Chicago's beef sandwich is quite distinct from its mainstream analogues in multiple aspects, including the cooking method, seasoning, manner of serving, and canonical condiments, and the dish's origins predate its commercialization as fast-food: Italian beef was a festive dish prepared by Taylor Street's working-class Italian-American families, a preparation most likely popularized by the neighbourhood's strong contingent of people from Naples and the surrounding towns who brought the recipe with them as part of their traditional cuisine. This view is confirmed by the attestation of an unambiguously clear antecedent of the Taylor Street recipe in a mid-nineteenth century work on popular Neapolitan cookery, recorded long before mass immigration brought Italians to Chicago and reflecting a long-standing method of cooking beef in southern Italy.

Beef, Dipped, Hot and Sweet...

To prepare the meat for Italian beef, one selects a large cut, typically a relatively inexpensive top or bottom round, which is partially trimmed of fat. There are multiple acceptable variations on the cooking method. Some cooks apply dry seasoning to the meat before browning it in a large pan with olive oil; others forego the browning and go directly to placing the seasoned meat in an oven; still others marinate the meat for a long time in a heavily seasoned broth before roasting in the oven. In all cases, it is absolutely essential that, when the meat goes into the oven, there is a substantial amount of seasoned broth in the roasting pan, as this liquid provides the ample amounts of 'gravy' that is a central feature of the dish.

Another key stage in the preparation of Italian beef comes at the end of the oven roasting: both the meat and the gravy are thoroughly cooled, which allows the cook to skim excess fat from the gravy and, more importantly, to slice the cooked meat as thinly as possible. Commercial preparations use large electric slicers, while domestic cooks must do as best they can with a knife. The gravy is then reheated and, preferably just before serving, the sliced beef is bathed in the gravy long enough for it to get thoroughly drenched and hot but not left so long as to break down.

Of central importance to the character of the dish is, of course, the seasoning, which always includes aromatic vegetables, especially of the allium family, and spices, though in the commercial setting there is a lamentable tendency to rely on powdered onion and garlic rather than actual fresh ingredients. In addition to garlic, onion, salt and pepper, both hot red chilli (*peperoncini*), usually in flaked form, and dried oregano are universally used in commercial kitchens, but many purveyors employ additional seasonings, and some regard this aspect of their recipes as a closely guarded secret. Some of these flavourings that I have encountered are rosemary, marjoram, fennel seeds, and paprika.

In addition, one finds a less common use of what Americans think of as 'sweet spices', something which I believe is an archaism and, in the commercial sphere, is a distinctive and much commented upon practice of the oldest continually operating beef stand in the Taylor Street neighbourhood, *Al's #1 Italian Beef*. Al's secret seasoning is thought by some to be allspice, but in my judgement it is surely a combination of cloves and cinnamon, perhaps also nutmeg, and, of course, the flavour combination of these three spices together with black pepper is what gave rise to the name 'allspice', a single spice originally produced only in Jamaica and generally alien to traditional southern Italian cookery.

The final preparation of Italian beef sandwiches naturally involves the union of bread, meat, and gravy. The bread for the sandwiches is traditionally a long loaf of a particular sort, called in much of America 'Italian bread' but referred to by Chicago's Italian bakers as 'French bread'; it is cut in sections and opened 'book' style. There are three options for the individual eater to choose: 1) 'dry', with the meat taken from the gravy and briefly drained before placement on the bread; 2) 'wet', with the meat placed on the bread without draining and with additional gravy spooned onto the sandwich;

3) 'dipped', with the composed sandwich entirely dunked for a moment or two into the gravy. Obviously, the bread of wet and especially dipped sandwiches becomes quite soft and the dripping gravy can make quite a mess.

The messy nature of an Italian beef sandwich is usually augmented by the two common additions. One can request the sandwich to be 'sweet', which indicates that roasted or fried ('sweet') bell peppers should be placed atop the meat; instead or in addition, one can request the sandwich to be 'hot', which calls for the addition of a style of *giardiniera* particular to Chicago's Italian-American community. This *giardiniera* is made by taking pickled vegetables, here typically cauliflower, celery, carrot and hot chilli peppers, mincing them and then further preserving them in olive oil, producing what is often a very piquant condiment. Thus, at a beef stand one might order as follows: 'a beef, dry and hot' or 'a beef, dipped, sweet and hot'. The best beef stands prepare these condiments in-house.

Variations on the basic Italian beef sandwich are several and include what is known as a 'combo', which combines the two main products of an old-fashioned beef stand, namely links of grilled southern Italian-style sausage, flavoured with fennel, and the roasted beef. 'Cheesy beef' is surely a more recent variant, with sliced low-moisture mozzarella or provolone topping the sandwich, reflecting the recent American predilection for adding cheese to almost everything. Far more interesting are the two inexpensive variants which do not contain any actual meat. These are known as 'gravy bread', which is simply the bread moistened or soaked with gravy, and the 'potato sandwich', which is gravy bread filled with French fries.

It should be noted that good beef stands typically produce excellent French fries, which are made with freshly cut potatoes and are twice-fried in the old Belgian fashion; they constitute the normal accompaniment to an Italian beef or combo.

Home Kitchens, Bakeries, and Taylor Street's Peanut Weddings

As mentioned above, food writers who have considered the origins of Italian beef all seem to assume that the dish arose in the States as a sort of peculiar Italian take on the Anglo-American roast beef sandwich. This assumption is manifested in the inclination, so common in popular food writing, to attribute the origins of recipes to an identifiable individual, an 'inventor' of the dish, and in this effort multiple owners of beef-businesses have been happy to cooperate, claiming that an ancestor of theirs, a couple of generations back, who founded their family business, was responsible for creating the dish.

Beyond such questionable claims, a couple of things are indisputable. First, the association of Italian beef with Italians is clear enough, from elements of the dish itself, from the location of the earliest beef stands in the once predominantly Italian Taylor Street neighbourhood, and of course from the name of the dish. Though the names and menus of hole-in-the-wall establishments such as beef stands are often without surviving written records – and, indeed, some, long-closed, exist only in the memories of old Taylor Street residents – it seems clear that the first such places arose in the years around the Second World War. Some local residents also insist that Italian beef appeared first as a commercial offering in existing small shops whose original and primary *raison d'être* was the making of sausage sandwiches, that is, both Italian sausage sandwiches and the particular Italian take on the Chicago-style hot dog.

That claims of the invention of the dish by forbears of one or the other owner of a surviving beef business are at best spurious is made clear by the occasional references in autobiographical and anecdotal documents related to the Taylor Street neighbourhood of an old local institution known as the 'peanut wedding', something which drew my attention many years ago because of its similarity to a parallel in my native New York area, the 'football wedding'. These jocular terms both refer to a style of wedding celebration once common among poorer, working-class Italian-American families in the early to mid-twentieth century, characterized in part by the humble fare served at the receptions. In New York, the main food was Italian submarine sandwiches, wrapped in paper, which could be tossed to guests like footballs. For Chicago's peanut weddings, the local folk etymology claims that roasted peanuts were featured, which is possible but not necessarily the full explanation of the name: 'peanut' here may have been more a jocular reference to the overall humbleness of these events, where, in fact, the main fare served was sandwiches of roasted beef with gravy, that is, Italian beef.

Some sixteen years ago, I became good friends with a baker, the late Frank Masi, proprietor of the last of the old neighbourhood Italian bakeries of Taylor Street. In exchange for helping out in his shop and filling in for absent employees, Frank taught me how to make all the traditional baked goods featured in his 'Italian Superior Bakery' – various kinds of bread, pan-pizza, *freselle, taralli* – and over the years we had many conversations about our shared Neapolitan foodways and about Taylor Street in the old days. Frank's parents came to the US from a small town just northeast of Naples in 1912 and opened their own bakery in Chicago in 1926, where Frank started working as a small child alongside his siblings. One of the most interesting and important parts of the history of Italian beef, something hitherto unnoted in any discussions of the dish, I learned from him.

According to Frank, for decades the bakery regularly was involved in the production of Italian beef for large social gatherings for the local Italian-American community, including events such as weddings and anniversaries, but did so in a now curiously old-fashioned way: the family or families organizing the events would prepare the beef in their home kitchens and bring the meat in oven pans for slow roasting in the bakery's large oven. Such events were good business for the bakery, which would be paid for the roasting and then also supply large numbers of long loaves of their 'French bread', with the beef and bread ultimately served as beef sandwiches at the event, which might be held in a large, rented venue or at a home.

Though today Italian beef is commonly thought of as a fast food, bought and consumed on the go for a quick meal, the dish remains for many Chicagoans, and especially those of Italian descent, also a quintessentially festive dish for large informal gatherings. Though the 'peanut wedding' is now a fading memory, Italian beef remains extremely popular for all sorts of social events, such as office parties, birthday celebrations, watching the Bears or other Chicago sports teams in important games, etc. If the gathering is not too large, some – especially Italian-American families – might well make their beef in the home kitchen, but virtually all commercial kitchens producing Italian beef offer both small and very large catering kits of beef, gravy, bread, peppers, and *giardiniera*, packed and ready for final assembly wherever customers care to have their celebration. Local grocery stores also commonly offer tubs of prepared beef and gravy for customers who can then purchase separately *giardiniera*, roasted peppers, and, of course, appropriate loaves of bread.

From Naples to Taylor Street

Despite its key role as a source of Italian influences on mainstream American cuisine and thence on global culinary trends, Italian-American cookery is remarkably misunderstood by both scholarly and popular food writers, who routinely confuse the highly adapted 'Italianoid' dishes of commercial kitchens and pseudo-Italian television chefs, aimed to satisfy the tastes and meal-structures of non-Italians, with the actual cookery of Italian-Americans.[2] This genuine style of cookery is, like the heritage language varieties of the old Italian-American communities, now moribund and in many places quite dead, as the process of cultural assimilation has transformed actual Italian-Americans – that is, Italian immigrants and their descendants who maintained to a noteworthy degree old world cultural traditions – into simply Americans of Italian descent, who perhaps still bear Italian names but are from a cultural standpoint thoroughly American. Members of this latter, fully assimilated group might develop a nostalgic appreciation of their heritage, but they cannot undo the break in the generational transfer of linguistic and culinary tradition, things only acquired through long exposure in an intimate, culturally homogeneous, and closed environment. Italian-American family and community structures long provided such an environment, from the period of the great diaspora, *c.* 1880-1924, up through the Second World War, but then, gradually at the community level but abruptly at the individual level, these structures began to break down. The ethnic mixing of the war itself, post-war educational and economic opportunities, and the urban upheavals and mass movement into ethnically mixed suburbs spread Italian-American influence into mainstream society but at the same time began the dissolution of both culturally tight-knit Italian-American families and communities, and ultimately led to the death of Italian-American linguistic and culinary traditions.

Chicago's Taylor Street neighbourhood started in the late nineteenth century as an ethnically mixed immigrant slum on the south edge of downtown, but in time its Italian population expanded westward in a narrow band around Taylor and Polk Streets as older ethnic groups moved out. By the 1920s, the core of the neighbourhood, stretching from around Halsted Street to Western Avenue, was predominantly working-class Italian and, though these Italians were of mixed regional backgrounds, the most prominent cultural group was from Campania, including Naples, with a strong contingent from the small city of Acerra and neighbouring towns just to the northeast of Naples itself. The Taylor Street neighbourhood served as precisely the sort of place where family and community structures were dense, supporting the maintenance of Old World cultural traditions to a considerable degree, up until a series of targeted urban 'renewal' projects replaced large swathes of residential properties with public institutions in the 1950s and especially the 1960s, forcing much of the Italian population to flee to various suburbs.

Italian beef, starting not as a commercial fast food but as a festive dish for working-class families, is a remarkable cultural relic from when Taylor Street was a poor Italian enclave with a strong Neapolitan cultural element. To understand the dish's history, its socio-economic and socio-culinary aspects are of paramount importance.

As the central point of a productive agricultural zone and capital of one of Europe's largest kingdoms for many centuries, Naples has a particularly rich culinary culture, which includes a style of elite cookery, with its international connexions, and also a traditional, non-elite style, firmly rooted in local food production but obviously open to influences from higher up the socio-economic scale, influences which logically would be manifested almost exclusively in the festive cookery of the non-elite.

If one considers the popular cookery of Naples and more generally of the *Mezzogiorno*, one sees that meat and especially beef plays a limited role, a fact which there, as elsewhere in Europe, reflects the high cost of meat and the poverty of the southern Italian masses until recent times. For most of these people, large pieces of muscle meats were at most to be enjoyed on special occasions, with meat's protein, fat, and flavour obtained more routinely in the form of organ meats, some of which were available as urban street foods. Here one thinks of *carnecotta*, beef offal and other less desirable parts of butchered animals, cooked so as to produce a great deal of broth, or *zuffritto* a.k.a. *zuppa forte*, an intensely seasoned stew made from pig's pluck; one could purchase just the broth or sauce with bread or, if one could afford it, also a portion of the cooked meats. Indeed, the vast majority of traditional southern Italian recipes for muscle meats have two central features: first, they typically involve preparations and cooking methods that render tough cuts of meat quite tender, and second, they all stretch the nutritional and flavour value of the meat by producing ample amounts of a broth or a sauce to be enjoyed with bread and pasta, the local staple foods. The two most prized meat preparations of Neapolitan popular cookery, the famous *ragù alla napoletana* and the less well-known but equally delicious *carne alla genovese* both illustrate the point. These dishes are decidedly festive and even in today's more prosperous times maintain an elevated status as the focus of Sunday or holiday meals. In both cases they yield delicious sauces (the tomato-based *ragù* sauce, the primarily onion *genovese*) which are used to dress a first course of special forms of pasta, with the meat(s) served separately as a second course.

With this in mind, let us consider the following recipe which appeared in one of the editions (1841) of the extraordinary book on Neapolitan cookery written by Ippolito Cavalcanti, a high nobleman keenly interested in his native city's cookery, both elite and non-elite. Most of the work describes the former and is written in Italian, but he also includes a section of popular, home-style cooking of the broader population which he wrote in the Neapolitan language. The recipe, simply called *Stufato*, opens with a brief discussion of appropriate top and bottom round cuts for the dish, and then continues:

Miettarraje dinto a nu tiano na fella de lardo pesato, na cepolla fellata, e ncoppa po nce miette la carne cu lo ssale, pepe, tutte spiezie, nu spicolo d'aglio, si te piace; lu farraje zuffrijere buono buono, e ogne ntanto nce mietarraje nu poco d'acqua, votanno sempe: quanno vide ca la carne s'è ffata rossa rossa, nce mietarraje l'acqua pe ffà lu brodo, che te pò servì pe li maccarune, pe la pasta menutola, pe na zuppa e pe nzò che buò.

You'll put into a casserole a slice of pounded *lardo*, a sliced onion, and on top then put the meat with salt, pepper, *all spices* [emphasis added], a clove of garlic if you like; gently fry it well, and every now and again you'll add a little water, always mixing; when you see the meat is well browned, you'll add water to make the broth, which can be served with macaroni, with tiny pasta, as soup, and for whatever you want.³

Very similar recipes for substantial pieces of meat cooked with liquid are known throughout Italy and beyond under different names – *stracotto, brasato, fricandò* – and this kind of dish often includes the sweet spices of cloves, cinnamon, nutmeg, alongside black pepper. Cavalcanti's inclusion of *tutte spiezie*, normally written together in Neapolitan lexicographical works as *tuttaspiezie*, surely refers to the combination of the three 'sweet' spices ground, a mélange that was available commercially in shops of the time, a practice also found in France, where the mélange was referred to as *tout-épice* or *quatre-épices*. Note too that in Cavalcanti's Neapolitan language recipe for *la genovese* of 1852, he includes as an ingredient *tutta spiezie* (p. 431). The use of these spices by Naples's lower classes was clearly reserved for special occasions.

In considering the relationship between Cavalcanti's *stufato* recipe and the variants which in Chicago's domestic and commercial cookery must all be considered genuine realizations of 'Italian beef', we note the following. Regarding the essential primary ingredients – top or bottom round of beef, water, onion, garlic – and the nature of the final product – a large, fully cooked piece of meat intended for slicing and copious amounts of flavourful liquid (*brodo* or 'gravy') to dress a starch, the *stufato* recipe and Italian beef recipes are identical. As for the cooking method, Cavalcanti's dish calls for browning the meat with the onion (and optional garlic) and gradually adding water with a final addition of a large amount of water before long, slow cooking on the stovetop. In Chicago, variation allows for optional browning and, though oven-roasting after the addition of the large volume of liquid is nowadays the norm, it seems quite likely that in the early days of the Taylor Street neighbourhood, when surely not all dwellings were equipped with ovens, the stovetop method was widespread; in any event, given the robust seasoning, the difference in the final taste profiles using the two finishing methods is small.

As for the seasoning, we call attention to the following. First, Cavalcanti suggests garlic as an option – *si te piace* 'if you like' – whereas in Chicago's beef recipes garlic seems now obligatory. and in many recipes, including all commercial versions, garlic is prominent, which accords with the rising use of this ingredient in Italian-American and especially pseudo-Italian cookery in the United States. Along similar lines, while Cavalcanti's *stufato* includes neither dried oregano nor *peperoncini*, these additions must now be viewed as canonical elements of Italian beef and perhaps represent accretions in the (Italian-)American culinary context, just as the increased use of garlic might be. Yet, we must also bear in mind that the culinary sensibilities of Cavalcanti himself, a nobleman, may have guided his formulation of the *stufato* recipe, and that his lower class contemporaries in the Naples region may have been adding to this dish more garlic, oregano or other herbs, and *peperoncini* from the garden for some piquancy in place of substantial amounts of store-bought black pepper. Finally, while today most domestic and commercial recipes for Italian beef eschew the addition of sweet spices, it is surely not coincidental that the oldest of Taylor Street's beef stands still employs them. Omission of these spices may, however, be an old (pre-emigration) practice among the poor.

All in all, Chicago's Italian beef is in essence the same dish as Cavalcanti's *stufato*, and the ways in which Taylor Street's versions deviate from the recipe recorded by the Duke of Buonvicino may reflect changes of seasoning brought about in the New World or they may simply continue practices of non-elite cookery in Naples and other towns surrounding Mount Vesuvius, practices which Cavalcanti was unaware of or rejected as unappealing.

Italian Beef in Broader Historical Context

This last point gives us cause to consider, albeit here only briefly, the intended audience of Cavalcanti's masterpiece. The literacy rate in the mid-nineteenth century Kingdom of the Two Sicilies (the *Regno*) was low, the vast majority of the population only poorly, if at all, educated and speaking habitually a local dialect of Neapolitan, Sicilian, etc. However, Naples itself was the capital of the Regno with a sizable educated population, comprised of noble families, clerics, elite bankers, merchants, etc., as well as a middle class of merchants, professionals, bureaucrats, etc., who were fully competent in Italian; a significant proportion of them were also able to speak Neapolitan. It is surely these elite and bourgeois sectors of local society which were Cavalcanti's primary target audience. His inclusion of the section in Neapolitan of local non-elite home-cooking (*cucina casareccia*) appears to reflect a desire to preserve and share this aspect of popular culture with the Italianising upper echelons of Neapolitan society and with interested readers from elsewhere in the Regno and beyond.

The Socio-Culinary History of Chicago's 'Italian Beef'

The accuracy of Cavalcanti's portrayal of popular Neapolitan cookery – a traditional cuisine passed down from generation to generation in intimate settings and most assuredly not through cookbooks – cannot be disputed; from my own standpoint, the aesthetics, methods, and many individual recipes accord completely with the cookery of my (non-elite) Campanian grandparents who emigrated to New Jersey around 1900. It would be absurd to think that the Neapolitans and other Campanians of Taylor Street did not bring this same cuisine to Chicago.

The *stufato* recipe that gave rise to Chicago's Italian beef must then be regarded as one of the traditional dishes of popular Neapolitan cookery and yet one may reasonably wonder why we do not find continuations of this preparation in other communities with Neapolitan culinary backgrounds. However, we do find this recipe widely diffused – but primarily, perhaps exclusively, in a variant form in which tomatoes replace the large dose of water for slow cooking the beef. It appears that by the nineteenth century, and surely already much earlier, this alternate version was winning out in popularity over the old, tomato-less *stufato*. There can be no doubt that this expanding use of tomato was much older in the cookery of non-elite urban and rustic circles, and in Cavalcanti's time we see it well established in the bourgeois and even elite cookery of the Regno's capital. Indeed, the recipe cited above, from the 1841 (third) edition of Cavalcanti's cookbook, and the essentially identical, tomato-less *stufato* recipe in the seventh edition of 1852 (p. 448) must be considered alongside the *stufato* recipe which appeared in the second edition of 1839 (p. 367-68) and which is identical to the others not only in name but in all aspects of preparation, including the addition of *tutta spiezie*, up until the addition of a large quantity of water, here replaced by fresh tomatoes (if in season) or *conserva de pommadore*. From an Italian-American perspective, the *stufato* with tomato described by Cavalcanti corresponds perfectly to what in my experience was commonly referred to as 'meat sauce' or 'gravy', i.e. *sugo di carne*, a typical Sunday preparation to be served with a special form of pasta.[4]

Thus, the Italian beef recipe reflects the older, 'pre-Columbian' Neapolitan *stufato* which, especially as a source for a sauce to dress pasta, was giving way to the post-Columbian version with tomato. We might conjecture that the reason this older recipe not only survived in Chicago but ultimately became established as an iconic local dish is because of its particular suitability to the institution of the peanut wedding. Home-catered in a rented venue, this form of celebration was not well suited to the serving of *maccarune e carne*, the classic celebratory meal of non-elite southern Italians, for the serving of pasta would have necessitated a further stage of cooking of the pasta and also the use of plates and utensils at the venue. The older *stufato* recipe, however, still offered a festive food – muscle meat seasoned with spices – but in a far more easily transported and served format. Once this style of celebration became established in the Taylor Street neighbourhood, the stage was set for the eventual commercialization of Italian beef alongside other Italian sandwich-style fast foods.

The relationship between the Neapolitan *stufato* and Chicago's Italian beef is clear, and while the family behind Al's beef stand can hardly be said to have invented the dish, they are to be commended for maintaining the old manner of seasoning that is otherwise receding not only from the commercial beef business but from local domestic cookery as well. The common assumption that Italian beef is in origin an Italian take on the Anglo-American roast beef sandwich is proved wrong not only by the very different manner of traditional seasoning but also by the central importance of the production of a large amount of thin gravy to serve, as Cavalcanti says, *pe nzò che buò* 'for whatever you want', which in Chicago – and probably already in Naples and Acerra – included flavouring large pieces of bread. Finally, we must note the socio-culinary role of Italian beef, which, though now regarded primarily as a portable fast-food, still retains its festive association as a food appropriate for large social gatherings, even ones of considerable importance. In this way, Chicago's Italian beef evokes the special place that such meat dishes held in the cuisine of the non-elite classes of Naples and more generally of the Mezzogiorno who were driven by poverty to emigrate to America where, at least in Chicago, they celebrated their humble peanut weddings with gravy-drenched beef sandwiches.

Notes

1. Aspects of this paper have been informed by decades of conversations with lifelong residents of Chicago's Taylor Street neighbourhood; special thanks to the late Frank Masi, Michael DiCosola, Joseph Assenato, Freddy Mancini, and the late Frankie 'Moon'. Further thanks to Zachary Nowak, Ernest Buccini, and Amy Dahlstrom. Appropriate disclaimers apply.
2. For further discussion of my heretical views on Italian-American cookery, see Buccini 2015 and 2021.
3. This recipe appears in the Neapolitan section of the 1841 (third) edition, reprinted in 2005: 22.
4. Of interest here are some recipes appearing in the Italian cookbook for Americans by Gentile (1919), who seems to have been an immigrant but of neither working-class nor southern-Italian origin. First, we note her recipe for 'Brown Stock (*Sugo di Carne*)' (p. 15), almost identical to Cavalcanti's tomato-less *stufato* and seasoned with cloves. Second, in a section presenting several *stufato* recipes there appears a *stufato alla francese* (p. 119), again very close to Cavalcanti's tomato-less version with both cloves and cinnamon (though also pieces of 'ham' and 'bacon').

References

Buccini, Anthony F. 2015. 'Italy', in *Ethnic American Food Today: A Cultural Encyclopedia*, ed. by Lucy Long (Lanham, MD: Rowman & Little), pp. 314–23.

———. 2021. 'Prejudice, Assimilation and Profit: The Peculiar History of Italian Cookery in the United States', in *Food, Social Change and Identity*, ed. by Cynthia Chou and Suzanne Kerner (n.p.: Palgrave Macmillan), pp. 73–90.

Cavalcanti, Ippolito. 1839. *Cucina teorico-pratica*, 2nd edn (Naples: G. Palma).

———. 1852. *Cucina teorico-pratica*, 7th edn (Naples: Domenico Capasso).

———. 2005. *Cucina casareccia in lingua napoletana*, ed. by Raffaele La Capria (Milano: Polifilo). [Excerpted from the third (1841) edition of *Cucina teorico-pratica*.]

Gentile, Maria. 1919. *The Italian Cookbook. The Art of Eating Well* (New York: Italian Book Co.; repr. London: Forgotten Books, 2012).

Achar: An Unexpected Stowaway on the Ships of the Dutch East India Company

Kathleen Burke

The Dutch East India Company founded a maritime empire in the Indian Ocean because of food, intent on monopolizing the supply of plant parts into Europe, called 'spices' by early modern Europeans. At the time, these 'spices' were cultivated almost exclusively in a handful of geographies in the Indian Ocean: nutmeg, mace, and cloves from the eastern Indonesian archipelago; cinnamon from Sri Lanka; and pepper from Sumatra and the Malabar coast of India. Despite these Indian Ocean geographies, most of the scholarly literature on 'spices' has focused on early modern Europe. Surprisingly few studies have examined the colonial food cultures that emerged in Company settlements in the Indian Ocean, reflecting disconnections between, on the one hand, gender and food history, and on the other the historiographical concerns of economic, and later global, history. This paper attempts to bridge some of these gaps by arguing that food matters to our understanding of the Dutch Company's empire in the Indian Ocean. It does so by examining how European colonial households reproduced a distinct series of cooking, eating, and drinking patterns on board ships as they travelled between Company ports in the Indian Ocean.

This paper argues that European colonial households sought to reproduce a novel cuisine on board ships, global Batavian cuisine, which had originated in these households in Batavia in the late seventeenth century. Batavia was the Company's headquarters, located in present-day Jakarta, Indonesia. The contours of this cuisine were produced in the negotiation of two principal actors: the 'mestiza' women partners of European men, and enslaved male cooks from India and Indonesia, who had been uprooted from their homelands and forcibly relocated to Batavia. Mestiza was a category of colonial governance, meaning 'mixed' woman, and, as such, framed from the gendered and racialized perspective of European men, who were the main migrants to the empire because of restrictive gendered ideologies that the Company used in its recruitment policies. This unequal exchange of knowledge resulted in new sets of culinary combinations, ingredients, techniques, and material culture which had never before existed in the same culinary context. It eschews any simplistic and linear categorizations based on geographical provenance or biological and cultural essentialism, but reveals the creative appropriation of food resources, labour, and knowledge that were assembled from across the Indian Ocean to create global Batavian cuisine in European colonial households.

By analyzing the movement of achar, an iconic dish of global Batavian cuisine, I argue that elite colonial households sought to reproduce global Batavian cuisine on board Company ships. Second, I argue that achar was characterized by a porosity of the boundaries between land and sea, drawing on Michael Pearson's concept of littoral societies.[1] This was facilitated by three main factors: common techniques for the preservation of food on land and at sea, including its preparation, storage, and transport; the mobility of material culture and of labour, especially enslaved cooks who were often forced to travel with elite colonial families; and the mobility of ideas about what constituted healthy food to reproduce colonial bodies. By examining a rare corpus of eighteenth-century travel journals, all of them written by Dutch-speaking women, I investigate how maritime provisioning shaped the contours of global Batavian cuisine and vice versa. The requirements of portability and durability, together with notions of which foods produced healthy bodies, were important factors that influenced foods prepared and consumed on board. These accounts also give us an insight into the porous boundary between official maritime provisions for the Company's crew and the cooking and eating practices of these elite families on board. Some of these synergies we can attribute to common technologies of preserving food over long distances.

This paper does not focus on the official provisioning of Company ships for its soldiers and sailors, for which an established literature already exists, and which, perhaps inadvertently, tends to reproduce the historiographical bias towards mobile men in the Company literature.[2] Instead, it focuses on neglected questions around how the colonial household was reproduced, transformed, and made mobile on-board Company ships. These questions necessarily involve examining the mobility of women, as part of a larger movement of their male partners or chaperones, enslaved people, and personal effects, between Company settlements. Just as women constituted a tiny minority of soldiers and sailors due to gendered ideologies that shaped Company recruitment policies, the numbers of these women were very small, and confined, almost exclusively, to partners or relatives of high-ranking, male Company employees. The Company devised specific rules for how many family members and personal effects a male employee could take with him on ships, prioritized according to his rank in the Company hierarchy.[3] As a direct result of this policy, the Company effectively limited the movements of whole households to the rank of merchants and above, meaning it was these colonial families who had the privilege of circulating between the Company settlements.

This paper starts by introducing the reader to the maritime world of the Dutch East India Company in the Indian Ocean, before briefly discussing methodology and sources. I then examine achar, an iconic dish of global Batavian cuisine, which travelled on Company ships throughout the Dutch Indian Ocean Empire, even as far as the Dutch Republic. Achar was produced by combining the technologies of salting, drying, and pickling, each of which had a long provenance in human history. We trace the journey of achar as colonial households transported it through the ports of empire, spotlighting its durability, portability, similarity to landed cuisines, and insertion into existing meal structures and digestive philosophies, all of which was underpinned by the infrastructure of the Company's maritime network.

Introducing the Dutch Company and Its Empire

The Dutch East India Company was founded in 1602 to control the supply of plant parts into Europe, called 'spices' by early modern Europeans. Starting with the Company conquest in Ambon in 1605, the Company sought to control the island geographies of 'spice' production. The brutal conquest and massacre of the inhabitants of the Banda Islands followed in 1621. Coveting the fragrant cinnamon bark, the best quality of which was cultivated in southwest Ceylon, the Company invaded the port of Galle in 1641, and later Colombo in 1654. But the Company's longstanding attempt to control the production of pepper, that other sought-after Indian Ocean 'spice', remained elusive. Cultivated in both Sumatra, the second-largest island in Indonesia, and Malabar, on the western coast of India, the Company struggled to maintain as fierce a grip on the production and supply of this plant part.

To achieve its objectives, the Company established a network of permanent settlements in port cities in the Indian Ocean, with its headquarters in Batavia. This network of port cities mapped onto the '*octrooigebied*' (charter area) defined by the Company's founding charter, east of the Cape of Good Hope across the Indian Ocean until the South China Sea. This contemporary geographical imagining from the perspective of the Company founders informs this paper's use of the category of the Dutch Indian Ocean Empire.

This network of port cities was imbricated in two important facets of the Company's trade: the intercontinental trade between Batavia and Europe, and what historians have called the 'intra-Asian' trade. While intimately related, these comprised two different trading circuits, which were governed by the monsoons. The Company sent around four ships annually from Batavia to Europe, departing between November and January.[4] These large vessels, called East Indiamen, were loaded with supplies of 'spices', the volumes of which were calculated to take account of expected demand and prices in Europe. But outside of this season, the Company sent a far larger number of smaller ships to ply the routes of the intra-Asian trade, exchanging 'spices' for textiles and other commodities, the profits of which they used to finance the costly intercontinental voyages to Europe.[5] This constituted a thriving and dynamic part of the Company's trade, like a complex, multi-player game with an almost infinite number of moving parts. These observations, together with Batavia as the heart of empire, have led some historians to describe the Company as an 'Asian' company.[6]

This is relevant when we consider that mestiza women and enslaved cooks borrowed many influences from the Indonesian archipelago in their shaping of global Batavian cuisine, as I explore elsewhere.

Sources and Methodology

To reconstruct the culinary worlds on board Company ships, I will rely on a small sample of travel journals written by Dutch-speaking women. These sources are extremely rare, given both the small number of women who made this journey and archival and collection practices which have, until recently, not prioritized the experiences of women in maritime travel and trade. While this wider contextualization makes their accounts even more valuable, we must be careful not to over-interpret them. As elite women, they all fit the pattern, described above, of being attached to prominent men in the Company hierarchy, which itself facilitated their travel over the oceans. But even then, their testimony offers us an opportunity to read their claims against the grain to illuminate something of the lives of enslaved cooks who frequently were forced to accompany them on their journey.

None of these women travelled alone, but among family members, at least one of whom was male. Johanna van Riebeeck travelled with her husband Joan van Hoorn on his repatriation voyage from Batavia to the Dutch Republic, accompanied by Joan's daughter, Petronella; several enslaved people including the enslaved cook Juni; and a Chinese doctor whom Joan entrusted with his care.[7] Sisters Helena and Johanna Swellengrebel travelled with their family, too, which included their father, retired Cape Governor Hendrik Swellengrebel, and the enslaved woman known to them as Susanna van Bengale, as they made their way from their hometown of Cape Town to the foreign shores of Europe. Meanwhile, another pair of sisters, Johanna and Maria Lammens, accompanied their brother to take up a position in colonial governance in Batavia.[8] Having departed from the Dutch Republic, Johanna and Maria are the only ones among these historical actors to have travelled without enslaved people. Slavery was banned in the Dutch Republic but not in its maritime empire. These observations reinforce notions of enslaved people belonging, against their will, to the colonial family.

What Is Achar?

Achar was a signature component of global Batavian cuisine. Widely used as an acrid accompaniment to meat, it fitted easily into the structure of meat as the centre of global Batavian cuisine, as I explore elsewhere.[9] It neatly encapsulated digestive philosophies around raising the appetite in the tropics, which European male observers widely thought suppressed the decoction of the stomach. But what are the origins of this mysterious dish? A word with Persian etymology, it may have been transported to Indonesia during the period of Arabic medieval trade in the Indian Ocean. While its precise origins remain elusive, European colonial households undoubtedly encountered it when they arrived on the shores of the Indonesian archipelago sometime after the Company's first conquest there in 1605.

Achar: An Unexpected Stowaway

Achar was made by combining three preservation technologies of salting, drying, and pickling in vinegar. We learn from George Rumpf, a German botanist on Ambon, that the defining characteristics of the category of achar rests on its peculiar method of preparation as well as the inclusion of three main ingredients: vinegar, green ginger, and chilli. The most common achar, he tells us, is made from a plant he calls the Indian crithmus.[10] The first step is to boil the plant in water, then let it dry in the sun, before immersing in briny vinegar, ginger, galangal, chilli, and papaya flowers. This comprises the three-fold preservation technique of drying, salting, and immersing in vinegar. The purpose of vinegar was at least partly to instil a sour taste, which may have been inspired by a digestive belief about raising the appetite.

Chilli was an indispensable part of achar. The chilli plant had been transported from the Americas to Indonesia sometime in the sixteenth century. George remarks that 'all kinds of fruits [...] are pickled in vinegar or brine, and which is never done without these pods'. The heating taste of chilli probably counterbalanced the sour taste of vinegar. 'They are so terribly hot,' he adds, 'or rather, have a biting and corrosive sharpness along with moderate heat.' This chimes with the achar that Johanna van Rieebeck records in her maritime provisioning journal, where she lists one of the provisions as '*aastiaar temperadoe*,' a transliterated form of the term *achar temperado* in Creole Portuguese. *Temprar* is a Portuguese word, roughly meaning to season or flavour. And indeed, from Johanna's description, this achar was highly seasoned with chillies. 'It was very strong on chillies, which was not pleasant, but otherwise it was reasonably good,' Johanna remarks, confirming George's suggestion that chilli pods were an essential ingredient in making achar.[11]

The most mysterious of all our references is to a specific kind of achar made with mangoes. In a separate series of letters to her parents in Batavia, Johanna asks them to send her 'a pot of mango achar' (*1 potje manges-aatsiaar*).[12] Curiously, while mango achar is a trope in a corpus of writings by European men, no recipe has been located for it in the eighteenth-century Dutch Indian Ocean Empire. Joan Stavornius, captain of the Dutch navy, tells us that mangoes, when they are still green, are pickled to make achar. We gain a deeper insight into this practice from George Rumpf, who singles out a peculiar Portuguese method of making mango achar. 'The Portuguese usually prepare them in a better way,' he writes, 'soaking them first for a long time in salt water, whereafter they stick green ginger or garlic in them or rather, fill the hollow left by the pit, and keep them in vinegar or ground mustard.' The Portuguese name for this dish, George tells us, is mango achar or *manga recheada*, which means stuffed mangoes in modern Portuguese, and may refer to removing the inner kernel and filling it with heating ingredients. The Portuguese method described by George, consisting of first salting the mangoes before immersing in vinegar, recurs with a striking regularity through our colonial corpus of achar recipes, including an anonymous manuscript, handwritten in the Dutch Republic in 1780, which contains the largest collection of achar recipes in the corpus.[13]

Signalling its value in global Batavian cuisine and as a gift exchanged among social networks, elites used their connections and capital to transport Batavian-made achars on Company ships to other ports of empire. For example, we find two achars listed in the auction lists of Company merchant Harmanus Frederick Kofferman in Ceylon on 15 January 1790. We encounter the first one of the list, entitled '*atjar bamboe*', the Dutch term for bamboo achar. We surmise that bamboo achar was in European colonial households in Batavia because it appears in the first printed cookbook in the Dutch Indian Ocean Empire, the *Kitab Masak Masakan India* (*Manual of How to Prepare All Kinds of Dishes in the Indies*), which parallels the Portuguese method of first drying, then salting, and finally boiling in vinegar that is allowed to cool.

Bamboo achar made in Batavia was transported to the other ports of empire, where it found a ready market among European colonial households scattered across the empire. At the Castle tavern in Colombo, bamboo achar was sold to two buyers, both with European names which suggest that they were part of European colonial society in Ceylon. An individual, called only Koetjaar, is recorded as purchasing a quantity of bamboo achar for two rijksdaalders, while a man called Hendrik Pieter purchased the same quantity for one rijksdaalder and twenty-five stuivers. A similar quantity of *atjaar bamboe* was sold, this time in the auction of the married woman Salomina Sophia Sperling, to a handful of buyers with European names including Boomgaard, Schubert, and Pedroe June Ondaatje in July 1793.[14] While the sums of money involved were small, these transactions do suggest that there was a private market for bamboo achar among the community of European colonial households in Ceylon.

Achar made in Batavia was a gift among the most elite in colonial society. In Cape Town on the tip of southern Africa, the Company merchant Coenraad Frederick Hoffmann received gifts from the Governor-General, Abraham Patras, which included nutmeg achar and bamboo achar from Batavia. These achars were delivered under the watchful eye of the first mate, Andries de Bruijn, aboard the Company provisioning ship the *Lage Polder* sometime before January 1737. Abraham made use of his superior connections, influence, and social capital as Governor-General to engage Andries in this task, for which he may have paid a small commission.[15]

As with bamboo achar, elites use their connections to transport other signature achars made in Batavia across the oceans. Harmanus Kofferman, who we encountered in Ceylon in 1790, has another achar in his collection entitled '*atjar roemeni*'. This curious term '*roemeni*' appears in a trilingual dictionary published in Batavia in 1780, with the term '*de roemenia boom*' in Dutch glossed as '*poon gandaria*' in Malay and '*alber roemenia*' in Creole Portuguese.[16] We know that '*poon gandaria*' probably corresponds to the tree, today known by its Linnaean name of *boea macrophylla*, which produces round, yellow ripe fruits somewhat akin to mango, which are used in sambal and achar in the Indonesian archipelago. The orthography of this plant places it firmly in geographies in the Indonesian archipelago and suggests that Harmanus valued this culinary knowledge, and probably acquired it on a ship from Batavia as an article of private trade, perhaps as a gift from a social contact there. Two individuals purchased Harmanus's *atjar roemeni* in January 1790: one a man called Philip de Waas, who purchased a pot for one rijksdaalder and thirty-six stuivers, while another individual, Goode Miller, is recorded as purchasing the same quantity at the same price. The sale of both bamboo and *roemeni* achars points to how Harmanus valued Batavian achars, which had probably travelled to Ceylon through private trade, and, furthermore, that other European colonial households placed similar value on Batavian achars which they were willing to purchase.

These oceanic exchanges of achar point to its mobility and portability on long ship journeys. Johann Stavornius hints at the value of mango achar in the Dutch Indian Ocean Empire, who says that it is 'sent to all parts'.[17] Johanna van Riebeeck, in asking her parents to send mango achar from Batavia, participated in these oceanic exchanges of achar as a mobile food among friends and family members.[18] This resonates with George Rumpf's comments that achar 'can be kept for a long time if it is tied very tightly so that no air can get to it', echoing the meticulous attention that Johanna pays to the way provisions are stored on board.[19] These technologies of preservation are mirrored in our corpus of colonial recipes, with the author of an anonymous manuscript in 1780 counselling her audience to seal the pot tightly with a wet clean cloth and store in a dry place, and noting that 'the longer it keeps, the better' in her recipe for lemon achar.[20] This durability was probably related to the three-fold preservation technologies of salting, drying, and then immersing in vinegar, which may have resulted in both tastier and longer-lasting achar.

But the other intriguing thing about achar is that it cut across the strict lines of social class. 'Seafarers are eager to take them on voyages and get great good from it,' Johann Wolfgang Hedyt, a Company soldier and surveyor informs us, suggesting that regular sailors appreciated the value of achar on long sea journeys as much, perhaps even more, than elite European colonial households. Johann goes further to note that bamboo achar 'is sent as a present to Holland', though here our records suggest that this was done by the merchant classes. Christoph Langhanß, another German soldier and sailor, intimates something about the use of achar among his peers which suggests, however, that the eating context was sharply different to the elite households we have examined. 'If one has nothing more than rice, achar is enough to make a meal out of it,' he writes.[21]

Achar even travelled as far as the Dutch Republic. Petronella van Hoorn asked her family to send her achar from Batavia. She was none too pleased about being wrenched from her childhood home in Batavia and relocating to the Dutch Republic in 1711.[22] She found herself swept up in the career trajectory of her father, Joan van Hoorn, Governor-General of the Dutch East Indies from 1704 until 1709, who was directed to repatriate to the Republic in 1710. At the age of only twelve, she accompanied her father Joan; his second wife Johanna van Rieebeck; the Chinese doctor, Thebita; and several enslaved people on the nine-month journey to the Dutch Republic. After arriving in the Republic in September 1710, Petronella was quick to notify her beloved grandfather, Willem, another former Company Governor-General, whom she had left behind in Batavia.

From this early correspondence, the young Petronellla missed the global Batavian cuisine of her childhood, and, most especially, achar. In response, Willem wrote to Petronella detailing the captains and officers to whom he had entrusted gifts of achar, and the way they were transported and stored among these officers' private provisions on ships. On 25 November 1711, he sent Petronella a pot of achar pickled in Batavia (*een pot met aatchiar hier ingelegt*), both under the watchful eye of Harman Stellingwerf, vice commander on the ship called *Generale Vrede*.[23]

Petronella's longings persisted long after her initial homesickness abated, and we see her emerge in her letters as a confident woman, who self-consciously positioned herself as an elite of empire. She continued to use her privileged set of connections with Company elites in Batavia to arrange for shipments of achar to Amsterdam. On 20 October 1719, she thanked her cousin, the Company merchant Rogier van Heyningen, for arranging a shipment of achar *leyley*, transported on the *Abraham Ram*.[24]

Elite Dutch women, like Petronella, exchanged multiple recipes for achar which spanned wide social networks, suggesting that this culinary practice became embedded in ways of eating in the eighteenth-century Dutch Republic. We surmise that this is because of its easy absorption into an existing meal structure as an acrid accompaniment to roasted meats. These achars usually reproduce the three-fold method we observed in global Batavian cuisine of first salting, then drying, and finally boiling fruits and vegetables in vinegar that is allowed to cool, showing the porosity between practices on land and sea. But we find intriguing adaptations to Republic-made achars from our corpus of global Batavian recipes. Instead of the signature ingredients of ginger and chilli, there are frequent substitutions of garlic and mustard seeds, probably because of the limited cultivation practices of ginger and chilli in the Republic and the already far wider spread cultivation of garlic and mustard plants in horticultural spaces there. But we also have evidence to suggest that ginger was omitted because at least one elite Dutch woman did not fancy its heating taste. Several of these recipes also make the intriguing addition of turmeric to achar, which we did not encounter in global Batavian cuisine.

We conclude our investigation of the travels of achar in the empire with its appearance in a handwritten cookbook from the eighteenth-century Dutch Republic. Johanna Henrietta Schorer (1757-1813) was born into a prominent regent family in the port city of Middelburg, in the province of Zeeland.[25] Zeeland, presently the second-largest province in the Netherlands, was one of the six chambers of the Dutch Company. Johanna's handwritten cookbook contains several intriguing adaptations to global Batavian achars. Johanna's recipe for quince achar starts with the familiar method of placing the quinces in a salt pickle for three days, and then removing them and cleaning them with a dry cloth. Omitting the step of drying in the sun, Johanna next instructs the cook to boil the quinces in wine vinegar with cardamon, cinnamon, and mace, before allowing the mixture to cool. The next step is the curious addition of crushed mustard seed and three cloves of garlic, after which Johanna suggests covering the pot with a cloth and leaving to rest for three weeks before it is ready to eat.[26]

This replacement of garlic and mustard seeds for the characteristic green ginger and chilli of global Batavian cuisine probably reflects the greater local availability of those ingredients in the Republic, partly due to the relative absence of growing cultures in ginger and chilli which had to be imported from Batavia.

But peering further into Johanna's achar collection, we find further transformations to the flavour profiles of global Batavian cuisine in her recipe for radish achar. The cook starts by peeling the radish, cutting it into long pieces, and then placing it in salt for twenty-four hours. After removing from the salt, Johanna tells the cook to place the radishes on a plate to dry in the oven – an adaptation of the method of sun-drying, probably because of the many rainy days in the Republic compared to Batavia. The next step is to boil it in wine vinegar and allow it to cool, mirroring the global Batavian method for achar, but then Johanna suggests adding a series of ingredients, including turmeric, mustard seeds, and garlic alongside the ginger which was a signature part of global Batavian achars.[27] We do not encounter turmeric in any of our achar recipes in the ports of empire. Its addition here is curious, even more so than mustard seed and garlic, because we have no evidence that turmeric grew widely in the Republic. But this combination of turmeric, mustard seeds, and garlic, often as a replacement for ginger and chilli, but sometimes in addition to them, recurs in our corpus of achar recipes in the Republic, suggesting that cooks and elite women authors made adaptations to global Batavian achars when they were prepared in kitchens in the Republic.

A final achar recipe provides another window into the dynamic character of these transformations. Johanna includes a recipe for achar *leyley*, which we encountered earlier in this paper as a gift that Petronella's cousin sends her all the way from Batavia.[28] Another version of this mysterious recipe, which translates as lily achar, appears in an anonymous handwritten cookbook produced by an elite woman in the Hague in 1780. The term 'lily' does not refer to lilies as an ingredient, but gestures to the variety of vegetables used to make the achar. Here the Hague author reproduces the methods and flavour profiles we encountered in Johanna's recipe for radish achar. 'Take savoy cabbage, cauliflower, Turkish beans, radish, yellow carrots, princess beans, and let them cook a little, then add cucumbers chopped into long pieces with the seeds removed,' the author begins. The next step is to place in a salt pickle for twenty-four hours before removing and letting the mixture dry in the sun. The following step is to fry two cloves of garlic in three spoons of olive oil on the fire, add two cans of ginger, shallots, garlic, turmeric, and crushed mustard seed, and pour over the dried vegetables, sealing tightly with a wet cloth and storing in a dry place.[29] In this recipe, we again encounter the curious addition of garlic, turmeric, and mustard seed, along with shallots, to the ingredients we encountered in global Batavian achars.

Conclusion

We have now reached the final port of call in our journey exploring how achar, used as an acrid accompaniment to roast meat in global Batavian cuisine, was transported across the Dutch Indian Ocean Empire. In combining three preservation methods of salting, drying, and immersing in vinegar, achar reveals the porous boundaries between foods consumed on land and sea. It also cut across social class, though this did not extend to consumption contexts, where elites signalled their superior class status to more numerous soldiers and sailors. European colonial households valued achar partly because of its portability and durability that allowed for its transport across long distances, even facilitating the incorporation of Batavian-made achars in the reproduction of global Batavian cuisine in other ports of empire. Colonial elites used their privileged connections to ship captains and officers to send achar as gifts to friends and family in other parts of the empire. Part of a digestive philosophy to rouse the appetite in the torrid climate of the tropics, achar also found its way to the Dutch Republic, where it was easily absorbed into eating structures there. Elite women made further adaptations to global Batavian achars to suit the ecologies and taste preferences of diners in the Republic. By following the travels of achar as it criss-crossed the oceans, we gain a richer, more complex picture into the way that food mattered in the Dutch Indian Ocean Empire, providing a revealing set of insights into the intersections between food, social hierarchies, and matters of taste.

Notes

1. Michael Pearson, "Littoral Society: The Concept and the Problems," *Journal of World History*, 17, no. 4 (2006): 353–73.
2. Blussé, Leonard. "John Chinaman Abroad: Chinese Sailors in the Service of the VOC." In Alicia Schrikker and Jeroen Touwen (ed.), *Promises and Predicaments: Trade and Entrepreneurship in Colonial and Independent Indonesia in the 19th and 20th Centuries* (NUS Press, 2015, 101–22).
3. Kathleen Burke, "The 'Pleasures of the Garden:' The Mobility of Plants, People, and Power in the Dutch Indian Ocean Empire," *Crossroads*, 19.1 (2020): 34–51.
4. Leonard Blussé, "On the Waterfront: Life and Labour Around the Batavian Roadstead". In Masashi, Haneda (ed.), *Asian Port Cities, 1600–1800: Local and Foreign Cultural Interactions* (Singapore: National University of Singapore, 2009), 119–138.
5. Leonard Blussé, "On the Waterfront: Life and Labour Around the Batavian Roadstead". In Masashi, Haneda (ed.), *Asian Port Cities, 1600–1800: Local and Foreign Cultural Interactions* (Singapore: National University of Singapore, 2009), 119–138.
6. Leonard Blussé, "On the Waterfront: Life and Labour Around the Batavian Roadstead". In Masashi, Haneda (ed.), *Asian Port Cities, 1600–1800: Local and Foreign Cultural Interactions* (Singapore: National University of Singapore, 2009), 119–138.
7. Collectie Van Hoorn-Van Rieebeck, Nationaal Archief, Den Haag, 1.10.15[19], *Verzameling van brieven*.
8. Marijke L. Barend-Van Haeften and Els S. van. Eyck van Heslinga, *Op reis met de VOC: de openhartige dagboeken van de zusters Lammens en Swellengrebel* (Zutphen: Walburg, 1996).
9. See Kathleen Burke, 'The "Pleasures of the Garden": The Mobility of Plants, People, and Power in the Dutch Indian Ocean Empire', *Crossroads*, 19.1 (2020), 34–51.
10. Georg Eberhard Rumpf (trans. Beekman), *The Ambonese Herbal. Vol. 5* (New Haven: Yale University Press, 2011), 317.
11. Collectie Van Hoorn-Van Rieebeck, Nationaal Archief, Den Haag, 1.10.15[19], *Verzameling van brieven*.
12. Collectie Van Hoorn-Van Rieebeck, Nationaal Archief, Den Haag, 1.10.15[19], *Verzameling van brieven*.
13. Koninklijke Bibliotheek, 's-Gravenhage, *Manuscript KW 135 E 43*, c. 1780.
14. Sri Lanka National Archives (SLNA), *VOC Venduboeken*, 1.3041.

15 Cape Town Archives Repository, South Africa, *Inventories of the Orphan Chamber of the Cape of Good Hope*, MOOC8/5.139 1/2.
16 *Nieuwe Woordenschat, Uyt Het Nederduitsch in Het Gemeene Maleisch En Portugeesch* (Lodewyk Dominicus, Batavia, 1780), 56.
17 J.S Stavorinus, *Reize van Zeeland over de Kaap de Goede Hoop, naar Batavia, Bantam, Bengalen, enz.: Gedaan in de jaaren MDCCLXVIII tot MDCCLXXI, door den heer J.S. Stavorinus* (Te Leyden,: bij A. en J. Honkoop, 1793).
18 Collectie Van Hoorn-Van Rieebeck, Nationaal Archief, Den Haag, 1.10.15[19], *Verzameling van brieven*.
19 Georg Eberhard Rumpf (trans. Beekman), *The Ambonese Herbal. Vol. 5* (New Haven: Yale University Press, 2011), 317.
20 Koninklijke Bibliotheek, 's-Gravenhage, *Manuscript KW 135 E 43*, c. 1780.
21 Christoph Langhanss, *Neue ost-indische Reise* (Rohrlach, Zürich, 1705), 383.
22 Gelders Archief, Huis Rosendael. Archiefblok 0525, Inventaris nummer 381.
23 Gelders Archief, Huis Rosendael. Archiefblok 0525, Inventaris nummer 381.
24 Gelders Archief, Huis Rosendael. Archiefblok 0525, Inventaris nummer 381.
25 Marleen Willebrands, 'Aan tafel bij de familie Royaards-Schorer: Koken en netwerken aan de Nieuwegracht in de achttiende en negentiende eeuw'. *Oud Utrecht Jaarboek* (2017), 150.
26 Johanna Schorer, *Het Recept Boek*, 1780. Private collection of Danny Jansen.
27 Johanna Schorer, *Het Recept Boek*, 1780. Private collection of Danny Jansen.
28 Johanna Schorer, *Het Recept Boek*, 1780. Private collection of Danny Jansen.
29 Koninklijke Bibliotheek, 's-Gravenhage, *Manuscript KW 135 E 43*, c. 1780.

Food for Travel, Travel for Food: The Story of *Ekiben*

Voltaire Cang

Beginnings

Ekiben is short for *ekiuri bentō* (*eki*=station, *uri*=sell, *bentō*=packed meal), that is, the meal packages that are sold at the train station of one's departure, destination, and most stops in between. *Ekiben* are often consumed by holidaymakers on long train journeys and by businesspeople going on and returning from business trips, though not by commuters on regular work or school train commutes. They are special meals for special train journeys.

Bento have an ancient history: a 2000-year-old intact but rock-hard *onigiri* ('rice ball') discovered at an archaeology dig in the western Noto Peninsula holds the claim to being Japan's oldest portable meal.[1] *Ekiben*, however, have a much shorter history, being necessarily linked to the history and development of train travel in Japan that began only in the late nineteenth century. The start date of commercial train travel in the country is well recorded and thus undisputed: 14 October 1872, when the train line connecting Shimbashi station in central Tokyo to the major port city of Yokohama began carrying its very first passengers. (This year, 2022, was celebrated as the 150th anniversary of railway travel in Japan.) Most of the travellers were well-off vacationers who paid the equivalent of 5000 Japanese yen in today's currency – more than 30 British pounds sterling – for the cheapest third-class ticket, which was the same price as a 10 kg bag of white rice at the time. (Today, the same trip costs 480 yen, or less than one-tenth the original price.) For those travelling first class, the ticket cost three times more.

Unlike the start date of commercial train travel, even the start year of the *ekiben*'s first appearance is undocumented and has therefore become the subject of vigorous – albeit friendly – dispute. Most of the relatively few studies that are concerned with or at least briefly discuss the history of *ekiben* declare that the first of such packaged meals were sold to travellers at Utsunomiya Station in Tochigi Prefecture, north of Tokyo, in 1885 during the station's opening. Shirokiya, a traditional inn (*ryokan*) in Utsunomiya, produced special portable meal sets for the occasion, each containing two pieces of *onigiri* stuffed with pickled plum (*umeboshi*) and sprinkled with black sesame seeds, with radish pickles (*takuan*) on the side. The *onigiri*-and-pickles sets were wrapped in a single bamboo leaf and sold by hawkers at the station to passengers headed to Ueno Station in Tokyo.[2] Aside from these prototype *ekiben* meal sets, the station hawkers also sold sweet buns with bean paste filling (*anpan*), hard bread, jellied sweets, tea, and a few other food items.[3]

Others who claim to have created the first *ekiben*, if not the original location in which they were sold, include Umeda Station in Osaka and Kobe Station some 40 km to its west (both assert an earlier date than Utsunomiya, in 1877), and Oyama Station (Tochigi Prefecture, 1885). The claims of Umeda and Kobe Stations are especially plausible, as they were two of the most important train stations used by soldiers and officials involved in the Seinan War (also known as the Satsuma Rebellion) that lasted from February to September of 1877. The two stations were major transportation hubs for the long journey to Kyushu in the south where the war was being fought.

Umeda and Kobe Stations would have been very good business locations for *ekiben* and food vendors: as food sociologist Takada Masatoshi writes, 'The demand for *ekiben* [in these stations] is more than conceivable.'[4] However, Oyama Station also has a rightful claim, since it is on the same line as Utsunomiya Station and opened at the same time; Oyama declares that it began selling sushi bento when Shirokiya was also selling its *onigiri* sets in Utsunomiya. The *ekiben* from Oyama contained at least three different types of sushi: one wrapped in nori seaweed, another topped with egg, and the third one consisting of sushi rice stuffed in fried tofu pockets (*inari sushi*).[5] (It is unclear if any of the three sushi included fish.)

Despite these competing claims, many consider Utsunomiya Station's *onigiri* sets as the first *ekiben* in the country. This is due to their being mentioned in company pamphlets issued by the then Nippon Railway (now part of JR East), Japan's first private railway company, soon after Utsunomiya Station began its operations. The pamphlets expressly describe the *onigiri* sets as 'Japan's first *ekiben*'.[6] While Nippon Railway could have only been flattering itself with the origin story declared in print and formally put on record, many people in Japan, including academics, rarely question Utsunomiya's stance. Nonetheless, '[t]he truth of which railway station sold the "first" *ekiben* in history remains a mystery [and] it is safe to say *ekiben* sales started in the late 1800s and grew with the expansion of the railway system in Japan'.[7]

Definitions

What is *ekiben*? If one goes by its literal meaning, it can be any packaged food item sold at the train station that to the purchasing traveller would constitute a complete meal. By this definition, an *onigiri*, a cheese sandwich, a snack bun, or even a granola bar could be considered *ekiben*, although a poor and banal version of the thousands of portable and often elaborate meals the traveller in Japan would readily find in most train stations.

Ekiben do have an official definition: a complete meal packaged as bento, containing rice and other food items, that is sold at the train station or onboard, and which ideally uses ingredients associated with the region, town, or city served by the train line and the stations selling the *ekiben*. *Ekiben* that meet the standards of this 'official' definition are readily identified: their wrappers invariably feature a patented logo in the shape of a square box containing the word *ekiben* rendered in Japanese and Roman alphabets and a red circle on the bottom left that evokes the red sun in the Japanese national flag (Figure 1).

Figure 1. Official ekiben *logo.*

This logo was created in 1988 by the *Nihon Tetsudō Kōnai Eigyō Chuōkai* (literally, 'Central Union of Businesses Within Train Station Premises in Japan', referred to below as the Station Businesses Union), which also conceived the definition supplied above. This definition, however, is not absolute: bento that use bread instead of rice have been granted 'official' *ekiben* status by the Station Business Union, such as the pork cutlet (*katsu*) sandwich *ekiben* and the egg sandwich *ekiben* both from Yamanashi Prefecture west of Tokyo, and the ham sandwich *ekiben* of Kamakura (Kanagawa Prefecture), which is the pioneering sandwich *ekiben* first sold in 1898. And while many *ekiben* use ingredients identified with their region, these may not have been necessarily procured or produced locally, even if they may be marketed as regional delicacies and are sold within local train station premises.

As for the design of the official logo, the Station Businesses Union explains that the square shape with cut corners evokes the thin wooden boxes (*kyōgi*, often from cedar or cypress) that are the traditional bento containers. The crossed vertical and horizontal lines represent the compartments found in the typical bento. The red circle stands for the *hinomaru* (literally, 'sun circle') bento – the red *umeboshi* pickled plum placed in the centre of the rice in a bento box, symbolizing Japan's national flag. The circle also represents 'warm human interaction', in the words of the Station Businesses Union. The Kanji characters for *ekiben* are written in the calligraphy of the Kabuki theatre, since it was during intermissions in Kabuki performances when *makunouchi* (literally, 'in between acts') bento were traditionally consumed.[8]

Makunouchi became popular among theatregoers since its contents could be eaten without making a big mess – there were no sauces or food items that were too wet or juicy – and none of the foods were crunchy or otherwise made too much noise that would have disturbed one's fellow audience members. *Makunouchi* were later appropriated by bento producers around the country to sell to travellers as *ekiben*, as they could also be consumed without mess and noise as one ate on a moving train.

Food for Travel, Travel for Food: The Story of Ekiben

Makunouchi is one of two major categories of *ekiben*, and is sometimes called by its other and blander name of *futsū bentō*, or 'regular/common bento', that is in stark contrast to the other *ekiben* category called *tokushu bentō*, literally 'special bento'. The reason for *makunouchi*'s 'common' bento status is mainly due its format, which is considered the most common for all kinds of bento, including *ekiben*. In this format, custom dictates for *makunouchi* bento to contain white rice, usually occupying half of the bento box and which is laid out in bite-sized pieces formed into the shape of rice bales. For the food accompaniments, the standard 'triumvirate' of fish, *kamaboko* fish paste, and *tamagoyaki* rolled egg omelette are included, along with other vegetables and seafood, all of which have been boiled, grilled, or fried and then cooled before packing so as to prevent any juices from dribbling or seeping out.[9]

The term '*makunouchi bentō*' first appears in historical literature from 1796.[10] The early versions consisted of *onigiri* and any number of vegetable or seafood accompaniments that were often packed in kyōgi boxes for convenient handling when brought to and consumed in the theatre. At present, *makunouchi* bento are no longer limited to the theatre, although audiences still eat and drink in between acts of classical Japanese theatre performances, which can stretch to several hours at a time. *Makunouchi* bento are also found in convenience stores, supermarkets, and takeaway bento shops, among many other different food shops in Japan, and sometimes in unlikely establishments such as pharmacies.

The other major category of *ekiben*, the so-called 'special bento', refers to any *ekiben* type that is not in the form of *makunouchi*.[11] These may come in unusually-shaped containers, such as cooking pots (*kamameshi*) or toy train boxes – which are the biggest bestsellers – and often contain a large serving of the specialty food of the region they represent, whether these be meat or seafood, fruit, vegetables, and even rice. ('Common', *makunouchi*-type *ekiben* may include various local specialty foods but in smaller portions.)

Ekiben Gatekeepers

The Station Businesses Union has all these different types of *ekiben* under its purview; its main concern is the *ekiben* sold in Japan's train stations. The organization was formed right after the end of World War II in 1946, and was initially a loose association of businesses that operated in a few stations of the then government-owned Japanese National Railways (JNR). Since then, it has grown into a federation of nearly ninety organizations that oversee the various businesses operating in all six passenger railway companies that form the JR (Japan Railways) Group today, which was created after JNR's privatization in 1987. The mission statement of Station Businesses Union sets out the purposes of the organization, namely:

1. To research and gather data that contributes to the development and improvement of businesses inside train station premises;
2. To research on food production technology and hygiene and quality improvement;

3. To conduct seminars and training programmes for employees of businesses located in train station premises so as to raise their levels of service and improve food hygiene practices;
4. To hold food product fairs and similar events;
5. To publish an organizational magazine; and
6. To pursue other activities as necessary to achieve the goals of the organization.[12]

This formal mission statement lacks detail concerning the organization's actual activities, but a visit to their website quickly reveals that almost all its work revolves around *ekiben*, be it research and educational activities, employee training, or publication and promotion. All the images in their website are of *ekiben* only, too, as are the main subjects or contents of their articles and news releases. The organization's web address (URL) is perhaps the most telling of all: ekiben.or.jp. *Ekiben* is very much a serious business.

Today, there are more than 4500 train stations in the JR Group nationwide. At the same time, it is estimated that more than 4000 types of *ekiben* are sold in Japan's train stations at any given period. One might then be prompted to conclude (on the spur of the moment) that a large percentage of train stations in Japan sell *ekiben* specific to their region. That, however, is not the case.[13] Many stations, particularly the bigger and busier ones, often sell multiple variations of the same kind and even name of *ekiben*, and the 4500 station count here does not include those operated by other private railway companies in Japan. (Latest statistics place the total number of train stations in Japan at 9024 as of March 2021, indicating the total to be almost equally split between JR and everyone else.)[14] In any case, the sheer number of *ekiben* available to the traveller in Japan does emphasize their popularity, as it also reflects the wide regional diversity of Japanese food traditions.

With these thousands of *ekiben*, it would be foolhardy to discuss or even list all the *ekiben* currently available in Japan. A discussion of Japan's most representative *ekiben* is nevertheless in order. For this we shall consider the *shūmai* bento, the most popular *ekiben* in Japan, and briefly discuss its history. We also briefly examine the role this *ekiben* plays in evoking and re-creating place, or creating it outright.

The *Shūmai* Bento, Japan's Quintessential *ekiben*

Shūmai is the Japanese name for *shāomài*, the Chinese pork dumpling popular as a dim sum nibble in Cantonese cuisine. It is the main food item in the *Shiumai bento*, Japan's bestselling *ekiben* that was first sold by the food shop, Kiyoken, at Yokohama Station in 1954.[15] The standard *shūmai ekiben* today contains five pieces of *shūmai*, aside from rice (eight bite-sized pieces shaped like rice bales and topped with an *umeboshi* pickle in the centre and black sesame seeds sprinkled on each 'bale'); one piece each of teriyaki grilled fish, *kamaboko* fish paste, *karaage* fried chicken, *tamagoyaki* rolled egg omelette, and dried apricot; boiled bamboo shoots, sliced *kombu* kelp, and slivered ginger (Figure 2). The contents have changed over the years, although the grilled fish and egg have always been included from the very first iteration, along with the essential *shūmai* and rice.

Food for Travel, Travel for Food: The Story of *Ekiben*

Figure 2. Kiyoken's Shiumai bento *(Wikimedia Commons).*

Like ramen and *gyōza, shūmai* is food with Chinese roots that has become a common fixture on Japanese household dining tables. Unlike ramen and *gyōza*, however, *shūmai* is rarely considered 'Japanese food' within and especially outside Japan. (One of the more blatant side effects of Japanese cuisine's global popularity in recent years is the 'Japanization' of ramen and *gyōza*, to the extent that these foods are often served in Japanese restaurants abroad, even in sushi restaurants, a practice that is not found in traditional sushi shops in Japan.) How, then, did this archetypal Chinese food make its way into one of Japan's most famous and popular *ekiben*?

Shiumai bento's originator, Kiyoken, started as a small food shop in Yokohama Station in 1908. In the beginning, the shop sold sushi and mochi alongside drinks such as cider and milk, all of which could be consumed quickly on the go. The journey from Yokohama to Tokyo is brief (about 25 minutes on the regular train, or 15 minutes on the *shinkansen* bullet train today), and passengers had no time, or perhaps even desire, to eat at their leisure. While Kiyoken's products sold well enough, its then president, Nonami Mokichi, decided to create a snack product that would distinguish his shop from the others who were all selling the same kinds of food and drinks.

Yokohama in the early twentieth century was home to Japan's largest enclave of Chinese, many from Canton, most of whom were engaged in trading or finance. The area occupied by these migrant Chinese was already called 'Chinatown' in the late nineteenth century by the international city's other foreign residents – who also had their own enclaves – although the Japanese called it 'Nanjing Town' (*Nankingai*). 'Nankin', that is, Nanjing, was then used by the Japanese as a label for all things Chinese, and sometimes for objects and places of foreign origin.

Yokohama also had the largest concentration of Chinese restaurants in Japan at the time, and these were patronized not only by the Chinese, but also by other Japanese and non-Japanese residents. Its Chinatown (i.e. *Nankingai*), however, had the reputation of being filthy and dangerous, and most Japanese visitors to the area were largely limited to the middle and lower classes who mostly came for business and sometimes for the food.[16] Nonami was among the frequent visitors to Chinatown, and it was on one of his visits when he first encountered *shūmai* and realized its potential as a distinct food product that he could appropriate and develop for his Yokohama-based business.

In the early 1900s, *shūmai* was already one of the two most popular dishes in Yokohama's Chinatown, along with *shina soba* ('Chinese noodles'), the wheat noodles served in broth that would later evolve into ramen. Like *shina soba*, *shūmai* is best eaten when it is piping hot, and Nonami had to find a means to make *shūmai* that did not lose its taste when it had become cold, as he planned to sell them with his other snack foods.

Nonami pirated a Cantonese chef named Wu Yusun from a Chinatown restaurant and employed him for the main task of inventing *shūmai* that would not lose its flavours when served at room temperature. After one year of constant experimentation, Wu finally succeeded: he added minced scallop to the ground pork base of his *shūmai*, which he found helped the pork mixture stay delicious even when cold, at the same time enhancing the umami flavour of the pork itself.

Kiyoken began selling its *shūmai* in boxes of 12 pieces each, at 50 sen per box, in 1928.[17] 50 sen was equivalent to the price of a decent meal at a regular Japanese restaurant in the 1920s. At this price, the *shūmai* box was relatively expensive, and sales were only moderately successful. Moreover, the impending Pacific War would soon adversely affect and interrupt Kiyoken's business in the succeeding two decades.

The end of World War II saw a Japan that would rebuild slowly at first, before rapidly developing into an economic giant from the late 1950s. Fast and widespread industrialization resulted in a vast improvement in train travel in the country, while rising incomes allowed more Japanese to travel for leisure. (Foreign travel was still prohibitively expensive, even after the deregulation of foreign tourism – which was restricted due to insufficient foreign currency reserves – in 1964, the year of the first Tokyo Olympics.) Kiyoken managed to survive the war and was soon able to continue producing and selling *shūmai* to travellers in Yokohama not too long after, in the early 1950s.

Nonami, who was still the president – a position he occupied until his death in 1965 – experimented with a new way of selling his company's *shūmai*: He hired young women, dressed them in bright red uniforms with sashes over one shoulder on which was written *shūmai musume* ('*shūmai* girl'), and fanned them out to the train platforms with baskets of *shūmai* to sell directly to train passengers. Nonami imitated the system of cigarette companies that were deploying their own 'tobacco girls' in many places at the train stations who sold their wares to mostly male patrons.

The *shūmai* girls became very famous after the novelist Shishi Bunroku published a serialized story in the nationwide *Mainichi Shimbun* daily that featured the love affair between a *shūmai* girl and a baseball player.[18] The story became even more famous when it was made into a hit film in 1953. As a result, *shūmai* became very popular all over Japan, with Kiyoken's *shūmai* (and its *shūmai* girls) lording over everyone else who also attempted to produce their own versions of *shūmai* in imitation of the originator.

Food for Travel, Travel for Food: The Story of Ekiben

In 1954, Mokichi's son and future heir, Yutaka, was already working alongside his father when he decided to further capitalize on *shūmai*'s popularity and develop it into *ekiben*. By then, leisure travel was on the cusp of becoming a boom among middle-class Japanese, and Yutaka was fully aware of the growing popularity of *ekiben* among travellers. However, most of the *ekiben* available at the time were of the *makunouchi* or 'common' type which, like the food products in train stations, all closely resembled one another. Yutaka and his co-workers innovated and created their own distinctive take on the *makunouchi* bento. Following the standard format, their *ekiben* contained rice, needless to say, as they also retained – or copied – the style of forming these into bale shapes. They also retained the *makunouchi* triumvirate of grilled fish, egg, and *kamaboko* fish paste, but replaced the rest with food not found in the regular *makunouchi* that were unmistakably local: Kiyoken's flagship *shūmai* (four pieces) and *fukujinzuke* chopped vegetable pickles made by Shuetsu, a manufacturer of preserved seafood and vegetables in Tokyo that has been in business since 1675.[19] (The *kamaboko* fish paste was also locally produced in Yokohama.)

In short, Kiyoken created a 'common' *ekiben* that became 'special', since it included local specialty foods, even if it did come in the standard and common bento box. The *shūmai* bento was a big hit from the beginning: *Shūmai* was already a popular food on its own, but including it – four pieces at that – in *ekiben* along with other local delicacies only enhanced its appeal to travellers. Priced at 100 yen – coffee then cost 50 yen in coffee shops – it was not luxurious food, but an affordable yet fancy treat.

Kiyoken's *Shiumai bento* today cost 860 yen (5 British pounds), which is about the price for a salaryman lunch or a regular bowl of ramen. They remain bestsellers, not only as food for travelling, but also as souvenir food gifts, picnic food, or even as a quick meal at work or at home. To bring its *shūmai* to as many consumers as possible, Kiyoken now sells its *ekiben* not only at train stations, but also in convenience and department stores,

Figure 3. The *ekiben* shop inside Tokyo Station.

supermarkets, and specialty food shops, as well as online. Most of its business operations are centred around Yokohama and the greater Tokyo area, although the company is expanding its reach into major cities in the western and northern regions of Japan. In 2022, Kiyoken sold an average of 25,000 *ekiben* boxes daily. At one recent festival event in Yokohama that celebrated the port's opening, a record of more than 30,000 boxes were sold in one day, from a single booth.[20]

Brief Discussion

Ekiben were first produced for people who travelled. In time, with their emphasis on using ingredients specific to and especially identified with its place of origin, *ekiben* evolved into food that represented the place one travelled to, as well as food that one travelled from. Most recently, as more region-specific *ekiben* have become available in places that one travels neither to nor from, many have become travel destinations in themselves, allowing one to visit places without the actual act of travelling, except perhaps with the mouth. (This vicarious form of travelling is on lavish display at the main *ekiben* shop inside Tokyo Station, which carries more than 200 varieties of *ekiben* from all over Japan at any given time. See Figures 3 and 4.)[21]

The history of Kiyoken's *Shiumai bento* tells the full story of *ekiben*, and its enduring popularity has made it Japan's quintessential *ekiben*. Kiyoken's *shūmai* boxes and the *Shiumai bento* are also perhaps the first to explicitly evoke place among all *ekiben*, or at least to use place explicitly to tempt travellers into buying them. The use of *shūmai* was a master stroke by *Shiumai bento*'s creators: not only does it reflect the cosmopolitan vibe of Yokohama, but it also represents the foreign and 'exotic' place that is China, which the consumer could now viscerally experience without the need to actually be there. It may well be for this reason – all the travelling made possible merely through the eating

Figure 4. The ekiben *shop inside Tokyo Station.*

of 'foreign' food – that has made *ekiben* among the popular and successful items to have come out of Japanese food culture.

But Yokohama is not China, as Kiyoken's *shūmai* is not Chinese *shāomài*. Chinese-American historian Eric C. Han, for example, considers *shūmai* more Japanese than it is Chinese, on the account that it is 'open-topped' and uses different ingredients such as shrimp and Japanese seasonings, aside from the obvious fact of its invention having taken place in Japan – albeit by a Chinese-born chef taking instructions from a Japanese food company entrepreneur.[22] Nonetheless, Kiyoken successfully riffed on this invention to produce a new food that soon enough became the main ingredient of Japan's most popular *ekiben* as well as an enduring symbol of Yokohama and, by association, China.

Chinese culture, too, is but only one of the many foreign cultures, especially 'Western' cultures, to have flourished and made its mark on Yokohama. This melting-pot character has made Yokohama unique from most other cities in Japan, including Tokyo. However, Japanese today associate Yokohama with Chinese culture (and its Chinatown) more than with any other non-Japanese culture. And in any discussion of Yokohama's food culture, Chinese cuisine is often the first thing that comes to mind among most Japanese, while the cuisine is also almost always the first to be mentioned, featured, and reviewed in all printed and online travel and food guides for the city. The *Shiumai bento* may not have singlehandedly caused the 'Sinicization' of Yokohama and its food culture, but the early adaption – or exploitation – of *shūmai* to represent the local food culture has not at all been insignificant. At the very least, Kiyoken was a very active collaborator, if perhaps an unwitting one, in the cause.

Final Thoughts

Unlike many food and restaurant businesses all over Japan, Kiyoken did not suffer big revenue losses during the recent global coronavirus pandemic. It did experience huge drops in sales at its transportation station shops during the pandemic's onset, as travelling and other non-essential movement were suddenly restricted. Kiyoken also faced a rash of cancelled orders from the many sports events, festivals, conferences, and other meetings for which it supplies *shūmai* bento meals. (Many of the several thousand staff and volunteers for the annual Tokyo Marathon, for example, are provided with *Shiumai bento* on the day of the event.)[23]

Kiyoken's losses, however, were soon recouped through online sales as well as from standalone and food retail shops. (Japan did not impose lockdown measures during the pandemic. While many schools, offices, and so-called 'non-essential' businesses were 'requested' to temporary cease their operations at the peak of the coronavirus crisis, most supermarkets, convenience stores, drug stores, and the like were permitted to stay open.) Although the pandemic prevented everyone from travelling, it was the *shūmai* that did the travelling for everyone; Kiyoken would continue to experience a sharp rise in sales for the rest of the pandemic season.[24] As with most other *ekiben*, *Shiumai bento* travels well, as it fulfils its *raison d'être* of enabling one to travel well.

Notes

1. 'Japan's Oldest Onigiri Unearthed: Sugiya Chanobatake Ruins', Nakanoto Town, Ishikawa Prefecture <https://www.town.nakanoto.ishikawa.jp/material/files/group/4/onigiri.pdf> [accessed 15 May 2022].
2. Kikuko Oda, '*Ekiuri bentō ni tsuite* (On *ekiuri bentō*)', *Journal of Cookery Science*, 9.1 (1976), 19–24.
3. Kikuko Yamazaki, 'Historical Changes of Ekiuri-bento in Japan', *Bulletin of the Institute of Modern Culture* (Showa Women's University), 778 (2005), 15–28.
4. Masatoshi Takada, '*Tetsudō no tabi to ekiben* (Railway Travel and *Ekiben*)', in *Tabi to shoku* (Travel and Food), ed. by Noritake Kanzaki (Tokyo: Domes, 2002), pp. 174–97, p. 176. In the main text, Japanese names are written in the conventional order, with the last name (Takada) first before the given name (Masatoshi).
5. Keizo Ono, '*Ekiben no hanashi no uchi: ekiben no ganso* (*Ekiben* Stories: Original *Ekiben*)', *Tabi* (Journey), 8 (Nihon Kōtsū Kōsha, 1930).
6. Takada.
7. Atsuko Hashimoto and David J. Telfer, '*Ekiben*, the Travelling Japanese Lunchbox: Promoting Regional Development and Local Identity through Food Tourism', in E. Park and others (eds.), *Food Tourism in Asia: Perspectives on Asian Tourism* (Singapore: Springer Nature, 2019), pp. 103–122 (p. 108).
8. *Nihon Tetsudō Kōnai Eigyō Chuōkai* (Central Union of Businesses Within Train Station Premises in Japan) <http://www.ekiben.or.jp/main/about/index.html> [accessed 15 May 2022].
9. Kan Sakurai and Jun Hayase, *Ekiben nyūmon* (Introduction to *Ekiben*) (Tokyo: Gentosha, 2014).
10. '*Makunouchi*', in *Nihon kokugo daijiten* (Grand Dictionary of the Japanese Language) (Tokyo: Shogakkan, 2001), XII, pp. 354–55.
11. Kan Sakurai and Jun Hayase.
12. *Nihon Tetsudō Kōnai Eigyō Chuōkai*.
13. Kan Sakurai and Jun Hayase.
14. '*Eki de-ta-be-su kaitei* (Station Database Revised)', Dream News 2021 <https://www.dreamnews.jp/press/0000233289/> [accessed 15 May 2022].
15. '*Shiumai bento*', instead of the conventional rendering of '*shumai* bento', is the official product name.
16. Eric C. Han, *Rise of a Japanese Chinatown: Yokohama, 1894–1972* (Harvard University Press, 2014).
17. '*Rongusera-kō: Shūmai bentō*' (In Consideration of Long Sellers: *Shūmai* Bento), *Comzine*: <https://www.nttcom.co.jp/comzine/no089/long_seller/index.html> [accessed 15 May 2022].
18. Han.
19. '*Rongusera-kō: Shūmai bentō*'.
20. IT Media Business, '*Korona ka demo fubunritsu yaburazu: Shūmai bentō Kiyōken ga kenji suru rōkaru burando* (Unwritten Law Undamaged by the Coronavirus Pandemic: Kiyoken *Shiumai Bento*'s Firm Hold as a Local Brand)' <https://www.itmedia.co.jp/business/articles/2106/04/news024.html> [accessed 15 October 2022].
21. Ekiben-ya Matsuri, '*Mainichi ga ekiben matsuri*' (Everyday Is *Ekiben* Festival Day) <https://foods.jr-cross.co.jp/matsuri/> [accessed 15 October 2022].
22. Han, p. 85.
23. IT Media.
24. IT Media.

The Raw and the Disordered: Portable Feasts, Blood, and Raw Meat Salads in Contemporary Political Movements in Thailand

Nattha Chuenwattana and Chonlatorn Wongrussame

Introduction

Portable food, and portable feasts, were an integral part of large street demonstrations and political gatherings in Thailand. Food was presented as a symbol of abundance, unity, and nourishment in many previous large-scale, well-organized movements.[1] However, the recent pro-democratic, youth-led movement (active since 2020) changed the meaning of food in its protests significantly.

Portable feasts, and the food served within their spontaneous mobile demonstrations, are a metaphor for political dissent, a clear representation of political identity politics, and an edible site of memory. Food is not only a representation of safety and nourishment, but also a unifying tool among the protesters. Over the years, raw meat salads consistently appeared in the middle of many political gatherings and public demonstrations against the government in Bangkok. This appearance was the first time in Thai political history that food from a minority culture assumed centre stage as both political tool and cultural outreach during political demonstrations.

Raw meat dishes, such as *Laab* and *Soi-Ju*, from northern and northeastern Thailand, rose to their prominence during the recent mass demonstrations. The Thai protesters transformed outdoor consumption of *Laab* and *Soi-Ju* into lively portable feasts. These portable feasts served as a transgressive food act that symbolized the resistance against the Thai government's ideology of public order. Raw meat dishes also served as a political metaphor that represents the identity of the opposition minorities, which includes many marginalized ethnic minority cultures. Raw meat salad preparation and consumption in open public settings, especially on the street in the middle of downtown Bangkok, was a deliberate method of protest.

Laab (ลาบ) is a popular ground meat salad seasoned with chilli and other spices. There are two variations in making *Laab*: raw and cooked. Blood is often added to season and emulsify ground meat in raw *Laab*. The *Laab* served during the protests was raw. *Soi-Ju* (ซอยจุ๊) is sliced lean raw beef, sliced beef liver, and sliced beef tripe, served with beef bile dip (*Jaew*, แจ่ว). Beef bile dip is seasoned with toasted rice, chilli, and fish sauce. *Brassica juncea* (*Pak-kad-hin*, ผักกาดหิน) is often served alongside *Soi-Ju* as its mild spicy flavour is complementary to *Soi-Ju*. While *Laab* can be cooked or raw, *Soi-Ju* is only served raw.

The protesters selectively used raw meat salads as a symbolic representation of minority groups and cultures within the movement. Eating raw meat, sometimes with blood, is a powerful message of dissent against the Thai government's ideology that eating raw meat is considered unhealthy and unsanitary.[2]

This paper will examine the changing meanings of food in the recent political demonstrations in Thailand. Food played a very important role as a symbolic representation in both the Red Shirts and the youth-led movement. It is important to understand the relationship between local culture, identity politics, and the process which makes raw meat salads a portable food (for the protesters) and also a delicacy in mainstream Thai culture, especially in Bangkok, where most political demonstrations occurred.

In the Name of Good Health and Cleanliness

Raw meat consumption is considered a way of life in the north and northeastern parts of the country. There are several raw meat dishes that are considered favourite local ingredients (*Pla-ra*) or ceremonial food, including *Laab* and *Soi-Ju*, served during important gatherings. However, the Thai government has long viewed raw meat consumption as an unhealthy eating practice. The anti-raw meat consumption campaign by the health ministry was established during the 1980s and 1990s.[3]

The campaign targeted the populations of the north and northeastern regions of Thailand. Historically, the north used to be Siam's colony, while the northeastern part of Thailand (or Isan) was always treated with prejudice as an internal colony. In 1984, the Ministry of Public Health started a liver fluke prevention and control programme in the northeast. From a survey in 1980-1981, the Ministry of Public Health found that the number of liver fluke cases was highest in the northeast, reaching 34.6% of the total number of cases across the country. From 1984 to 1991, the Ministry of Public Health organized the project to treat patients with liver flukes. The government implemented a massive public health education campaign to change raw meat-eating habits. Raw fish consumption is a major cause of parasite infection (fluke) that in turns leads to liver cancer.[4]

Even after the heavily broadcast anti-raw meat consumption campaign, *Laab* appeared at the protest after the Black May Massacre (พฤษภาทมิฬ) in the north in 1992. The gathering of progressive artists, organized by Tasnai Setthaseree and Mit Jai-in, in Chiang Mai was called the 'Chiang Mai Social Placement'. *Laab* was served and eaten at Tha Phae Gate, Chiang Mai.[5]

The anti-raw meat and raw fish campaign was deemed successful.[6] It was able to reduce cases of liver disease, but it deepened the historical bias against these regions. As a result of this campaign the perceived image of raw meat consumption was viewed as unclean and the food of the inferior, which resulted in the racist bias against these raw meat-eating cultures. Just as raw meat salad was rejected and continuously devalued, so was the politician most highly favoured by the north and northeast populations: Thaksin

The Raw and the Disordered

Shinawatra, the Thai prime minister, was overthrown in a *coup d'état* in 2006.[7]

After the 2006 coup, the Red Shirts political movement was established. The Red Shirts positioned themselves as pro-democracy, anti-dictatorship, and anti-coup. The Red Shirts movement not only consisted of labourers, but also other pro-democracy Non-Governmental Organizations and many in the media.[8] The majority of the Red Shirts' members came from lower socio-economic backgrounds and those excluded from the socio-political order. Most of them were Thaksin's supporters from the north and northeastern parts of Thailand.[9] The Red Shirts assumed the *Phrai* identity, which is the opposite of the *Phu Di* or the 'gentleman'.[10] It is important for the upper-class Thai to educate and conduct themselves as a *Phu Di*. There are many personal qualities used to separate *Phrai* from *Phu Di* based on the bourgeois ideal imported from Victorian England with one of them being 'orderly'.[11]

According to this notion, *Phrai* was perceived as uneducated and disorderly. *Phrai* is the lower-class status of the commoners or labouring class located just above the slaves, according to the traditional Sakdina hierarchical system. They were often exploited and oppressed by the upper class Thai traditional leadership. Different levels of personal freedom separated slaves from *Phrai*. Therefore, the Red Shirts intentionally assumed this *Phrai* identity to gain political legitimacy for their movement for equality and true democracy.[12]

In 2008, Two Red Shirts poets, Krit Luelamai and Mai Neang K. Kunathi, both intentionally used *Laab*, blood, and other cultural references to the commoners from the north and the northeast as part of their protest poems:

[…] Come, come eat Laab-koi together from the cutting board
Together, we rolled fermented tea leaves
Let's fill the blood with our blood
Make it angrily boiled more than ever
Butchered meat, fueled the meat with great force
Let them know that we will not back down anymore!
Cattle gave their lives to us
So we have the energy to fight the elites and all the angels […]
                                    Krit Lualamai, 'In the Big Eating Circle'[13]

'[…] Commoners rolled sticky rice dipped into Laab-koi
Eat teeny fish, tiny shrimps plunged in Jaew
Nourished the flourishing political awareness
Citizens are the precious, are the dearest […]'
                                    Mai Neang Kor Kunati, 'Kaysone Phomvihane'[14]

During the same year, the protesters from the northeastern region cursed and poured blood in front of the residence of the privy councillor (Gen. Prem Tinsulanonda) as they believed he was responsible for the coup. The Red Shirts movement was perceived even more as savage and unclean after they conducted the blood-pouring ceremony. This tainted image of the Red Shirt movement legitimized and justified the violent crackdown by the government to end their demonstration in downtown Bangkok in 2010.[15]

In March 2010, at least 98 Red Shirt protesters were killed and approximately 2000 were wounded as a result of the militarized operation that dispersed the protesters in downtown Bangkok.[16] The 'Big Cleaning Day' activity was held in downtown Bangkok the very next day. The event was organized by the mayor of Bangkok. Downtown highstreets, where the Red Shirts protesters were shot and killed, were power-washed by municipality workers and volunteers. The event was covered by mainstream media across the country.[17]

Laab

Even though Thaksin Shinawatra's cabinet was overthrown by the coup, his sister, Yingluck Shinawatra, was elected to become the Thai prime minister on 5 August 2011. Like Thaksin, Yingluck got overwhelming support from the Red Shirts movement and north and northeastern voters. Yingluck's cabinet faced strong opposition protests. The pressure from the protests forced her to call for an early election on 9 December 2013. On 10 January 2014, the People's Democratic Reform Committee (PDRC) moved to prevent elections from taking place in Thailand.[18]

Tasnai Setthaseree and Mit Jai-in, artists behind the Chiang Mai Social Placement event, organized a protest to support an election. The event was titled 'Eat and Wait for an Election', and once again they lay down mats around the Tha Phae gate in the shape of a cross and invited people to join and eat *Laab* together.[19]

The coup finally happened on 22 May 2014 after many months of tension and protests by the PDRC. Anon Nampa, a lawyer, and Pansak Sritep, an activist whose son was shot dead during the Red Shirt massacre, decided to organize a protest against the coup in front of the Bangkok Art Gallery on 23 May 2014. They brought *Laab* to eat and posted their *Laab* eating on social media. The protest was subsequently dispersed, and many protesters were arrested.[20]

Other protests followed. On 10 December 2014, Pansak, Phayao Akhad, whose daughter was shot dead during the Red Shirts massacre, and other students organized the activity 'Eat Laab, Bow Down to the Constitution, Blow Cake' at the Democracy Monument. *Laab* was the main dish of this event, and it was served at this New Year party to all the attendees. Phayao said that if her daughter had lived, her family would have celebrated their birthday party together at the end of every year.[21]

The most powerful *Laab*-eating protest took place on 31 August 2017. Phayao, Pansak, and a group of relatives of the massacred Red Shirts organized the 'Seven Years Empty, Waiting in Anguish' event. This came after the Supreme Court dropped the criminal and disciplinary cases concerning the Red Shirts' crackdown against Prime Minister Abhisit Vejjajiva and Deputy Prime Minister Suthep Thaugsuban.[22]

Phayao was wearing a light blue nurse vest stained with the blood of her daughter,

Kmongade, at the event. As at the protest in Chiang Mai, food was served to attendees on a mat. *Laab* with blood, steamed sticky rice, and fresh vegetables were carefully arranged on the mat. Phayao then lit incense and said to her daughter's spirit that 'Everyone will not die for nothing'. Pansak, whose son was shot on the street during the Thai government's crackdown in 2010, then proceed to eat *Laab* with blood.[23]

At all these events mentioned, *Laab* served with blood connected the identity of the *Phrai* or the commoners with raw meat-eating cultures from north and northeastern Thailand. *Laab* eaten communally empowered their political movement, while challenging the centralized narrative of the Thai government that repeatedly and systematically ostracized their minority cultures and their political decisions. Today the Red Shirts movement is considerably much smaller than it was in 2010. *Laab* is still considered a traditional dish of the north and northeast, while *Soi-Ju* entered the raw meat-eating protest scene. *Soi-Ju* managed to captivate the youth protesters' palates, and both *Laab* and *Soi-Ju* became trendy dishes in Bangkok a few years later.

The Rise of *Soi-Ju*

Anon Nampa, a human rights lawyer, is responsible for the popularity of *Soi-Ju* in contemporary Bangkok. As a proud northeasterner from Roi-et province, Anon publicly ate *Laab* and *Soi-Ju* at most of the youth-led political protests. He frequently posted his *Laab* and *Soi-Ju* eating activities on social media. Countless pictures of red raw meat with spicy bile dip have been featured to Anon's over 300,000 followers. That these raw meat salads are Anon's favourite dishes is common knowledge among Thai political activists.[24]

Anon often used *Soi-Ju* with spicy bile dip to challenge his younger, Bangkok-based, and highly urbanized activists. According to Anon, northeastern children are not allowed to eat raw meat, especially *Soi-Ju*, when they are young. *Soi-Ju* and its bitter bile dip are considered adult, grown-up food. Anon treats *Soi-Ju*'s eating and the bitter taste of spicy bile dip as a rite of passage, a transition to adulthood, for the younger political activists. He often prepared the dish to offer to other activists during many of the youth-led protests that have happened since 2020.[25]

On 21 September 2020, the *Laab* Group or 'คณะลาบ' was created. Their philosophy is that the '*Laab* shop' is an important social area where all classes of people gather to enjoy *Laab* without discrimination. *Laab* Group social media and their gatherings served as a place to produce ideas and innovations as a force to support Thai democracy.[26]

On 14 October 2020, *Laab* Group organized an event inviting people to eat *Laab* at the Democracy Monument. They invited people to bring steamed glutinous rice, vegetables, *Laab*, and a local boiled liquor: 'Let's eat spicy salad for democracy, come on my brothers and sisters.' This gathering by the *Laab* Group served as a picnic that united the experienced activists and the younger generation of protesters together through *Laab* and *Soi-Ju* eating.[27]

Soon after, on 19 October 2020, *Laab* Group gathered in front of the Bangkok Special

Prison. The activity 'Eat *Laab*, Eat *Koi* with the *Laab* People' took place in front of the prison's entrance. *Laab* Group members and other protesters joined together in a circle to make and distribute *Laab* and *Koi* (another northeastern raw beef dish), which was served with steamed sticky rice on a mat which was laid out on the concrete floor close to the prison's entrance. Three protesters then hugged their friends' necks and shouted, 'Let's come out to eat *Laab*!' According to *Laab* Group's social media post on that same day, *Laab* Group activists considered eating *Laab* by hand as an act that could overthrow the Thai Dictatorship. Their *Laab* eating act in front of the Bangkok Special Prison served as a reminder that *Laab* lovers like Anon and other activists were still imprisoned.[28]

After Anon was released on bail, he organized a political protest to support the 'People's Draft' of the Thai constitutional amendment on 17 November 2020. Anon invited people to eat *Moo Grata* (หมูกระทะ) or meat barbecue with soup prepared on hot charcoal just in front of the Thai parliament. He also intended to sell *Soi-Ju* in front of the parliament.[29] Anon intended to price his *Soi-Ju* at 112 bahts, as 112 is the number of the *Lèse-majesté* law which he strongly opposed. The Royal Thai Police released a warning that it is illegal to set up charcoal stoves to cook meat and grill shrimp on the street.[30]

On 27 November 2020, the largest and most important *Laab* Group's eating event was held at Lad Phrao Intersection. This event was held as part of a larger protest with more than 10,000 protesters. The response to this 'Eat *Laab* against the Coup' event was overwhelmingly positive. The Group gave away all 112 sets of *Laab*, *Koi*, *Somtam* (papaya salad). Anon also gave a speech against the *Lèse-majesté* law and criticized the current government at the protest.[31]

But it is clear that *Soi-Ju* eating is synonymous with Anon Nampa and other protesters. As the democratic movement became mainstream, *Soi-Ju* also became a trendy food in Bangkok. After the push for raw meat dishes by both Anon and the youth protesters, *Laab* and *Soi-Ju* is no longer just the food of poor minorities. As of 15 May 2022, Wongnai.com's search results list around 9900 restaurants in Bangkok that include *Soi-Ju* on their menus.

Conclusions

Laab has been used to symbolize the identity of the politically marginalized voters from north and northeastern Thailand. However, eating *Laab* and *Soi-Ju* during recent the mass demonstrations changed its meaning. Raw meat-eating on the street was not only an act of dissent against the Thai government's order and ideology, but also a unifying medium that bridged the Red Shirts and the youth-led movement together as one. Raw meat salad eating fostered unity within the movement.

Raw meat salads played an important role among the Red Shirts and the youth-led movement in recent years. For traditional Thais, the act of standing and walking while eating is considered impolite. In this case the protesters often passed around the food bowl, even while eating on the ground, acts scorned by traditional Thais as practices of the poor and uneducated. *Laab* and *Soi-Ju* served during the protest was often eaten by hand with steamed sticky rice, making this raw meat salad highly portable, allowing activists to move easily and quickly if the Thai riot police moved in to disperse the protest. Therefore, the systematic act of cooking, walking, and eating previously unwanted raw meat dishes at illegal feasts on the street became acts of defiance and thus a preferred method of protest.

Notes

1. David Sutton and others, 'Food and Contemporary Protest Movements', *Food, Culture and Society*, 16.3 (2013), 345–66.
2. Christina Sunyoung Kim and others, 'Role of Socio-Cultural and Economic Factors in Cyprinid Fish Distribution Networks and Consumption in Lawa Lake Region, Northeast Thailand: Novel Perspectives on *Opisthorchis viverrini* Transmission Dynamics', 2017, 170, *Acta Trop*, pp. 85–94; Department of Disease Control, Ministry of Health, *Manual for Disease Control: Fluke. For Health Officer* (Bangkok: Sam Charoen Panich, 1993), preface.
3. Christina Sunyoung Kim and others; Department of Disease Control.
4. Christina Sunyoung Kim and others; Department of Disease Control.
5. 'Chiang Mai Artists and Activists Organized, Eating while Waiting for the Election', *Prachatai*, 10 January 2014 <https://prachatai.com/journal/2014/01/51080> [accessed 12 November 2022].
6. Department of Disease Control.
7. David M. Engel, 'Blood Curse and Belonging in Thailand: Law, Buddhism, and Legal Consciousness', *Asian Journal of Law and Society*, 3 (2016), 71–83.
8. Michael Volpe, 'Frame Resonance and Failure in the Thai Red Shirts and Yellow Shirts Movements' (unpublished doctoral dissertation, George Mason University, 2015), pp. 167–68, 297–302.
9. Engel.
10. Salisa Yuktanan, 'Ritualizing Identity-Based Political Movement Challenging Thailand's Political Legitimacy through Blood-Sacrificing Rituals', *The Michigan Journal of Asian Studies*, 1:2, 2012, pp. 89–110; Patrick Jory, 'Thailand's Politics of Politeness: Qualities of a Gentleman and the Making of "Thai Manners"', *South East Asia Research*, 23.3 (2015), 357–75.
11. Jory.
12. Yuktanan.
13. Krit Lualamai, 'In the Big Eating Circle', *Prachatai*, 27 March 2010 <https://prachatai.com/journal/2010/03/28533>[accessed 12 November 2022].
14. Mai Neang Kor Kunati, 'Kaysone Phomvihane', *Prachatai*, 13 December 2010 <https://prachatai.com/journal/2010/12/32260> [accessed 12 November 2022].
15. Engel; Kevin Hewison, 'Royalism and Democracy: Rebellion, Repression and the Red Shirts', *East Asia Forum Quarterly*, April–June 2010, pp. 14–17; Yuktanan.
16. Engel, pp. 71–72.
17. '12 Years Big Cleaning Day Washes Away the Deaths from the Protests, Leave Only Memories', *Thairath*, 23 May 2022 <https://plus.thairath.co.th/topic/speak/101551> [accessed 12 November 2022].
18. Volpe.
19. 'Chiang Mai Artists and Activists Organized'.
20. 'Amnesty Issued a Statement Pointing Out that the Arrest of the Protesters Created "Dangerous Norm" Urges Thai Military to Show Caution', *Amnesty International Thailand*, 24 May 2014 <https://www.amnesty.or.th/latest/news/oldweb15/> [accessed 12 November 2022].
21. 'The 'Birthday Cake Cutting' at the Democracy Monument', *Prachatai*, 11 December 2014 <https://prachatai.com/journal/2014/12/56942> [accessed 12 November 2022].
22. 'Phayao Akhad Burns an Incense to Tell Her Daughter after Supreme Court Supreme Court Acquitted Suthep-Aphisit Lawsuit Against the 2010 Protests', *Prachatai*, 1 September 2017 <https://prachatai.com/journal/2017/09/73048> [accessed 12 November 2022].
23. 'Phayao Akhad Burns an Incense'.
24. Nattha Chuenwattana, Chonlatorn Wongrussamee, and Anusorn Tippayanon, 'Anon Nampha and The

Laab Group: Resistance on a Plate of Raw Meat', *The Momentum,* 18 October 2021 <https://themomentum.co/feature-revolutionaryfood-arnon-nampa/> [accessed 12 November 2022].

25 Chuenwattana, Wongrussamee, and Tippayanon.

26 See for example 'The Laab Group Was Born', *Facebook,* 21 September 2020, <https://www.facebook.com/kanalab2563/photos/pb.100059436827356.-2207520000./104286314767583/?type=3> [accessed 12 November 2022].

27 คณะลาบ2563, '14 October, Let's Go to Eat Laab at the Democracy Monument', *Facebook,* 4 October 2020 <https://www.facebook.com/kanalab2563/photos/pb.100059436827356.-2207520000./117681470094734/?type=3> [accessed 12 November 2022].

28 'I want to eat here! The Mob Organized Eating Laab-Koi in Front of Klong Prem Prison', *Daily News,* 19 October 2020 <https://d.dailynews.co.th/politics/802011/> [accessed 12 November 2022].

29 'The Mob, Nov. 17, Surrounded the Parliament by "Anon's Soi Ju"', *KomChadLuek,* 16 November 2020 <https://www.komchadluek.net/scoop/449228> [accessed 12 November 2022].

30 'Police Warns, Setup Tables to Eat Grilled Pork, and Shrimp, on the Street Are Illegal', *The Standard,* 16 November 2020, <https://thestandard.co/police-point-setting-table-to-eat-pan-fried-pork-on-the-road-is-illegal/> [accessed 12 November 2022].

31 'Anon Joined the Laab Group to Eat Sticky Rice and Papaya Salad Against the Coup in the Middle of Lad Phrao Intersection', *Matichon,* 27 November 2020, <https://www.matichon.co.th/politics/news_2461794> [accessed 12 November 2022].

Savour the Flavour: A Portable History of Instant Coffee

Julia Fine

In early 2020, a new pandemic trend came into vogue alongside sourdough bread and feta pasta: whipped instant coffee. Also called dalgona coffee, this Korean-inspired trend was touted by stalwart American media outlets like *The Wall Street Journal*, *The New York Times*, and *The Washington Post*. But even as dalgona coffee was extolled, the very same pages decried its central ingredient, instant coffee: 'Long reviled by coffee snobs, instant coffee is having its moment,' *The Wall Street Journal* wrote. 'There's just one issue: People are spending an awful lot of time trying to make it taste good.'[1]

Yet despite the old-guard American media's censure of the drink, around the world, instant coffee is incredibly popular. As of 2019, for instance, 72% of all coffee consumed in China was instant coffee (as compared to 18% freshly ground coffee and 10% ready-to-drink coffee).[2] Similarly, in a survey of UK coffee drinkers, the largest segment of the population (33%) enjoyed instant coffee, with Nescafé as the most beloved brand.[3] As of June 2021, the plurality of South Koreans drank instant coffee every day.[4]

Scholars have suggested various reasons why instant coffee is so despised in the US but beloved worldwide, pointing to its convenience and low cost. But this understanding suggests that instant coffee is consumed around the world despite, not because of, its taste. Rather than beginning from the assumption that instant coffee tastes inherently bad (or good), here, I draw on Christy Spackman and Jacob Lahne's argument that the body is not merely a 'molecular interface, a reduction that locates sensory experiences squarely in physical objects that can be measured and quantified'.[5] Instead, I follow Sarah Besky's analysis of tea production and consumption to ask how ideas about quality are created and 'what quality *does*' in the case of instant coffee – 'what claims about it are made, by whom, and with what consequences'.[6] I ask how instant coffee rose in popularity and why its qualities developed in different ways in different spaces. Through this globetrotting exploration, I ultimately hope to demonstrate that if taste, as some scholars have pointed to, is a 'social agent', it is critical to recognize that the social context itself is a 'taste agent', influencing how a variety of foodstuffs are experienced and understood in different cultural and geographical environments.[7]

A Moveable Feast

Before delving into the history of instant coffee, it is critical to delineate the history of coffee more generally and its intimate relationship to imperialism. The arabica coffee plant, which makes up the majority of coffee consumed today, is indigenous to Ethiopia, although it is unclear exactly when humans began to cultivate the plant. By the sixth century, Ethiopians occupying Yemen may have set up coffee plantations, thereby spreading the beverage to the Arab world. Ottoman Turks began to occupy Yemen in 1536, bringing the beverage back to their own empire. From there, smugglers brought the coffee seed to Southern India, and the Dutch began to cultivate coffee in Ceylon, Sumatra, and other parts of their burgeoning empire, which, together with Portuguese colonialism in the Indian Ocean, popularized the beverage in Europe.[8]

Within the context of early modern Britain, as historian Brian Cowan argues, 'virtuoso travelers and, even more so, the virtuoso readers of travel literature and accounts of the commodities of exotic cultures' popularized coffee within the British Isles as an appealing and avant-garde drink.[9] This 'exotic' drink became firmly entrenched in British culture as a 'new ingredient to the seventeenth-century English pharmacopeia', both as a 'medical cure and a newly desirable consumer pleasure [which] could maintain the oriental mystique of other exotic drugs while also remaining mostly free from the negative associations attached to more powerful psychotropic drugs'.[10]

Coffee's connection to empire thus spurred its adoption in early modern Britain. And British imperialism also spurred the creation of novel types of coffee, including early prototypes of instant coffee. While soluble instant coffee, or dried granules rehydrated with water, can be dated to the early twentieth century, portable, near-ready-to-drink coffee has a much longer history in the west. For instance, while early modern British coffee culture is often defined by sedentary consumption of the novel beverage in coffeehouses and the concomitant creation of a new public sphere, as far back as 1771 the British government granted a patent for a 'coffee compound', an easily transportable coffee essence.[11]

There is very little information accessible about this 1771 British coffee compound. What is clear, however, is the way this sort of portable coffee compound was, from its origin, indelibly related to imperialism. This relationship is immortalized in later discussions of instant coffee. In a 1964 cookbook published by Maxwell House, one of the most popular American instant coffee brands, the writer describes the origins of their brew in the context of the informal British Empire:

> One day, just after the turn of the century, an Englishman living in Guatemala was waiting for his wife in the garden of their home. A servant had just set out the impressive silver service which – in the English tradition of keeping home-away-from-home standards high – the wife had brought with her. Idly, the man savoured the delicious aroma of fresh coffee as he watched the steam rise from the lovely silver pot. Then his attention focused on the spout of the pot. He saw a fine, dark powder forming there. The powder, it seemed to him, must be the condensation of coffee vapours. That gave him an idea. He began experiments by which he hoped eventually to produce for commercial consumption coffee which could be made in the cup, simply by adding hot water to the dark powder. He was successful and, in 1909, put the first instant coffee on sale.[12]

Savour the Flavour: A Portable History of Instant Coffee

It is telling that this apocryphal story of instant coffee took place in Britain's informal colonial periphery. Further, that its genesis was allegedly inspired by the 'English tradition of keeping home-away-from-home standards high' gestures to how instant coffee – more easily portable as it would keep fresh for longer than beans – allowed coffee produced in the colonies but packaged and sold in the metropole to be sent back to the colonies.

A Soluble Solution

Soluble instant coffee's genesis also lay in military and imperial expeditions. It is unclear exactly when and how the technology for producing soluble instant coffee debuted. Some scholars point to the development of soluble instant coffee by the G. Washington Coffee Company in the early 1900s, others ascribe its rise to Nestlé's processing techniques developed in the 1940s, and others point to the development of soluble 'Kato Koffee' by a Japanese inventor in 1899 or a New Zealand inventor in the 1890s.[13] Whatever the exact genesis of the beverage, instant coffee emerged sometime around the early twentieth century, using a 'dry hot-air process' to remove moisture.

This new drying technology and the soluble coffee it produced proved incredibly useful during World War II, when the US government insisted upon instant coffee as a 'standard component' of US troop rations.[14] As far back as the US Civil War, coffee had been an 'important and vital part of the soldier's diet' due to its role in encouraging 'cheerfulness', 'comradeship', and hard work among men 'who had to march or work by day or watch by night'.[15] By the First World War, soluble instant coffee was employed particularly in the trenches on the front line, given that it required little-to-no equipment to brew.[16]

However, it was only during World War II that instant coffee became a standard part of 'every ration in which it was practicable and possible to use', including in life raft and lifeboat rations, individual combat rations, and in-flight lunches for long-range missions.[17] This deeply invigorated demand for instant coffee among former soldiers and stimulated new manufacturers' development. Before World War II, only Nestlé, G. Washington, and a handful of other manufacturers produced instant coffee. These companies could not meet the massive demand of the US army, thereby catalyzing 'about ten new manufacturers' to open during the war.[18]

Spurred on by wartime demand, the rate of instant coffee consumption in the US continued to explode after the war. By 1960, instant coffee accounted for 20-25% of the national coffee consumption.[19] Much of the marketing around instant coffee continued to target young men who may have served – the papers of American advertising executive Jean Rindlaub hold letters she wrote instructing a major US advertising firm to promote instant coffee with a 'masculine image', suggesting slogans like 'It's a man's cup of coffee – strong and rich and robust'.[20] Beyond former soldiers, instant coffee companies also worked to reach a new market: middle- and upper-class American housewives. According to a nationwide survey of American housewives conducted in 1955 for the National Coffee Association (NCA), half of all households surveyed used instant coffee. Today many

Americans think of instant coffee as a 'lower-class' beverage, but in 1955 the NCA survey found just the opposite. The highest proportions of instant coffee use were found in the top-income group (57%) and white-collar households (60%).[21]

Instant coffee brands worked to cultivate this loyalty by presenting the beverage as more standardized and purer than other forms of the beverage. Food historians have demonstrated how, in the face of rising income inequality, the 'ideal of purity' in food was heavily valued by urban middle-class white women as a means to 'distinguish [themselves] from the lower classes and claim moral authority over industrialists and corporations that were eclipsing them in wealth and power'.[22] Many advertisements and brand-produced books emphasized the purity of instant coffee as compared to regular coffee. A 1964 Maxwell House cookbook describes the process of producing instant coffee: 'The fundamental difference between regular and instant coffee is that the percolator or extraction step in making the brew is done in the plant, under scientifically controlled conditions, instead of in the home.'[23]

Moreover, instant coffee also allowed a growing number of office workers to drink the beverage on the go. As *Redbook Magazine* notes, while the vast majority of instant coffee (91%) was consumed at home, the 'remainder being used in offices, in factories, or purchased from vending machines', suggests the growing importance of portability in the postwar world.[24] Historian Mark Pendergrast describes how the invention of the coffee break – an idea promoted by the Pan American Coffee Bureau in 1952 – increased the consumption of instant coffee, buoyed as well by the invention of vending machines dispensing hot instant coffee.[25] One 1956 pamphlet published in Minneapolis contained a comic strip titled 'From Coffee House of Yesterday to Coffee Break of Today'. One panel shows a work-time coffee break in which two men in suits are served coffee in an office kitchen by a blonde woman. Over their heads, the words '62 per cent of industrial managers say coffee breaks increase office efficiency'.[26] This strip depicts instant coffee – specifically the fact that this energizing and productivity-inducing beverage could be easily and systematically brewed outside the home – as bolstering the American workforce.

Instant Coffee's Failure to Launch

Despite this vision of instant coffee's future primacy, by the 1970s the tides began to fully turn against instant coffee. While statistics from this period are scant, in this decade it is clear that 'consumption of instant coffee had levelled off or begun to decline' in countries in the Global North, including the United States.[27] Why did instant coffee lose its sheen in American markets?

Some scholars point to the poor taste of instant coffee as the reason Americans moved away from the drink in the 1970s. This poor taste is often ascribed to the fact that instant coffee is made with the stronger and more acerbic robusta beans, rather than the milder arabica beans. In the late-nineteenth century, many coffee-producing regions in Asia were struck by the coffee rust fungus, *Hemileia vastatrix*, which destroyed the sensitive arabica plant. While some places like Ceylon turned from coffee to tea, others began to look for a hardier coffee plant that could withstand the vagaries of the fungus. As Stuart McCook details, from 1905 onward coffee farmers and scientists worked to develop robusta coffee as a 'commercially viable replacement for arabica coffee' across Asia and the Pacific.[28]

In the same period, Brazil's production of arabica heavily outpaced supply. By the 1930s, Brazilian producers were 'accumulating such huge coffee surpluses that she had to destroy part of the harvest in order to prevent a further fall in prices, which were considered already to be too low'.[29] While Brazilian producers and governmental figures did this in order to stabilize coffee prices to their own benefit, there was a major unintended consequence: buyers in the United States and Europe began to purchase robusta coffees as a more economical brew than arabica, with its artificially buttressed prices.[30] Further, in addition to Brazilian price stabilization, robusta coffee's low price was influenced by decolonization, as newly independent colonies in Africa and Asia were 'eager for dollar infusions to their war-devastated economies', as well as political instability spurred on by US forces in arabica-growing countries in Central America.[31] Robusta was a prime candidate for use in instant coffee as well in order to keep costs down: by the 1950s, instant coffee categorically contained a minimum of 50% robusta coffee, with many brands using 100% robusta.[32]

So, despite historical scholarship alleging otherwise, robusta had been used in instant coffee for over two decades with little apprehension – American housewives even cooked with instant coffee, which was marketed as a 'delicious, new wa[y] to dress up desserts' and a way to make main dishes that would 'earn a reputation as a gourmet cook'.[33] And the advent of freeze-drying technologies in the late 1960s actually meant instant coffee began to taste better: freeze-drying preserved the bean's aroma, allowing producers to forgo the 'rancid'-tasting coffee oil meant to evoke the scent of a freshly brewed cup.[34] Thus, flavour was not the whole or even most salient issue undergirding the fall of instant coffee. What was?

One answer lies in the rising class anxieties associated with changing income patterns in this era in the United States. Paul Krugman described the economic crisis of the 1970s as leading to a 'Great Divergence' in the United States, in which average income rose substantially 'mainly because a few people have gotten much, much richer', while 'median income, depending on which definition you use, has either risen modestly or actually declined'.[35] Scholars have shown that this stagnation of the middle class gave rise to a 'contemporary food revolution'. In this 'revolution', middle-class Americans began to favour 'foods constructed as elegant, virtuous, and exotic' in place of the 'mid-century preference for the familiar, plain, and ample' to display their class aspirations and alleviate class anxiety.[36]

Instant coffee was a food that had long been explicitly advertised as familiar and accessible. For example, advertising executive Jean Rindlaub, in a memo to Luzianne Coffee, suggested they use the slogan, 'Use half as much coffee – Luzianne's so packed with flavor you'll still get the good taste you like'.[37] Market research bore out Rindlaub's approach in the 1950s: the plurality of all households surveyed in a 1955 National Coffee association survey viewed the coffee-making industry as 'unfavorable', mainly due to coffee being 'too high-priced'. Instant coffee was positioned as a cheap alternative: 70% of those surveyed who changed their coffee-making habits due to high price switched their brew to instant coffee.[38]

But by the 1970s, coffee became a convenient way to reinforce class aspirations catalyzed by the 'Great Divergence'. While, as we have seen, before this period instant coffee was presented as a 'modern' brew, in this era advertisements instructing how to 'brew the perfect cup' of 'gourmet' coffee began to proliferate. One pamphlet from the Coffee Development Group distributed in the 1980s instructed housewives on what exact type of grind to use with what brewing method; here, coffee was presented as a 'simple art' rather than a science.[39] Another publication by the Pan-American Coffee Bureau released in 1974 contained a guide 'all about buying coffee'. It begins by noting that '*Arabica*, grown chiefly in Central and South America, has a fine, full flavour. *Robusta* is a hardy variety from Africa that is not as fragrant as *arabica*. You won't find either of these names on a coffee can, but the price will give you a hint'. Here, the flavour of the coffee is intimately related to purchasing power.

In the face of these rising class anxieties, new technologies emerged that allowed American housewives to both remain frugal and alleviate their class anxieties. While, as demonstrated by the Pan-American Coffee Bureau pamphlets, range-top percolators and vacuum-type coffeemakers were in vogue in this period, in 1972 the first home electric-drip coffee maker was introduced by North American Systems. This machine, dubbed 'Mr. Coffee', effectively undercut the popularity of instant coffee. Like instant coffee, Mr. Coffee was presented as a frugal investment: the built-in 'coffee-saver' effectively 'raises the coffee grounds to the perfect brewing height ... automatically', and it continued to present a standardized cup. However, this frugal method allowed housewives to take control over brewing more 'gourmet' coffee: as historian Rebecca Shrum notes, 'the increasing availability of 100 percent arabica coffee beans during the 1970s made an easy-to-use brewing device in which the taste of better-quality coffee could be noticed and appreciated.' It is telling that this device producing allegedly 'better-quality coffee' was gendered as male. By the late 1960s, Shrum argues, 'women even found themselves blamed for the poor taste of instant coffee.'[40]

At first, it may seem contradictory that one standardized, packaged brew was replaced by another standardized, packaged brew albeit with a different production method. However, as Margot Finn reminds us, there is often a lack of ideological coherence in the food revolution, often with contradictory ideals.[41] And this period also saw the rise of specialty coffee houses in the 1970s which sold whole-bean coffee. As George Howell, the head of Harvard Square's Coffee Connection remembers, 'We were an overnight success. Customer enthusiasm covered us. They were like parched people coming out of a desert and finding an oasis.'[42] This rise of specialty coffee allowed the declining middle class to differentiate themselves from the lower classes based on their artisanal know-how and specialized coffee vocabulary. Instant coffee became shunned by the influential middle class, and marked as a low-grade, poor-tasting concoction.

Beneath and Beyond American Hegemony

But even as middle-class American consumers turned instant coffee into a receptacle for their class anxieties, other regions embraced the brew. For one thing, as John Talbot notes, 'instant coffee seems to be the most acceptable coffee product in traditional tea-drinking countries', in particular, the UK and Canada. Part of this may be its ease of preparation. Much like the tea bags so beloved in the UK and its former colonies, instant coffee requires little in terms of ingredients, technical know-how, or equipment.[43]

In addition to this similarity to tea, multinational corporations explicitly marketed instant coffee as a symbol of American modernity – even as many American consumers shunned the beverage. Susie Khamis has detailed how Nestlé, between the 1950s and 1970s, worked to increase Australian consumption of instant coffee by appealing to 'seminal shifts in the nation's cultural identity', including 'the growing influence and appeal of the United States, the nation's deepening engagement in advanced consumer capitalism, and the postwar arrival of non-British immigrants'.[44] More recent advertising of Nescafé in Bangladesh has mimicked this model, presenting the drink as the harbinger of Western-centric modernity. Scholars have documented Nescafé's 2012 television campaign, aired in Bangladesh, with the slogan 'My first cup' and the tagline 'Time you started'.[45]

But this is not to say that the only reason instant coffee is embraced in parts of the Global South is due to US hegemony or prejudiced visions of modernity. Instant coffee's robusta flavour may have been more familiar in many Asian and African markets, leading to its greater popularity. As Yannis Apostolopoulos, CEO of the Specialty Coffee Association, reminds us, 'Coffee is the epitome of colonialism, one way or another. The proximity to Central or South America is what makes the US appreciate coffees from the Americas, which are not available in other geographies.' While the taste for arabica versus robusta in the US was, as we have seen, culturally constructed, there was deep familiarity with and availability of arabica in the nation. Instant coffee tends to be more popular in places where robusta is grown, both because of taste proximity, as Apostolopoulos points out, and price.[46] In 2019-2020, Vietnam produced 29.1 million bags of robusta, a 10% rise spurred on by what Nikkei Asia describes as 'Asians' love' for instant coffee. Likewise, Indian production of robusta coffee rose 8.1% to 4 million bags in 2019.[47]

And so we return again to Korea's dalgona coffee. Rachel Laudan, in her work *Cuisine and Empire*, argues that although McDonald's was presented as a symbol of American tastes, diners throughout the world 'adapt[ed] McDonald's to their own purposes, which are quite different from those in its American homeland', and 'everywhere [...] American fast food chains have gone, they have provoked local competition'.[48] Similarly, while corporations like Nescafé may have positioned instant coffee as a harbinger of Western modernity, these countries created their own forms of instant coffee consumption – like the dalgona so heavily decried in *The Wall Street Journal*.

Conclusion: An Instant Future?

Even as old-guard American media outlets condemned instant coffee in their pages, in 2021, a different sort of shift was happening at the vanguard of cuisine in the United States. A plethora of US-based companies – including Gen-Z superstar Emma Chamberlain's own coffee brand – have launched instant coffee products appealing to both quality and convenience. Other brands, like the start-up Molecule Coffee, appeal to instant coffee's convenience as well as the wellness industry to offer instant brews with 'bioavailable' ingredients like Ashwagandha root and lemon balm.[49]

Why has instant coffee come back into vogue in the present-day United States, helmed by Gen-Z influencers and TikTok trends? Finn, in her work *Discriminating Tastes,* details the logical evolution of the American food revolution: while instant coffee was once snobbishly eschewed by middle-class consumers for its cheapness, plainness, and availability, today middle-class consumers are eschewing snobbishness itself.[50] As one sociological study argues, using omnivorousness to signify variegated cultural consumption rather than literal consumption of both meat and non-meat, 'Highbrows are more omnivorous than others and they have become increasingly omnivorous over time.'[51] This 'performance of omnivorousness', which 'itself communicates privilege' has effectively rehabilitated the image of instant coffee among middle- and upper-class American consumers, who are imbibing an upscale, cosmopolitan version of the drink that serves to erase the way these same groups once deemed the brew as low-class.[52]

This is not necessarily to wholly castigate middle-class consumers for this 'soft bigotry of taste' – taste, after all, is a product of structural acculturation. But, as Finn reminds us, consumers should be 'more critical about the narratives they embrace when it comes to what kinds of foods are better or worse'.[53] This need for greater awareness becomes clear when discussing robusta coffee: despite the rehabilitation of instant coffee in the eyes of middle-class consumers, robusta has not received the same treatment. When I spoke to the co-founder and former CEO of Molecule Coffee, an instant coffee brand that blends the brew with bioactive compounds, he sheepishly disclosed the fact that his brand used robusta beans, rather than arabica. 'We plan to launch single-origin arabica instant coffee,' he said. 'But we started from a simpler version that was robusta, because of cost.' He noted that many consumers decried the 'cheaper' taste of the coffee: 'People expect really great taste from our coffee. They don't care about the quality of reasons,' he says, referring to their explanations as to why the company used robusta instead of arabica coffee.[54] Thus, while instant coffee is being reappropriated by elite American brands, robusta continues to be denigrated.

However, US brands and consumers may have to embrace robusta wholeheartedly amidst a changing climate. Robusta, as a hardier plant, can withstand higher temperatures while producing more fruit, and it is resistant to fungal and pest pressures that are increasing in the current climate crisis. Already, Brazil – which traditionally focused on arabica – has increased its production of robusta by over 20%.[55] And, between 2021 and 2022, total arabica output decreased from 102 million bags to 88 million bags.[56] The global history of instant coffee suggests that American preference for arabica is neither innate nor preordained; American consumers may have to step out of their 'othering' of robusta and interrogate their class anxieties in order to continue to enjoy a cup of joe during our current age of climate emergency. Miguel Meza, the owner of Paradise Coffee Roasters in Minneapolis, put it simply: 'This has to be part of our future.'[57]

Notes

1. Frances Yoon, 'People Are Spending a Long Time Making Instant Coffee', *Wall Street Journal*, 7 June 2020, section A <https://www.wsj.com/articles/people-are-spending-a-long-time-making-instant-coffee-11591553888> [accessed 22 December 2022]. See also Julia Fine, 'Plant of the Month: Robusta Coffee', *JSTOR Daily* (blog), 18 August 18 2021 <https://daily.jstor.org/plant-of-the-month-robusta-coffee/> [accessed 22 December 2022].
2. 'Analysis of China's Coffee Market Status and Development Trends 2020', *Forward Intelligence* (Qianzhan), May 2020 <https://bg.qianzhan.com/trends/detail/506/200525-1cf1000e.html> [accessed 22 December 2022].
3. 'Food and Hot Drinks in the UK 2019', *Statista*, February 2020, <https://www-statista-com.ezp-prod1.hul.harvard.edu/study/66624/food-and-hot-drinks-in-the-uk/> [accessed 22 December 2022].
4. 'Frequency of Drinking Instant Coffee in South Korea as of June 2021', *Kantar Media*, September 2021 <http://www.kantarmedia.co.uk/businesses/tgi/> [accessed 22 December 2022].
5. Jacob Lahne and Christy Spackman, 'Introduction to Accounting for Taste', *The Senses and Society*, 13.1 (2 January 2018), 1–5 (p. 4) <https://doi.org/10.1080/17458927.2018.1427361>.
6. Sarah Besky, *Tasting Qualities: The Past and Future of Tea* (Berkeley: University of California Press, 2020), p. 5.
7. David E. Sutton, 'Food and the Senses', *Annual Review of Anthropology*, 39 (2010): 217.
8. Mark Pendergrast, *Uncommon Grounds: The History of Coffee and How It Transformed Our World* (New York: Basic Books, 2019), p. 24–25.
9. Brian Cowan, *The Social Life of Coffee: The Emergence of the British Coffeehouse* (New Haven: Yale University Press, 2008), p. 14.
10. Cowan, pp. 31–32.
11. Frances H. Martin, 'A History of Coffee Prices in the United States, 1840-1954', *Monthly Labor Review*, 77.7 (1954): 765–67. For more on early modern British coffeehouses and their role in social life, see Cowan and Troy Bickham, *Eating the Empire: Food and Society in Eighteenth-Century Britain* (London: Reaktion Books, 2020), pp. 77–87.
12. *The Maxwell House Coffee Cookbook*, ed. by Doris McFerran Townsend (New York: Pocket Books, 1964), p. 239.
13. See John M. Talbot, 'The Struggle for Control of a Commodity Chain: Instant Coffee from Latin America', *Latin American Research Review*, 32.2 (1997): 117–35; 'Regular and Instant Coffee, Substitutes and Other Coffee Beverage Products: Notes on Consumption and the Household Market', *Redbook Magazine*, January 1957; Kieun son and Heewon Shin, 'Academic Manual: History', *F Magazine*, n.d., 46.
14. Talbot, p. 120.
15. Franz A. Koehler, *Coffee for the Armed Forces: Military Development and Conversion to Industry Supply* (Washington DC: Historical Branch, Office of the Quartermaster General, 1958), pp. 11–12.
16. Koehler, p. 16.
17. Koehler, p. 18.
18. Talbot, p. 120.
19. Talbot, p. 120.
20. Cambridge, MA, Schlesinger Library, Papers of Jean Rindlaub, MC693 Box 8 Folder 5.
21. 'Survey of Consumer Attitudes on Coffee for National Coffee Association', Princeton: Benson & Benson Inc., 12 October 1955, p. 26.

22. S. Margot Finn, *Discriminating Taste: How Class Anxiety Created the American Food Revolution* (New Brunswick: Rutgers University Press, 2017), pp. 62, 9.
23. *The Maxwell House Coffee Cookbook*, p. 240.
24. 'Regular and Instant Coffee', p. 28.
25. Pendergrast, p. 189.
26. *The Magic Bean: The Story of Coffee* (Minneapolis: Tadlock, 1956).
27. Talbot, p. 121.
28. Stuart McCook, *Coffee Is Not Forever: A Global History of the Coffee Leaf Rust* (Columbus: Ohio University Press, 2019), p. 90.
29. Jean Heer, *World Events, 1866-1966: The First Hundred Years of Nestlé* (Chicago: Harshe-Rotman & Druck, 1966), p. 164.
30. McCook, p. 103.
31. Pendergrast, pp. 188, 194–95.
32. Pendergrast, p. 262.
33. *25 New Coffee-Flavored Desserts Starring Nescafé* (New York: Nestlé Co., 1948), p. 2; *The Maxwell House Coffee Cookbook*, p. 43.
34. Rebecca K. Shrum, 'Selling Mr. Coffee', *Winterthur Portfolio*, 46.4 (Winter 2012), 272–298 (p. 272) <https://doi.org/10.1086/669669>.
35. Paul Krugman, *The Conscience of a Liberal* (New York: Norton, 2009), p. 167.
36. Finn, pp. 10–11.
37. Papers of Jean Rindlaub, MC693 Box 8 Folder 5.
38. 'Survey of Consumer Attitudes', pp. 54, 64.
39. *Secrets for a Perfect Cup* (Washington DC: Coffee Development Group, 1986).
40. Shrum, pp. 271, 286, 282.
41. Finn, p. 31.
42. Pendergrast, p. 226.
43. Talbot, pp. 122, 121.
44. Susie Khamis, '"It Only Takes a Jiffy to Make": Nestlé, Australia and the Convenience of Instant Coffee', *Food, Culture & Society*, 12.2 (June 2009), 217–33 (pp. 218–19).
45. Duncan Barnes, Danielle Fusco, and Lelia Green, 'Developing a Taste for Coffee: Bangladesh, Nescafé, and Australian Student Photographers', *M/C Journal*, 15.2 (May 2012) <https://doi.org/10.5204/mcj.471>.
46. Yannis Apostolopoulos, Interview with CEO of the Speciality Coffee Association, *Microsoft Teams*, 1 June 2022.
47. Sachiha Kurose, 'Asians' Love for Instant Coffee Fuels Robusta Cultivation', *Nikkei Asia*, 6 October 2019 <https://asia.nikkei.com/Business/Markets/Commodities/Asians-love-for-instant-coffee-fuels-Robusta-cultivation2> [accessed 22 December 2022].
48. Rachel Laudan, *Cuisine and Empire: Cooking in World History* (Berkeley: University of California Press, 2015), p. 311.
49. Denis Simonov, Interview with Co-Founder of Molecule Coffee, *Zoom*, 24 May 24 2022.
50. Finn, p. 196.
51. Richard A. Peterson and Roger M. Kern, 'Changing Highbrow Taste: From Snob to Omnivore', *American Sociological Review*, 61.5 (October 1996), 900–07 (p. 900).
52. Finn, p. 196.
53. Finn, pp. 216-217.
54. Apostolopoulos.
55. Zac Cadwalader, 'In Response to Climate Change, Brazilian Producers Are Switching to Robusta', *Sprudge*, 17 August 2021 <https://sprudge.com/in-response-to-climate-change-brazilian-producers-are-switching-to-robusta-180686.html> [accessed 22 December 2022].
56. Son and Shin, p. 57.
57. Brooks Johnson, 'Coffee Roasters Look to Long-Snubbed Robusta Bean as Climate Changes', *Star Tribune*, 4 January 2022.

Mobile Gastronomies: A Case Study of Food on Wheels Culture in Hyderabad, South India

Kashyapi Ghosh and V. Vamshi Krishna Reddy

Portability in today's world has attained new meaning: we are in the midst of a technological transition that has changed our lives in many ways. With the burgeoning of technology and the shrinkage of the world into a 'global village', portability has taken new forms (McLuhan 1964). Among all other forms of portability, portable food has grown immensely in the last two years owing to the effects of the global pandemic. A total of 2.2 million daily food orders were placed through an online medium, 16% of which came from one of the fastest developing cities in India, Bangalore. The dominant space of portable food in India has two sources: one, that of food software delivery applications, known as 'gastro apps', like Swiggy, Zomato, Dunzo, and UberEats, which had smaller predecessors like FoodPanda, Tinyowl, and others (Srinivas 2021:18). At the same time, brick-and-mortar restaurant kitchens have been evolving into cloud kitchens, which are different from restaurant kitchens as temporary spaces for preparing food and easing the delivery process. In this paper we aim to delineate a form of portable food that does not fit easily into the usual categories: the temporary mobile vans that serve India's burgeoning tech hubs.

Portable food has had a long and interesting history in the Indian subcontinent. According to David Burton, the earliest known form of portable food, colloquially called the 'tiffin', was introduced in India by the British in the nineteenth century (Burton 1994: 17). Around the 1880s, the colonizers introduced a light snack in between lunch and dinner to beat the severe tropical heat. From colonial times to now, portable food has acquired different forms. India has always had the culture of travelling and had long fascinated travellers from foreign lands, and food remained an integral part of that travelling. The food away from the traditional dining table began with the offerings available at roadside eateries. Chitrita Banerjee praises road food 'from a kaleidoscopic panoply of eateries offering signature regional cooking' (2007: 48). This regional cooking ranges from rice, vegetables, and fish in the east to *chapatis* (flat bread roasted over fire), *dal* (lentils), and *paneer* (cottage cheese) when one ventures towards the west. However, the food for religious travellers, Buddhist monks, Hindu saints, and Sufi mystics, was simply rice and *dal* cooked in a vessel over open fire (Banerji 2007: 50). Road food is an institution that promotes the mobility in the country. When Sher Shah

Suri built the highway covering the Golden Quadrilateral, he also constructed inns and hostelries to make travelling along the highway convenient. The palate underwent a transformation as one travelled from the east to west. Rice was replaced by flat breads made in clay ovens and kilns. Such has been the history of portable food in India.

However, there is another form of portable food that needs to be introduced before engaging with the main argument. Many conversations about portable food in India have revolved around the *dabbawala* service in metropolitan Mumbai. Premiere institutions like the Harvard Business School and the Indian Institute of Management, Ahmedabad, have studied their business model. As Sara Roncaglia explains, 'The Mumbai *dabbawalas* are food deliverymen who connect homes and workplaces – messenger boys, urban servants who are fast and precise, trustworthy and discreet, clean and punctual' (2013: ix). They are like the food delivery applications except they work without smartphone apps on a strategic internal system. The *dabbawalas* date back to 1890, when Mahadeo Havaji Bachche started a tiffin delivery service for migrant workers. From 1930 onwards he attempted to make it into a union, and it was not until 1956 that the first registered charitable trust, called the Nutan Mumbai Tiffin Box Suppliers Charity Trust (NMTBSCT), was established. This tiffin box supplier delivered 200,000 dabbas to people working across the length and breadth of metropolitan Mumbai (Roncaglia 2013: xiii). This robust network of suppliers delivering lunch to office-goers, students, and others was extensively featured in Ritesh Batra's award-winning movie, *The Lunchbox* (2014). The *dabbawala* service is the starting point of many conversations about portable food which we mentioned formerly has attained variegated dimensions in today's world dominated by technological innovations. This is a cursory glance of the evolution of portable food and how a South Asian country like India went from eating food while travelling to food travelling to our homes during the pandemic.

Portable Food: The Beginnings

In this paper, our aim is to unpack an alternative form of portable food: food far away from the table, different from the roadside mobile hotels or the *dabbawalas* that have been an important part of the cosmopolitan spectrum. A lot of research has been conducted on food trucks as a form of portable food by social science scholars (Anenburg and Kung 2015). It has been said the first food truck in India was started during the 1970s, the Emergency era, by a young adult, Aroon Narula. It was a bright and yellow mobile van called Hawker parked at the Faculty of Arts, Delhi University, and later shifted to pavement in front of Delhi School of Economics. Since then, food trucks and bustaurants (bus restaurants) have staggeringly increased their presence over the years. Food trucks are an attractive business set-up with low investment and fewer resources involved especially with the pop-up restaurant trend in the 2010s. This decade saw the emergence of the middle class, a by-product of the liberalization of the economic market, and India's eating-out culture gained a momentum in urban and semi-urban areas with increasing globalization and the greater disposable income of the middle class. The entry of the first

American fast-food chain, McDonald's, in the capital city of New Delhi, was a game changer in the Indian food scene. The success of McDonald's dovetailed Domino's, KFC, Pizza Hut, Dunkin' Donuts and other quick service restaurants (QSR) to the Indian foodie scene and subsequently the concept of the fine dining experience was introduced. However, there is an alternative portable food culture that has grown in parallel with the burgeoning of Information Technology (IT) hubs and parks in cities like Noida, Gurgaon (now Gurugram), Pune, Bangalore, and Hyderabad.

The form of portable food that we want to discuss are the mobile vans which essentially do not come under the high culture of food trucks. These mobile vans are the makeshift vehicles strewn around financial districts and IT hubs across India. Before we begin to explore the trajectory of these food vans and how they came into the Indian food scene, a short background on the evolution of the IT industry will help us to understand the context better. This background about the IT sector is important because our form of portable food is positioned at the heart of that evolution. The IT industry is one of the most booming economic sectors in India. India's IT industry made a mark after the recession of the early 1990s slowed the US economy. Then the National Association of Software and Service Companies (NASSCOM) reported an annual revenue of $40 million. By 2011, as Balakrishnan points out, 'The number of people directly employed in the industry was reported as exceeding 2 million and those indirectly employed as many as 7 million. The industry accounted by then for 5.2 percent of India's gross domestic product (GDP), and its annual expenditure in the domestic market was reported at $16 billion' (2011: 3). The nature of such massive employment created flux in the sectors of housing, transportation, and food as young people moved away from the comforts of their homes to new cities. Food became a ubiquitous aspect of the resulting 'work hard, party harder' lifestyle: 'According to FICCI, the average Indian used to eat out around 2-4 times in a month in 2010. However, the frequency increased to an average of 6.6 times in a month a decade later' ('How India Eats Out?' n.d.). This data charts the immense growth of the food industry in India. Chef Suchit Garg explains, 'Eating out trends have not only evolved with the new restaurants in the market but has also become a regular form of entertainment – especially in the metros, mini-metros and Tier-I cities – that is driven by higher incomes, greater number of nuclear families and working women and urbanisation' (Nusra 2013).

Having traced the evolution of the portable food in India from the colonizers' light snack between meals to the *dabbawala* community to the present gastro apps which deliver food at the doorsteps, we arrive to our investigation of the makeshift mobile vans. The temporary mobile vans are a representation of the alternative forms of infrastructure that dominate city life. Modern life has been overtaken by technology at every step, forcing humans to migrate in search of better opportunities and better living standards. These mobile vans that appear just before lunch and close their business before dusk make an important contribution to the urban geography of the new financial districts. These vans are the in-between, the 'intersectional spaces' where people from all classes converge. The vans stand as an important metaphor for modern life, where everything is transitioning. They are neither a big business enterprise like the 'gastro apps' funded by millionaires, nor are they the quick fix start-up venture of the food trucks. These mobile food vans suggest an alternative and inclusive conversation around mobile gastronomies. One might question the importance of such a study: why do these mobile vans comprising the unorganized labour sector in a financial district in South Asia matter? Toby Miller notes that '[a] key lesson of cultural studies is the value of making the everyday central to politics rather than epiphenomenal or dilettantish' (2013:141), and here the idea is to bring this overlooked everyday aspect of portable food into academic conversation. Therefore, these mobile food vans will be the starting point of our conversation about alternative forms of portable food in India.

Positioning the Portability of Food

In 'Disjuncture and Difference in the Global Cultural Economy', Appadurai talks about how the global cultural economy is no longer simple, unilinear, and cannot be 'understood in terms of existing centre–periphery models' (2006: 588). Hyderabad, the city of the Nizams, known for its *dum biriyani* (a famous dish made of rice and meat cooked together which originated in Mughal cuisine but has been adapted in different parts of India), is a symbol of the complex global cultural economy that has grown significantly in terms of infrastructure and population. Hyderabad became a melting pot of cultures, professions, and economies during the 1990s. In 1994, the chief minister of the state of Andhra Pradesh (presently Hyderabad is a part of Telangana after a bifurcation of the state in 2014) had the vision of transforming Hyderabad into a metropolitan city with the creation of 'knowledge enclaves' akin to Malaysia's Multimedia Super Corridor. His goal was to create growth-driven reform in a state where agriculture was the dominant form of livelihood. This reform grew along the lines of major urban restructuring and creation of IT hubs. A number of foreign and native software companies like Microsoft, Google, Infosys, and Wipro established offices in these 'knowledge enclaves'. Das notes that 'in conjunction with the name Hyderabad, this new cyber enclave has been named "Cyberabad"' (2015: 54). Cyberabad is a technology township also known as HITEC City (the Hyderabad Information Technology and Engineering Consultancy City), which was inaugurated in 1998 with the Cyber Towers, 'a 10-storey "intelligent" building with a 580,000 ft^2 area fitted with dedicated fibre-optic internet links and uninterrupted power supply' (Das 2015: 237). It is this HITEC City that forms the ground for our narrative on portable food. HITEC City includes the suburban areas of Madhapur, Gachibowli, Kondapur, Manikonda, Nanakramguda, and Shamshabad and hosts a number of small and big multi-national companies including Facebook (now Meta), Amazon Global Campus, and Microsoft. These sites are significant to our narrative of mobile food vans that are symbolic of city life.

To understand and analyze the mobile food vans strewn across the campuses of these companies, we visited in and around Gachibowli and Madhapur to observe the workings of the mobile food vans. These mobile food vans cater to a wide range of metropolitan citizens from people who offer housekeeping, maintenance, and security services to the people who work as white-collar employees in these multi-storeyed office spaces (Figure 1). What is striking about these vans are how they just appear an hour before lunch and disappear well before evening. These mobile vans offload their items onto small makeshift tables and serve their food on disposable plates (Figure 2). The food is cheaper than at the average restaurant, making it a viable option for the everyday meal. We used non-participant observation, walking around the financial districts of Gachibowli and Madhapur to understand as well as experience the multi-faceted dynamics of the mobile food vans. As Pink observes, 'to understand everyday life, we need to acknowledge that it is neither static nor necessarily mundane' (2012: 14); these mobile food vans are a synecdoche of everyday life in the financial districts and a witness to the dynamics of city space. The vans move around in time and space, depicting the vagaries of city life which are often unrepresented.

Mobile Food Vans: Platforms of City Life

As Dhiraj Barman opines, 'Third World urban conditions are full of contradictions and complex layers of contestation, which is articulated in many ways' (2020: 62), and the same can be applied to the financial district of Hyderabad. These mobile food vans are metaphorical of the 'complex layers of contestation' that dominates urban life in the Global South. These vans do not have fixed places of service, nor do they have proper licences and documentation to run a business, yet they thrive for an odd hour or two during lunch time, reasserting the 'spatially blurred and mixed hierarchies' that mega-cities are comprised of (Castells 2010: 436). People from all sections of society come to these food vans as a matter of convenience as well as availability. They offer a more limited, pocket-friendly range of food items compared to food delivery applications and restaurants. These food vans can be read as 'platforms' which are 'emerging as a focal point of diverse literatures' (Barns 2019: 2).

Figure 1 (left). The employees of financial districts crowding the mobile vans in Nanakramguda, May 2022. Figure 2 (right). Makeshift tables and disposable plates: quintessential parts of mobile gastronomy.

The food vans are emerging platforms that change the socio-spatial experience of city life. Stationed at street corners, they are ubiquitous pictures of the mega-cities that are 'discontinuous constellations of spatial fragments, functional pieces, and social segments' (Castells 2010: 436). These mobile vans essentially do not fit into the markers of the global city like skyscrapers, shopping malls, and technology-subsumed spaces. They form another aspect of the burgeoning field of platform urbanism. Platform urbanism, as Barns suggests, converses with a set of ideas about 'how the increasing ubiquity of platform ecosystems is reshaping urban conditions, institutions and actors' (2019: 19). These vans are part of the greater 'platform ecosystems' which reshape the urban space, the emerging global city that everyone wants a piece of. These 'platform ecosystems' not only reshape urban conditions but also alter the way we look at the 'actors' of these ecosystems. The vans in the financial district allow the common citizen to engage with these platforms promoted by city life.

These 'platform ecosystems' are also intersectional spaces where an entire gambit of people gathers. The mobile van as a platform ecosystem has a complex layer of components. They provide food for people offering housekeeping services to the multinational companies and people who offer maintenance services like plumbers, electricians, and lift men, but also to the people who have white collar jobs in these companies. The whole corpus of employees who benefit from these urban structures gather around these vans as an important part of their everyday life. Now, the contention is to position these mobile vans in the mainstream of activities. The owners of these vans do not own the means of production to have a bigger business set-up; the vans could also be a secondary source of income for many. But this 'platform ecosystem' gives them an opportunity to be a part of the urban infrastructure and benefit from it. This what Barns claims is 'reshaping urban conditions, institutions and actors' (2019b: 19). It allows the ordinary middle-class individual to become a part of the urban socio-spatial experience. It gives them a chance to stand at the centre of the discursive space and create a diversity of their own. By selling their food in makeshift vans, they not only engage with urban conditions, but they also reshape the existing spaces.

Portability of Mobile Food Vans

When we started looking for different kinds of portable food, the obvious point of reference became the food delivery applications that Srinivas calls 'gastro apps' that have proliferated in global metropolitan cities as well as in small towns. Young men and old going around the city in orange (Swiggy) and red (Zomato) coloured t-shirts with a bag in pillion, calling themselves 'hunger saviours', riding their two-wheelers from one neighbourhood to another. However, we decided to explore the emergent space of these mobile food vans instead of the dominant space taken up by these 'gastro-apps'. Of course, food delivery applications have become the dominant conversation about portable food in India, but these small makeshift food vans can also be an important conversation starter about portable food in the rich socio-cultural fabric of India.

Food on Wheels Culture in Hyderabad, South India

As we reached the financial district of Cyberabad, to look closely at the functioning of these mobile food vans, a lot that met our eyes couldn't be theorized in academic jargon. A flurry of events marked these urban transitional spaces just before lunch. We took turns to visit Knowledge City, Gachibowli, and observe closely the workings of these mobile food vans. These mobile food vans offload their comestibles at small temporary 'platforms' with a multi-coloured garden umbrella to provide respite from the scorching May summer (Figure 3). The nature of these vans is transitionary: they do not have a fixed spot nor do they have a seating facility. They found unused blocks of concrete or stones to provide temporary seats for customers; most of the time customers are seen standing and eating. The range of comestibles offered by these food vans is limited; mostly lentils, rice, and *sambar* (South Indian lentil stew made of *brinjal*, drumsticks with a zest of tamarind), a minimalist version of the famous Hyderabadi *dum biriyani,* curd rice (a South Indian fare where curd and rice are mixed together over a simmer of spices), and lemon rice (made with a spicy mixture added to white rice, giving it a yellowish tinge). The price is kept in an affordable range to cater to all classes of people. The food is usually stored in big stainless-steel containers of different shapes and sizes. Pickle is a mandatory compliment with most of the food. The food mostly caters to the south Indian palate, but people who work in these financial hubs come from different regions in India. They prefer to eat at these vans owing to the pricing and quick service.

Just outside the HSBC Park at Gachibowli, we see an old worn out Maruti Omni (one of Maruti's earliest models) parked at the corner, and food is being brought out and arranged on plastic tables just before the lunch time rush hour begins (Figure 4). Right around 1:00 pm, we observe people flocking from the HSBC Park towards these corner street 'platforms'. Grabbing their lunch here is more convenient than ordering from the gastro apps as there is little transit time. Ordering online is sometimes an expensive affair with restaurant packing charges, distance fees, and high demand charges being piled up on the customer. We observe a range of people flocking to these food arenas. There are

Figure 3 (left). The garden umbrellas that give respite from the summer heat in HITEC City. Figure 4 (right). Maruti Omni parked outside HSBC Park in Knowledge City, Gachibowli, May 2022.

Figure 5. The QR code scanner: a ubiquitous technological intervention in the Madhapur suburb of HITEC City, May 2022.

casually dressed youngsters probably working as interns or junior developers in the IT companies, people in security uniforms, and women in housekeeping aprons. There is a whole gambit of people who extend their services to the IT hubs in all capacities, and they are the ones who take advantage of this form of portable food.

As we walk around and note down our observations from the field, one thing that strikes us is the usage of the digital form of payment that has grown rapidly with the country's vision of a Digital India. At the extreme corner of one of the food vans, we observe a Quick Response Code (QR code) scanner to make swift payments and reduce currency transactions (Figure 5). Positioned within the IT hub, these small business owners clearly do not want to lose out on customers and have aligned themselves with the latest technological infrastructures. They have been forced to 'think infrastructurally' (qtd. in Barns 2019a: 2) to keep up to the emerging platform urbanism. This whole set-up of the mobile vans coming into this technology dominant spaces, offloading their items and serving it to the citizens, is also a part of the intersectional spectrum. Intersectionality as 'a theoretical and methodological paradigm' chooses to intervene and interrogate in and around social situations. In this social situation, the vans are the intersectional spaces which bring out people from all sectors and classes to have a meal. It doesn't segregate based on their profiles or the kind of services that they do. As Cho, Crenshaw, and McCall note, 'Intersectionality was introduced in the late 1980s as a heuristic term to focus attention on the vexed dynamics of difference and the solidarities of sameness in the context of antidiscrimination and social movement politics' (2013: 787); from the 1980s intersectionality has come a long way to bring out alternative narratives and conversations that do not depict the dominant section of society but gives equal representation to the emerging faces of the techno-social space. Here, the spaces around the mobile vans can be read as intersectional owing to the fact that they give voice to an unorganized sector trying to align to the global city life. These people with their small business set-ups truly use the platform and make it intersectional.

The portable food vans are entangled in this idea of intersectionality where power relations are not always clear: 'Intersectionality has travelled into spaces and discourses that are themselves constituted by power relations that are far from transparent' (Cho, Crenshaw, and McCall 2013: 789). We do not clearly understand whether they are included in the dominant discursive space of city life or whether they have been thrown to the margins. The power relations in these situations are convoluted; these vans do not run on government certification or norms, and they do not adhere to road or safety rules. They are often at the brink of elimination from their spaces by the state apparatuses. As Castells observes, 'Mega-cities' functional and social hierarchies are spatially blurred and mixed, organized in retrenched encampments, and unevenly patched by unexpected pockets of undesirable uses' (2010: 436). These are the unexpected pockets where people from the unorganized sector create an economy of their own. They take ownership of the city space that clearly hasn't included them in its dominant narrative. This form of portable food in a greater conversation about mobile gastronomies brings visibility to this intersectional space.

Conclusion

Portable food in India has undergone a long and meandering journey from the roadside eateries to the *dabbawalas* to the recent trend of food trucks and finally the food delivery applications. A lot has changed in the last five years in terms of portable food, especially food away from the table. Restaurants and pop-up restaurants have taken over city as well as country life. We can see the food delivery people in small towns in India riding their two-wheelers, acting as the 'hunger saviours' of the modern world. The pandemic has also changed the way we look at portable food. The visibility of food delivery applications has intensified. The mobile vans that we discuss are an alternative form of portable food, limited to the city space. As Barns says, 'Cities are constantly being disrupted and rebuilt with every passing wave of technology innovation' (2019:12), and, with every disruption, there will be a realignment to new and different social practices. City life is dynamic and bound to change with every fresh wave of technological advance.

These mobile food vehicles might completely be dismantled in the upcoming technology wave; however, there is a chance that they could realign to the latest technology and transform themselves. This work is an experiment in tracing the alternative forms of portable food that marks the city scape of India. Of course, it is limited to the financial districts of Hyderabad, Southern India and delves into the socio-spatial experience of that region. The everyday life of these mobile food van owners show how they come to terms with the urban social space and the platform infrastructures that it creates. This work can be extended to other financial hubs of India like Bangalore (Karnataka), NOIDA (New Okhla Industrial Development Authority, Uttar Pradesh), Gurugram (Haryana), Pune (Maharashtra), and Kolkata (West Bengal). The forms of portable food in these regions will be an interesting space to add to the greater narrative of mobile gastronomies in India. The engagement of different forms of portable food with the urban social space would make a reasonably good case study. The future of portable food is extremely dynamic if the factors of urban migration, globalization, and technological innovation are considered. The food scenario in India is seeing reforms like never before. With the growth of slow food, veganism, and healthy eating, it will be interesting to see how these mobile food vans survive in the ever-changing social geography and the dominance of social media fads. Whether they would survive the ravages of technological waves or realign themselves to something novel would be something that would also make a new pathway of research.

The mobile gastronomies represented by the food vans is one among the many forms of food portability that the mega-city witnesses. Documenting other forms of portable food like the smaller vendors selling street food and observing how they add to the tapestry of city life would be the way forward for the study of mobile gastronomies.

References

Anenburg, Elliot, and Edward Kung. 2015. 'Information Technology and Product Variety in the City: The Case of Food Trucks', *Journal of Urban Economics*, 90, 60-78

Appadurai, Arjun. 2006. 'Disjuncture and Difference in the Global Cultural Economy', in *Media and Cultural Studies: Keyworks*, ed. by Meenakshi Gigi Durham and Douglas Kellner (Oxford: Blackwell), pp. 584-97

Balakrishnan, Ajit. 2011. 'India's IT Industry: The End of the Beginning', *Social Research*, 78, 1-20

Banerji, Chitrita. 2007. *Eating India: An Odyssey into the Food and Culture of the Land of Spices* (New York: Bloomsbury)

Barman, Dhiraj. 2020. '"Global"-izing Indian Cities: A Spatial Epistemological Critique', in *City, Space and Politics in the Global South*, ed. by Bikramaditya K. Choudhary, Arun K. Singh, and Diganta Das (Oxon: Routledge), pp. 61-82

Barns, Sarah. 2019a. 'Negotiating the Platform Pivot: From Participatory Digital Ecosystems to Infrastructures of Everyday Life', *Geography Compass*, 13

——. 2019b. *Platform Urbanism Negotiating Platform Ecosystems in Connected Cities* (Singapore: Palgrave Macmillan)

Burton, David. 1994. *The Raj at Table: A Culinary History of the British in India* (London: Faber and Faber)

Castells, Manuel. 2010. *The Rise of the Network Society*, 2nd edn (Oxford: Blackwell)

Cho, Sumi, Kimberlé Williams Crenshaw, and Leslie McCall. 2013. 'Toward a Field of Intersectionality Studies: Theory, Applications, and Praxis', *Signs*, 38, 785-810

Das, Diganta. 2015a. 'Hyderabad: Visioning, Restructuring and making of a high-tech city', *Cities*, 43, 48-58

——. 2015b. 'Making of High-Tech Hyderabad: Mapping Neoliberal Networks and Splintering Effects', *Singapore Journal of Tropical Geography*, 36, 231–48

'How India Eats Out? Is a Food Evolution Happening?' n.d. *The Restaurant Times* <https://www.posist.com/restaurant-times/features/indian-in-a-food-evolution.html> [accessed 8 March 2023]

McLuhan, Marshall. 1964. *Understanding Media: The Extensions of Man* (New York: Signet Books)

Miller, Toby. 2013. 'Cultural Studies and Food Sovereignty', *Cultural Studies Review*, 19, 138-42

Nusra. 2013 'The changing culture of Eating in India', *RestaurantIndia.in*, 7 June <https://restaurant.indian-retailer.com/article/The-Changing-Culture-of-Eating-in-India.6076> [accessed May 20,2022]

Pink, Sarah. 2012. *Situating Everyday Life: Practices and Places* (London: Sage)

Roncaglia, Sara. 2013. *Feeding the City: Work and Food Culture of the Mumbai Dabbawalas* (Cambridge: Open Book Publishers)

Srinivas, Tulasi. 2021. '"Swiggy it!": Food Delivery, Gastro Geographies, and the Shifting Meaning of the Local in Pandemic India', *Gastronomica*, 21,17-30

An Exploration of Food Mobility through the Cultural Practice of *Koseli* in Contemporary Nepal

Binti Gurung

Historically, the movement of food in Nepal has manifested through its trade relationships. In *Medieval History of Nepal*, Dilli Raman Regmi points out that until the eighteenth century Nepal dominated a strategic highway between Tibet and Nepal, one of the two convenient links along the Himalayas.[1] As pointed out by economist Yadav Prasad Pant, a treaty signed between Nepal and Tibet (as part of China) in 1872, considered to be the country's earliest trade document, initiated trade relations between the two.[2] Even though the volume of trade was minimal, the treaty enabled communities along the Himalayan frontiers to exchange food and other goods that could not be produced locally. For centuries, salt and grain had been the basis of such trade between Nepal and Tibet. The high altitudes meant that Tibet could not produce its own grain but had surplus salt, so the treaty opened up a channel for both countries to exchange produce and goods. Some of the food items such as salt, medicinal herbs, and yak tails have been recorded as imported to Nepal, while agricultural produce including rice, chillies, and onions lacking in the Tibetan region were exported from Nepal.[3]

An oral interview with a Gurung elder in his 80s reveals that trade treks from his village in lower Manang of central Himalayas to the border areas of Tibet were common right up to the 1970s. Trekking for several days to make food exchanges at the Tibet-Nepal border, the travelling party sheltered at the bordering village of Larke in Gorkha. Locally grown grains such as wheat and naked-barley exchanged for salt would be brought back to the villages to supply the traders further down in the valleys.

If this type of trade served as the basis of food portability in Nepal, a cultural practice of food movement can be seen through the exchange of *koseli*. *Koseli*, a culture of food exchange performed to express and strengthen feelings of generosity and reciprocity, can be seen in many contexts. Though it will not be the focus of this paper, one recent form that has emerged within the contemporary travel and tourism context is the practice of passing food and local goods as souvenirs.

Even though the cultural practice of *koseli* is a prominent aspect of Nepali society, a recorded history of its origin is difficult to trace. The existing caste structure creates more complexity as to what can be exchanged as a *koseli*. Added to this challenge, many indigenous groups in Nepal have had no written records, and the customary practice of transmitting knowledge has been the oral traditions. Within this context, I will attempt to explore the culture of *koseli* through the exchange of *raksi* (an alcoholic drink), foraged foods, and meat products within the social environment of Nepal with reference to the indigenous cuisine, and how these foods linked with caste, religion, and ethnicity contribute to, as well as differentiate, the way food moves in Nepal. Furthermore, in view of the indigenous knowledge informing this paper, I will attempt to explore the ideas presented, where appropriate, through an indigenous methodology of the Talking Circle.[4]

The Diverse Food and Cuisine of Nepal

There is no dispute that the tradition of *koseli* finds its place in the many communities of Nepal. Nepal has well over a hundred different castes and indigenous groups, of which fifty-nine are formally recognized by the State. Over the centuries, unique food habits, practices, and cuisines have developed as a result of the cultural interactions from the south and north, and, in some cases, also from the challenges imposed on local mountainous communities by the isolated and rugged terrain. The varied elevation and ecological condition of the southern plain fields, the hills, and the Himalayas contribute to a diverse food culture. Indigenous groups such as the Gurungs, Thakalis, and Bhotes in the northern highlands have nurtured their distinct food cultures through foraging and pastoral practices. Traditional animal husbandry involving the rearing of yak, sheep, and mountain goats provides a variety of meats along with milk and milk products of *chhurpi*, dried cheese, and buttermilk. Native crops such as barley, buckwheat, and amaranth are specific staples that reflect the cuisine of the mountain region. In the mid-hills, where crops such as maize and finger millet are plentiful, corn and millet-based *dhido* (thick porridge), foraged wild tubers, asparagus, and fiddlehead ferns make part of their cuisine. The relatively stable climatic conditions also allow the local communities to anticipate seasonal cycles. During the dry winter months, when wild and green vegetables are not readily available, indigenous communities return to age-old fermenting techniques to create *gundruk* (fermented green spinach) and *sinki* (fermented radish leaves). In the fertile land of the south, indigenous Tharu, as the fishing group of Majhis of the Ramechhap, consume a water-based cuisine of fish, *ghongi* (snails), and mussels.

Over the centuries, many of these regional communities have found innovative and locally sustainable ways to create distinct food identities. Along with the foods mentioned above, traditional food preservation techniques are used. Because raw meat does not tolerate higher temperatures due to its high moisture and protein contents that spoils the meat quickly, the locals in the Himalayas process the meat by smoking, drying, and fermenting.[5] In the month of August, when it is harvest time, unripe maize is gathered just before its full maturity to retain the sweet flavour of the maize. Later, in the dry months, when food is scarce, maize and dried meat that had been processed and stored are prepared to make several food items. One such dish is *phalghi*, a hearty slow-cooked stew prepared by the indigenous Sunuwar and Sherpa communities using meat, corn, local radishes, and varieties of beans.[6] This is also one of the examples that demonstrate how indigenous mountain people invent and innovate to address their local food challenges and adapt to their specific environment. Similar examples come from the way sausages are prepared. Smoked dried sausage of *Ghimti, Gyurma*, or *Se Kru* in the mountain region is prepared with *rakti* (yak blood) and offal. Despite the ingenuity involved in such food preservation techniques, very little recognition has been given to the local cuisine of this region outside of their community.

Caste, *Raksi*, and Food Relations

The historical narrative of the caste system in Nepal reflects considerable complexity when it comes to food culture. Caste and religion have long dominated and shaped the local cuisines of Nepal. The cultural acceptance of food imposed by the religious and caste considerations continue to exert a great influence in contemporary Nepali society. Regmi informs us that the early movements of people entering into Nepal Valley from the Gangetic Plains may have facilitated this caste process.[7] Even though a caste-based structure had previously existed in ancient Nepal, it was not until the reign of Jaya Sthitimalla, a Newar king of Nepal Valley, when it was firmly established.[8] By borrowing the Brahmanical classical framework of the *Dharmasastra*, which stratified the Newar community into four varnas and thirty-six *jatis* (castes) and many sub-castes, Jaya Sthitimalla formalized a caste-based system based on occupation.

This system introduced in the fourteenth century standardized what duties one could perform in the society. The Brahmins who were placed at the top of the hierarchy greatly benefitted from their position as priest and scholars, as this enabled consolidation of political powers. In the second order were the Kshatriyas, who continue to wield power even today through participating in administrative and ministerial work. The social groups that remained were classified in various varnas of Vaishyas and Sudras. The agrarian and peasant group of Jyapu, including meat sellers, butchers, oil extractors and the many food merchants and producers, found themselves at the lower rung of the hierarchy. The caste hierarchy is deeply embedded in Nepali society; even today discriminatory practices exist towards specific food producers and their communities.

Very little changed after the unification of Nepal in the late eighteenth century. In fact, the Muluki Ain of 1854, Nepal's first legal document drafted at the initiation of Prime Minister Janga Bahadur Rana, further strengthened the dominance of a Hindu-based legal system by institutionalizing social customs on purity and impurity. The document elucidates how indigenous food such as pork and *raksi* are positioned within the Hindu-based policies. The slaughter, sale, and consumption of cow meat were banned, and Brahmins, referred to as a sacred-thread wearing caste, caught committing specific crimes would be fed with pig meat accompanied by *raksi* and have their caste degraded. The food regulations set a negative precedent for the food culture of indigenous people. In fact, some of the policies even enforced what sort of food and drinks could be accepted by a person and from which caste. In the following excerpt, consumption of *raksi* by a high-caste Brahmin is also expressed in this way:

> If someone belonging to a Sacred-Thread wearing (to indicate the Brahmins) or an alcohol drinking caste (of indigenous groups), [...] knowingly, by force or for no reason makes a person belonging to a Sacred Thread wearing caste consume alcohol etc. or any other forbidden substance which leads to his caste degradation, his share of property shall, in accordance with the Ain, be confiscated and he shall be imprisoned for a year.[9]

By implication, indigenous groups classified in the alcohol-drinking category of *matuvali* are designated as an impure caste, while legitimizing the position of the high-caste Hindu Brahmin.[10]

In contemporary Nepal, the cultural treatment of food practices is deeply embedded within this caste system. The cuisine of minority groups such as the Dalits, considered untouchables, and many indigenous groups are considered inferior, and, in some cases, as Daniggelis has noted from her fieldwork, the forager Rai women of eastern Nepal are referred to as *jangali* by the high-caste Hindu Brahmins and Chhetris, a term implying that someone is primitive and backward.[11] Scholars such as Ganesh M. Gurung, Thomas Cox, and Hannah Rauber have explored the social process of caste within the indigenous groups of Nepal. While none of them has addressed its direct impact on food culture in detail, they have discussed some aspects in passing. For example, in *Far Out: Countercultural Seekers and the Tourist Encounter in Nepal*, Mark Liechty does not mention this connection, but he does briefly comment on the caste-based living arrangements in Kathmandu Valley, where the lower caste groups are physically dislocated from area political institutions.[12] However, highlighting M. N. Srinivas's theory of Sanskritization, Gurung notes that social groups at the lower end of the caste hierarchy attempt to improve their social status by giving up their own cultural norms and rituals and adopting the cultural values of the higher-caste Hindus. In some cases, as has been documented in the Khyampas, nomadic traders of western Nepal, the group has entirely given up eating their cultural foods of yak and beef meat in order to assimilate with the Hindus.[13] These social processes have reinforced significant restrictions on the cuisines of the indigenous people, while enhancing the food values of the higher-caste.

Furthermore, the high-caste Hindu standards forbid meat such as pork and buffalo and grains such as millet and foraged food on the grounds that they are impure, yet for many indigenous groups in Nepal these foods go beyond the question of daily sustenance. For example, in the Gurung oral and cosmological narratives of *Pe*, locally distilled *raksi*, *jaad* (fermented beer), *pa* (fish), and parts of the animal such as the liver are ritual items necessary to offer the ancestral deities. According to a Gurung shaman, *po-ju*, the teachings contained in *Pe*, goes even further to recite the origin of *marcha*, a starter used in indigenous alcohol-making process to ferment alcoholic beverages. Therefore, when the policies contained in Muluki Ain categorize *raksi* and meat as polluting foods, it minimizes the larger spiritual and cultural meaning of indigenous food knowledge, in order to perpetuate Hindu-based ideology.

Furthermore, when millet-based *raksi* and meat items like pork are exchanged in *koseli*, it becomes evident how factors such as caste and religion continue to perpetuate meanings of particular foods, sometimes at the cost of minimizing another food culture. In one context, *koseli*, as a vehicle of food, is a '*maiti jada lagne khana*', local food gifts prepared by married women for their maternal families. Indigenous Rai of the eastern Nepal use pork meat in particular to express affinity as food gifts. Existing customs within the group dictate that specific parts of the meat are selected, which is determined by one's respective affiliation within the lineage, but most often either pork thighs or trotters are boiled and placed in the parcel, along with boiled eggs. *Titepati*, mugwort leaves, are placed on top of the food parcel to ward off bad spirits as the parcel is carried past dense forest and rivers. For the Rais, as with other indigenous groups, nature is never far away; rather, it is integrated in their daily lives as foragers as well as in *koseli*. The customary practice of what food products are involved in *koseli* has regional variations within the indigenous groups, yet what remains common is the inclusion of *raksi*, which is mandatory, *sel-roti* (deep fried ring-shaped bread), and meat items. In many ways, a traditional *koseli* with locally produced food items serves as a vessel of food movement with multiple cultural layers and meanings.

Ghimti (Blood Sausage) and the Customary Institution of Ghampa

As stated in the previous sections, the historical progression of caste structure set in place a discourse that defined the place of food within the caste hierarchy, including culturally important foods for indigenous groups like the Rai, like *sargyangma* and *ghimti* (types of blood sausages) and most of all *raksi*; distinctive features of many indigenous cuisines occupy a low rung on this caste-based hierarchy.[14] A discussion conducted through a Talking Circle with the elders of Lower and Upper Manang on food processing knowledge and the *Ghampa* system offered a space for better understanding of key aspects of the regional cuisine and food habits.[15]

From the Talking Circle, it became clear that the indigenous mountain communities have been using traditional strategies to cope with their local food demands. Because of the short farming season, crops are harvested once a year, so processing and storing of food are a crucial part of the process of living in the rugged terrain. While there are different types of food such as *pote*, dried and ground young buckwheat leaves, added in lentil soup, and dried yak meat, the processing of *ghimti* forms the defining feature of mountain cuisine. As explained during the discussion, the preparation of *ghimti* in Manang, and in the wider Himalayan region, is never a solitary activity. The effort is shared not only because of the scale of the preparatory work involved, but also because it allows community members to come together in an activity that serves to strengthen bonds and share feelings of generosity and reciprocity. In Lower Manang, *ghimti* is prepared in two ways, one in which *rakti* is mixed with offcuts of the meat and other in which buckwheat or corn flour is mixed with *rakti* to form a thick paste for filling. Once this step is complete, each filling is encased in the intestine with the ends securely tied, then the sausage is left in simmering water before hanging over the fireplace for several weeks. Nothing is wasted in the process, as all parts of the meat are used in the minced filling. *Ghimti* prepared in this way can be used for six to seven months, and also travel as *koseli* to the relatives in the city. In the villages, though, during the winter months, locals enjoy it as a dish accompanied by *raksi*.

In this process, *Ghampa*, the customary institution, also plays a part. Due to the interconnected nature of the food production process within the community, members of *Ghampa* are actively involved in managing the yak in the region. A study conducted in the village of Nar and Phu in Upper Manang by Gurung and McVeigh explains the role of *Ghampa*.[16] According to the study, *Ghampa* maintain a significant responsibility in the management of the village activities. *Ghampa* are structured in a way that enables all community households to participate, either as decision makers or implementers, and creates opportunity for all to serve in the committee. Decisions regarding grazing rights, timing of the yak movement, areas for grazing, and dissemination of community information are all conducted by the members of *Ghampa*. However, one of the elders in the Talking Circle mentioned that they are a disappearing institution; since the formation of political institutions, the institution no longer exists in Lower Manang, and any decisions relating to yak management are entirely taken up by the shepherds. In the village of Nar and Phu, near the border of Tibet in Upper Manang, the yaks are grazed upon an area abundant in medicinal herbs and wild garlic; as a result, yak meat of this region is considered to be flavourful. As *koseli*, dried yak meat and *ghimti* are transported to the cities, most often to members from one's community as part of larger cultural identity. At the same time, the process serves to facilitate an important aspect of transmitting intergenerational knowledge, where the younger generations are connected to the food of their ancestors.

Conclusion

Up to the twentieth century, the process of food mobility was facilitated through trade agreements along both frontiers along the Gangetic Plains and the Himalayas. Trade arrangements until the late 1970s were motivated by a prime necessity to secure food that could not be locally produced. In contemporary Nepal, though, a customary practice of *koseli* acts as a carrier of food between people and places. Many communities embody this practice to convey a sense of generosity and reciprocity, and members express these feelings through food gifts. More specifically, indigenous women prepare traditional food of *sel-roti* and brew local raksi for *koseli*, and meat is always involved. These food gifts are exchanged during social occasions and festivals, when a visit is expected from the daughters at their maternal home. Even though it is a widely practiced culture, the acceptance of *koseli* and the selection of what constitutes *koseli* remain highly contentious. The caste-based system introduced since the medieval times and its historical progression have heavily contaminated the food culture in Nepal, as the food one ate began to take distinct cultural meanings tied to caste status and hierarchy. The disappearance of specific food cultures as a result of this process is also a cultural loss that helps us understand the diverse food meanings within the multicultural environment of Nepal.

Notes

1. Dilli Raman Regmi, *Medieval Nepal: A History of the Three Kingdoms 1520 A.D. to 1768 A.D.* (Delhi: Rupa & Co., 2007), pp. 527–28.
2. As highlighted by Dirgha Man Gurung, there was a trade treaty made between British India and Nepal, and the first that was signed was in 1792. The trade relationship between the two countries was tumultuous until Ranuddip Singh succeeded Jang Bahadur in 1877 (Yadav Prasad Pant, 'Nepal – China Trade Relations', *The Economic Weekly*, 14 April 1962, 621–24).
3. Regmi, p. 528.
4. Government of Alberta, *Indigenous Pedagogy: Talking Circles Protocol* <https://www.learnalberta.ca/content/aswt/indigenous_pedagogy/documents/talking_circles_protocol.pdf> [accessed 18 March 2023] (excerpt from *Contemporary Issues Teacher Resource* (Toronto: Nelson Education, 2006), pp. 445–46).
5. Arun K. Rai, Uma Palni, and Jyoti Prakash Tamang, 'Traditional Knowledge of the Himalayan People on Production of Indigenous Meat Products', *Indian Journal of Traditional Knowledge* 8.1 (January 2009), 104–09.
6. Part of an oral interview conducted with Shova Sunuwar, Chairperson of Sunuwar Women Indigenous Group, March 2022.
7. Nepal Valley refers to Kathmandu as it exists today. Prior to the unification process initiated by Prithivi Narayan Shah in the eighteenth century, there was no concept of Nepal. Instead of Nepal as a nation, there were twenty-four kingdoms of Chaubise Rajya.
8. Regmi, pp. 641–709.
9. Rajan Khatiwoda, Simon Cubelic, and Axel Michaels (eds), *Muluki Ain of 1854: Nepal's First Legal Code* (Heidelberg: Heidelberg University Publishing, 2021), p. 34.
10. The cultural implication of *Matuvali* referred to the indigenous groups of Tibetan-Burmese origin in a derisive term that expresses a state of 'getting out of control' after consuming alcohol. Introduced in Muluki Ain 1854, the cultural connotation of *matuvali* remains today.
11. See for example E. Daniggelis, 'Women and Wild Food: Nutrition and Household Security among Rais and Sherpa. Gender Relation in Forager-Farmers in Eastern Nepal', in *Women and Plants: Bio-Diversity Management and Conservation*, ed. by P.L. Howard (London: Zed Books, 2003), pp. 83–97.
12. Mark Liechty, *Far Out: Countercultural Seekers and the Tourist Encounter in Nepal* (Chicago: University of Chicago Press, 2017).
13. Ganesh M. Gurung, 'The Process of Identification and Sanskritization: The Duras of West Nepal', *Kailash: Journal of Himalayan Studies*, 14.1-2 (1988), pp. 41–56.
14. Kamal Maden, Ramjee Kongren, and Tanka Maya Limbu, 'Indigenous Knowledge, Skill and Practices of Kirat Nationalities with Special Focus on Biological Resources', in *Identity and Society: Social Exclusion and Inclusion in Nepal*, ed. by Joanna Pfaff-Czarnecka, Kristen Stokke, and Mohan Das Manandhar (Kathmandu: Mandala Book Point, 2009), pp. 123–48.
15. A discussion following the protocols of Talking Circle took place on 29 May 2022; four members of elderly men and women from Lower and Upper Manang region participated.
16. G. Gurung and C. McVeigh, 'Pastoral Management and Yak Rearing in Manang's Nar-Phu Valley', in *Yak Production in Central Asian Highlands: Proceedings of the Third International Congress on Yak held in Lhasa, P.R. China, 4–9 September 2000*, ed. by H. Jianlin and others <http://agtr.ilri.cgiar.org/sites/all/files/library/docs/yakpro/SessionA10.htm#TopOfPage> [accessed 18 March 2023].

Fair Food To Walk With

Peter Hertzmann

A few years back I found a photograph of my six-year-old father with his parents at the 1915 Panama-Pacific International Exposition. They were posing in front of a backdrop looking very serious. My grandfather, who was a haberdasher at the time, was in a natty suit. My grandmother wore a sombre day dress. My father wore a sailor-style outfit that was popular for boys at the time. None of them smiled (Figure 1). From what I read about the Exposition, it was great fun for the fair goers. Why my relatives didn't seem to be enjoying themselves will remain a mystery.

Did they visit The Palace of Food Products, more popularly called The Palace of Nibbling Arts?[1] As one fair visitor described, 'I spent so much time in the Palace of the Nibbling Arts partaking of the Scotch scones with raspberry jam that I was forced to leave [after three days] without visiting the majority of Foreign Buildings.'[2] Did my grandparents try the scones?

Figure 1. Sigmund, Flora, and Walter (age 6) Hertzmann posing for a photograph at the 1915 Panama-Pacific International Exposition. (Author's collection.)

American Fairs, Expositions, and Festivals

The Panama-Pacific International Exposition occurred near the middle of a century of world's fairs or expositions in the United States. Starting with the New York Crystal Palace Exhibition in 1853 and 1854, a commercial disaster; followed by the Centennial International Exposition of 1876 in Philadelphia, the first commercially successful fair of the genre; and ending with the financial wreck of the New Orleans World's Fair in 1984, fairs captured the public's imagination in a time before Disney's Epcot Center and other permanent exhibition centres.[3] The foreign exhibits at early fairs were a chance to view foreign peoples as exhibits, as in a zoo, rather than a chance for the fair-goer to truly experience a foreign culture. That content is now accessible to any traveller, whether in-person or sitting on an armchair, with fewer overlying layers of racism and exploitation that the expositions provided.[4]

Although termed 'World's Fairs', these large events were, in fact, American national, or in some cases regional, fairs with foreign exhibits provided for entertainment or promotional purposes. As Robert W. Rydell notes, 'World's fairs performed a hegemonic function precisely because they propagated the ideas and values of [America's] political, financial, corporate, and intellectual leaders and offered these ideas as the proper interpretation of social and political reality.'[5]

Food at world's fairs can be roughly divided into two broad categories: food served in restaurant-like settings and food eaten while walking around the fair. Walk-around food was purchased from kiosks or stands or from roaming vendors, mostly working on consignment. Additional walk-around food was at hand in the form of free samples provided by food processors extolling their products. The restaurant settings could be substantial. The Centennial International Exposition of 1876 boasted a German restaurant with 1,500 seats and a substantial French restaurant to compete with it.[6] Walk-around food requires disposable packaging, much of which is not invented until the first decade of the twentieth century and doesn't become common until later.

In the two hundred years since America's first 'fair' in 1810, world's fairs comprise only a tiny fraction of America's fairs.[7] Not to be outdone and often for similar commercial purposes, smaller political units throughout the nation have also sponsored fairs. The first state fair was held in Syracuse, New York, in 1841.[8] Other states followed. California had its first state fair in 1854, just four years after it became a state.[9]

Unlike world's fairs, which are fashioned as international events, state fairs attract exhibitors mostly from within the state or from companies that have a significant presence in the state. Most state fairs were started by state agricultural authorities, and their emphasis still follows an agricultural theme. Rural county fairs also follow agricultural themes, whereas urban county fairs have had to adapt to the changes by urban life.

State and county fairs may offer barbecue contests and chilli cookoffs, but these themes also exist as stand-alone festivals. Every type of vegetable or meat has a festival, from artichokes to garlic to lamb testicles to zucchini.[10] Each of these festivals offers a plethora of dishes made with the honoured ingredient plus plenty of other fair fare.

Another form of food festival happens in large urban areas with loads of fine-dining establishments. The largest is Taste of Chicago with about 3 million visitors each year.[11] The food at these festivals consists of small bites provided by each restaurant. In some cases, food trucks augment the variety. There are often special events of multi-course dinners. Wine and beer are essential elements of these festivals. Depending on the festival, you may find a place to sit to enjoy your nibbles, or you may just need to eat and drink while you walk.

In areas of America where immigrant groups have enough members to support a festival, festivals often include the foods of the ethnic group.[12] These tend to be annual events, but in areas with large ethnic populations, separate groups may sponsor multiple festivals in a year.

The last type of modern fair or festival is the non-food event that also features food and wine as part of the attraction. Each suburban town in my area has an 'art and wine festival' one weekend each year. The sponsor is usually the local merchants' association or Chamber of Commerce. The festivals are organized by professional companies that coordinate all the vendors – artists, crafts people, and food sellers – arrange for all the permits, recruit the entertainment, and do the set-up and tear-down. These festivals are coordinated so dates do not conflict from town to town in the same area.[13] Within a large city, various neighbourhoods host art and wine or wine and food festivals like the suburban versions.

Large, multi-day music festivals also make food available for the attendees. One of the many music festivals in the San Francisco Bay Area, Outside Lands, has over ninety food vendors serving a variety of ethnic and local food to the more than 200,000 attendees.[14]

All the above festivals are just as commercial as the state and county fairs, but they are less promotional of specific industries. In most cases, the purpose of the festival is to raise money for a single charity, a group of charities, or maybe the promoter.

On the smallest scale are schools and churches that sponsor an annual fair or festival. Most are small, but some churches invest in midway-style rides and support outside food vendors.[15]

Fair Foods

The variety of foods served at modern fairs and festivals is as varied as the venues. Yet some foods are traditionally associated with American fairs; some are even thought to be created for the fair that made it famous, although most existed long before the fair.

The scones of the 1915 Panama-Pacific International Exposition existed in some form at least a century before the fair and are still a common item in San Francisco bakeries.[16] Little is known about these scones. Were they based on wheat like modern scones or oats like traditional Scottish scones?[17] One newspaper article laments the end of the fair and thus the end of 'sconing', or as the writer describes it, the 'most exciting and perilous of indoor sports'. The author tells us only 'the scone[,] which is a triangular product, [is] crusty and corrugated without, softly absorbent of butter and jam within'.[18] Hardly a recipe.

Fair Food To Walk With

Some reports of fair food carry much detail while other reports remain cloaked in gauzy details. Most early fair food remains with us today, although few require a trip to a fair to partake in it.

Beer

Beer has been a part of fairs and festivals since Friar Tuck rode through Sherwood Forest on an ale wagon.[19] In the early days of fairs in America, beer was ever present, even when objected to by the local temperance league, but it was not yet a walk-around food. By the time of the 1876 Centennial Exposition in Philadelphia, large breweries were established in America and together they were able to occupy a brewery hall where beer was produced daily.[20] Fair-goers could sample the beer on the premises by taking their schooners or mugs to the first-floor balcony or the roof-top observatory.

At the World's Columbian Exposition in Chicago in 1893, conditions were better. German lager-style beer was now produced all year because mechanical refrigeration had advanced enough to allow breweries to control the temperature of production.[21] Milwaukee breweries began selling beer in glass bottles in 1880, although no evidence seems to exist that beer was sold in bottles at the fair.[22] America was still a decade away from the invention of paper cups.[23] Most likely, the beer was still served in glass schooners or mugs.

As America inched towards the Volstead Act taking effect in 1920, the various temperance leagues became increasingly effective in limiting the sales of beer and other alcoholic beverages at fairs around the country.[24] A decade later, ten months before the end of prohibition, the Cullen-Harrison Act allowed people to drink low-alcohol content beer.[25] By the 1960s, beer in cans became more popular than beer in bottles.[26] Today, at least one concession at most fairs and festivals sells beer. Regions with many craft breweries may have several beer sellers. The beer is served in a plastic or paper cup – perfect for walking around – or simply in a can.

Bel-Gem Waffles

The Belgian-style waffle was introduced to America in 1962 at the Seattle World's Fair.[27] When introduced in Seattle it was called either a Brussels or a Belgian Waffle.[28] When reintroduced in 1964 at the New York World's Fair, the snack was sold as a Bel-Gem Waffle by a different family, although some articles referred to it as a Belgian Waffle.[29] Considering that Americans want their waffles at breakfast, selling 2500 waffles a day was quite a feat.[30]

Many Belgian waffle recipes claim to be authentic, but the wife of the 1964 proprietor asserts that her husband's original recipe was licensed only once, in 2002 to the family that sold the waffles at the New York State Fair with the condition the recipe could be used only at the fair.[31]

Today, Belgian waffles are more likely to appear on the menu of an American coffee shop than at a festival or fair.

Cotton Candy

Modern cotton candy was developed at the turn of the twentieth century in Nashville, Tennessee. Like most mechanical developments, the exact date of their first working machine is unknown, but the inventors, William J. Morrison and John C. Wharton, filed their patent in late 1897.[32] Their first machine melted sugar in a gas-flame-heated chamber that spun at high speed. The new candy was called Fairy Floss. It was packaged in a wooden box and sold for a quarter.[33]

The combination of a hand crank and a gas flame must have been unacceptable because four years later, the pair filed a new patent for an electric-powered and electric-heated candy machine.[34] When Morrison and Wharton introduced their Fairy Floss at the 1904 Louisiana Purchase Exhibition, they benefitted from the fair's ultra-modern electric power generation plant.[35] Using their improved machine, which still had balance issues, they introduced Fairy Floss to the world. Fair goers purchased almost 70,000 two-bit boxes of the candy.[36]

The name 'cotton candy' was the creation of Joseph Lascaux, a dentist in New Orleans. He invented his version of a fairy floss machine and marketed it as a cotton candy machine. The doctor and the machine weren't successful, but the name was a winner.[37]

Modern cotton candy machines no longer have a balance issue.[38] These newer cotton candy machines are inexpensive enough for church and school groups to purchase or rent.[39] Entrepreneurs can purchase a machine built onto a cart that was suitable for sales in a larger festival or fair.

All the supplies needed for producing and selling cotton candy are readily available. Sugar is available pre-coloured and pre-flavoured, or the flavouring can be purchased separately for quantity mixing with granulated sugar. The flavour list includes pink vanilla, blue raspberry, birthday cake, and piña colada plus more mundane flavours of apple, cherry, watermelon, grape, and orange. All the ancillary product needs, such as paper cones, bags, boxes, holders, and a product tree are available.[40]

Cotton candy may be the iconic walk-around fair food.

Cracker Jack

Although sources often claim that Cracker Jack was introduced at the 1893 World's Columbian Exposition in Chicago, the facts may be a bit sticky.[41] For twenty years prior to the fair, brothers Frederick and Louis Rueckheim made various candy confections.[42] For the fair, the brothers produced a gooey confection of molasses, peanuts, and popcorn called simply Candied Peanuts and Popcorn. Compared to their later product, the one we know as Cracker Jack, '[t]he basic ingredients were the only similarity; the stickiness of the molasses proved uninviting to consumers'.[43] Although some histories claim that the fair was a huge success for the brothers, company history says 1896 was when Louis Rueckheim perfected a means of drying the molasses so the final product wasn't sticky.[44] 1896 was also the year the mixture was renamed Cracker Jack.[45]

The company history was rewritten when it was purchased by Bordon in 1964. There's no record, including participating in the fair, available prior to then. Cracker Jack is found at circuses and baseball games as well as corner stores, but there is no true record of it being sold at fairs and festivals.

Cupcakes, Cookies, and Brownies

Church and school festivals and fairs have always relied upon the good will of parishioners or parents to provide, at their own expense, items of food that can be sold or auctioned off to profit the stated beneficiary of the event. The most popular items of food are baked goods, especially those that can be sold in single serving sizes.

Since the 1950s, even the busy parent with no baking skill can produce adequate results using a boxed mix. For one brand of brownies, everything is in the box except two eggs, a half cup of vegetable oil, and three tablespoons water.[46] Another brand even throws in a disposable baking pan.[47]

Occasionally, professionally baked goods can be found at street festivals and outdoor music events run by professional organizers, but these goods are more common in farmers' markets with a regular schedule.

Ethnic festivals may have their own specialty baked goods such as *malasadas* at Portuguese festivals or *andagi* at Okinawan festivals. In Hawaii, where both groups have a long history, both items may appear at the same festival.

Soda and Other Drinks

By the time of the 1876 Centennial Exposition in Philadelphia, soda water, both plain and flavoured, was a common product available by the glass and by the bottle, from soda fountains and in the home.[48] Fair records show that 1,900,000 bottles of lemonade,

Figure 2. Arctic-brand soda dispenser by the James W. Tufts Co. at the 1876 Centennial Exposition in Philadelphia. (Hagley Museum and Library, Wilmington, DE. Used with permission.)

soda water, and ginger beer were sold during the six-month run of the fair.[49] A single company, James W. Tufts, paid $52,000 for the rights to set up a number of fountains throughout the fair (Figure 2). The *Visitors' Guide* promised, 'Soda-water fountains have been located at convenient points within the principal buildings, and in pavilions erected for the purpose on the grounds. Charge per glass, 10 cents'.[50]

Eighteen vendors signed up to sell 'pure and mineral waters, natural and artificial' at the 1893 World's Columbian Exposition in Chicago.[51] Numerous other vendors sold the equipment to produce and bottle the water. Even the bottled soda was not for walking around since the bottles had to be returned to the vendor when emptied.

Soda water and all the flavoured sodas that followed have become the ubiquitous beverage for Americans. New varieties are developed each year to compete in the marketplace. Americans no longer must go to a fair or a drugstore to be refreshed. Some people even have a soda-fountain dispenser in their homes.[52] In many locales, the nearest bottler will bring a dispensing truck to a church or school fair to provide a variety of soft drinks.

Funnel Cakes and Deep-Fried Everything Else

Modern state and county fairs would not be the classic events they've become without a plethora of non-conventional deep-fried foods. In the 1980s, any food that could be coated with pancake batter and deep-fried was.[53] The date and location of the first deep-fried Twinkie is unknown, but the root of this class of foods is funnel cakes. Fried dough is common in many cultures, some ancient. The current fair food, the funnel cake, is thought to be derived from similar preparations by seventeenth- and eighteenth-century German immigrants known today as the Pennsylvania Dutch.[54] A funnel cake is simply a batter drizzled in a random pattern into hot oil. Once cooked and drained, the 'cake' is served with a thick application of powdered sugar.[55] Other toppings such as fruit, chocolate sauce, and whipped cream are also available to increase the fat and calorie count.

Twinkies, Oreo cookies, Snickers candy bars, Cadbury Eggs, Reese's Peanut Butter Cups, cheesecake squares, s'mores (a toasted marshmallow and chocolate confection), and chocolate-chip cookie dough are all commonly deep-fried at fairs. The process is the same for all: thoroughly freeze the food to be fried; coat the food in a thick batter with more than normal amount of baking powder; fry in 190 °C (375 °F) oil until golden-brown; drain well; and add toppings of choice.[56]

Single Ingredient Themes

From 1979 to 2019, the small California farming town of Gilroy (the self-labelled 'Garlic Capital of the World') held the Gilroy Garlic Festival. Most of the activities were garlic themed. All the food served, and that's what most festival-goers came for, was garlic-based. Over 1800 kg (4000 lb) of fresh garlic was consumed each year.[57] The Gilroy Garlic Festival may be the most famous, but the United States hosted twenty garlic festivals in 2021.[58]

Garlic is just one of the foods with its own festival. Onions have at least five festivals ranging from the Vidalia Onion Festival in Vidalia, Georgia, to the Walla Walla Sweet

Onion Festival in Walla Walla, Washington, and the Maui Onion Festival, in Lahaina, Maui, Hawaii.[59]

Seafood festivals celebrate a variety of sea creatures: crabs, crawfish, frog (legs), lobster, mussels, oysters, salmon, and shrimp. Fruits have festivals. Nuts have festivals. Animal meats and offal have festivals.[60]

Other Fair Fare

The list of common fair foods is too large to describe each in detail, but each should be acknowledged:

- Peanuts and all the tree nuts that have been featured through the years.
- Fruits, like bananas that sold for a dime each at the 1876 Exposition, and pineapples, that were still exotic in 1915.[61]
- Popcorn, which was sold dusted with sugar at early fairs, and now can be had at fairs in multi-coloured, multi-flavoured bricks or giant bags.[62]
- It's difficult to determine when the modern hot dog got its start. The bun-covered sausage at 1893 Fair was not new to the German immigrants that sold it.[63]
- Corn dogs, sometimes regionally referred to as Pronto Pups, seem to be around since the 1930s and are still quite popular at fairs.
- The hamburger of 1885 is not today's hamburger, and not the twenty-first-century ramen burger served at music festivals in San Francisco.[64]
- The Louisiana Purchase Exposition claimed to have served America's first ice cream cone.[65] Today, ice cream is sold as 'soft serve' and is available in a plethora of flavours and in a cup or a cone.
- Competing with ice cream is the snow cone, sometimes called shaved ice; crushed ice with one or more flavoured syrups.
- Lemonade is common, either made of lemons or sweetened-lemon concentrate.
- Grilled sausages of various types.
- Grilled corn is so popular that there are special large cookers designed for fair use.
- Pizza, either barbecued or baked in a mobile pizza oven.
- Ethnic foods of too many varieties to mention.

Bringing the Fair to a Close

All fairs and festivals come to an end, whether for the world or sponsored by a church club. Attendees rush to see one more exhibit, ride one more ride, and, in 1915, eat one more scone dripping with raspberry jam.[66]

Now, more than a century-and-a-half since America's first world's fair, the large, national fair is no longer financially feasible.[67] Today, state fairs are less common, and many counties have stopped sponsoring fairs. Fairs compete with amusement parks, outdoor music festivals, and community festivals and events.[68] Much of the walk-around food from the early fairs is found either in a modern grocery store or has changed to the point that earlier fair-goers wouldn't recognize the food. Even deep-fried Twinkies in two flavours can be found now in a grocer's freezer compartment, ready to bake or fry.[69]

Notes

1. Bonnie M. Miller, 'The Pure Food Exhibits in the "Palace of Nibbling Arts"', *Southern California Quarterly*, 100 (2018), 150–82 (p. 156) <https://www.jstor.org/stable/26499875>
2. 'Philately at the Fair', *Mekeel's Weekly Stamp News*, 4 December 1915, p. 420.
3. David Gerard Hogan, 'World's Fairs', in *The Oxford Companion to American Food and Drink*, ed. by Andrew F. Smith (Oxford: Oxford University Press, 2007), pp. 636–37.
4. Robert W. Rydell, *All the World's a Fair: Visions of Empire at American International Expositions, 1876-1916* (Chicago: University of Chicago Press, 1984), pp. 142–43, 2.
5. Rydell, p. 3.
6. *Visitors' Guide to the Centennial Exhibition and Philadelphia* (Philadelphia: J.B. Lippincott & Co., 1876), p. 20.
7. America's first 'agricultural' fair occurred in Pittsfield, MA, when Elkanah Watson 'staged, in conjunction with his neighbours in 1810, the celebrated "cattle show" that evolved into an American institution known as the county fair' (Fred Bassett, 'Elkanah Watson Papers, 1773–1884, SC13294 and SC12579', New York State Library, n.d. <https://www.nysl.nysed.gov/msscfa/sc13294.htm > [accessed 26 March 2022]).
8. 'State Fair History', *The Fair, a Division of the New York State Department of Agriculture and Markets* <https://nysfair.ny.gov/about/fair-history/> [accessed 26 March 2022].
9. Carson Hendricks, *California State Fair* (Charleston, SC: Arcadia Publishing, 2010), p. 8.
10. Kathie Fry, 'U.S. Food Festivals by Name', *Do It in the Americas* <http://www.doitintheamericas.com/us/events/food-festivals-by-food.htm> [accessed 12 April 2022].
11. 'Chicago's Largest Festivals', *Chicago Business*, Crain Communications, Inc., 2007, archived from the original on 28 September 2007 <https://web.archive.org/web/20070928045556/http://chicagobusiness.datajoe.com/app/ecom/pub_viewhtml.php?listid=2022&year=2007&htmlkey=maM91S4hq44x.> [accessed on 18 April 2022].
12. The term 'immigrant' here is used to refer to anyone who identifies with the ethnic group, not just first-generation immigrants.
13. Rob Hard, 'How to Plan a Community Food Festival', *The Balance Small Business* <https://www.thebalancesmb.com/how-to-plan-community-events-and-food-festivals-1223700> [accessed 18 April 2022]
14. Janelle Bitker, 'Outside Lands: 5 Standouts from Festival's Myriad Food Options', *San Francisco Chronicle*, 31 July 2022, p. G14.
15. *St. Pius X Summer Festival* <http://www.nkybestfestival.com/> [accessed 18 April 2022].
16. Christine Ingram and Jennie Shapter, *The Cook's Encyclopedia of Bread* (New York: Lorenz, 1999), p. 54.
17. 'A Brief History of the Scone', *Foodways* <https://www.freshways.co.uk/a-brief-history-of-the-scone/> [accessed 18 April 2022].
18. Helen Dare, 'Now Sconing, With the Exposition, Also Ends: And "Sconing," as We All Know Who Have Been Participants or Spectators of the Game, Is Most Exciting and Perilous of Indoor Sports', *San Francisco Chronicle*, 5 December 1915, p. 19.
19. *Robin Hood: Prince of Thieves,* dir. by Kevin Reynolds (Warner Bros. Pictures & Morgan Creek Entertainment, 1991).
20. The beer produced in the exhibit may not have been the same beer consumed since then: before mechanical refrigeration, lager-beer production was a wintertime activity, as German lagers required a lower fermentation temperature than English ales.
21. Mark Benbow, 'German Immigrants in the United States Brewing Industry', *Immigrant Entrepreneurship: German-American Business Biographies, 1720 to the Present* <https://www.immigrantentrepreneurship.org/entries/german-immigrants-in-the-united-states-brewing-industry/> [accessed 18 April 2022].
22. Dirk Hildebrandt, 'They Brought Their Beer', *Wisconsin Magazine of History*, 102 (2018), 14–27 (pp. 22, 23).

23 'Company History', Hugh Moore Dixie Cup Company Collection, 1905-2008 (website) <https://sites.lafayette.edu/dixiecollection/company-history/> [accessed 18 April 2022].
24 U.S. Pub.L. 66-66, 'National Prohibition Act', 28 October 1919.
25 'Five interesting facts about Prohibition's end in 1933', *Constitution Daily: Smart Conversation from the National Constitution Center* <https://constitutioncenter.org/blog/five-interesting-facts-about-prohibitions-end-in-1933> [accessed 18 April 2022].
26 Brendan Byrne, 'The Rise of the Beer Can', *The Atlantic*, 27 May 2016 <https://www.theatlantic.com/technology/archive/2016/05/rise-of-the-beer-can/484406/> [accessed 18 April 2022].
27 'Waffle Different', *Spokane Daily Chronicle*, 24 April 1962, p. 2, which also notes that the Belgian waffle was introduced at Expo 58 in Brussels by Walter Cleyman; Stefanie Stuart, 'The Best Tips to Make Belgian Waffles', *ALD Professional Kitchen Equipment* <https://ald.kitchen/blogs/news/the-best-tips-to-make-belgian-waffles> [accessed 31 March 2022].
28 'Waffle Different'; Alan Stein, 'Belgian Waffles Are Introduced in America at the Seattle World's Fair on April 21, 1962', *HistoryLink.org* <https://historylink.org/File/10092> [accessed 31 March 2022].
29 Ula Illnytzky, 'She Sold Belgian waffles at the 1964 World's Fair', *yahoo!life*, 21 April 2014 <https://www.yahoo.com/lifestyle/tagged/health/she-sold-belgian-waffles-1964-145828573.html> [accessed 31 March 2022]. I ate a Belgian Waffle at the 1964 World's Fair. It was good, but at the price of one dollar, it was equal to the price of four packs of cigarettes. I was sixteen years old; do you think I had another waffle?
30 Adam Chandler, 'Patriotic Waffle House Takes a Strict Stance Against Belgian Waffles', *The Wire*, 1 July 2014 <https://www.theatlantic.com/international/archive/2014/07/waffle-house-takes-a-strict-stance-against-belgian-waffles/373787/> [accessed 31 March 2022].
31 Casey Barber, 'World's Fair Belgian Waffles', *Good. Food. Stories.* <https://www.goodfoodstories.com/worlds-fair-belgian-waffles/> [accessed 31 March 2022]; Illnytzky.
32 William J. Morrison and John C. Wharton, 'Candy-Machine', U.S. Patent 618,428, 31 January 1899.
33 A quarter is a coin in the U.S. equal to one-fourth of a dollar, or twenty-five cents – two bits in slang – which in 1900 was equivalent to about $8.56 in 2022. All money conversions are made with 'CPI Inflation Calculation', U.S. Official Inflation Data, Alioth Finance (website) <https://www.officialdata.org/> [accessed 16 April 2022].
34 William J. Morrison and John C. Wharton, 'Candy-Machine', U.S. Patent 717,756, 6 January 1903.
35 Jackie Finch, 'It's the Centennial for St. Louis World's Fair', *The Herald Times*, 11 April 2004.
36 Pamela J. Vaccaro, *Beyond the Ice Cream Cone: The Whole Scoop on Food at the 1904 World's Fair* (St. Louis: Enid Press, 2004), p. 122.
37 Gerard Paul, 'The History of Cotton Candy – A Dentist's Summer Gift', *Many Eats*, 13 September 2020 <https://manyeats.com/history-of-cotton-candy> [accessed 4 April 2022].
38 'The History of Cotton Candy', *Spun Paradise*, 4 November 2018 <https://www.spunparadise.com/blog/the-history-of-cotton-candy> [accessed 5 April 2022]. The company was founded in 1872 (The *Confectioner's Journal*, 48.65 (1922), p. 126a).
39 Many party rental companies rent cotton candy machines, and, if required, electric generators, and sell the supplies needed to make and sell the candy.
40 *Gold Medal*, Gold Medal Products Co. <https://www.gmpopcorn.com/> [accessed 5 April 2022]; *Standard Concession Supply* <https://www.standardconcessionsupply.com/> [accessed 5 April 2022].
41 'Cracker Jack Popcorn', *Chicago Inventions* <http://websites.umich.edu/~eng217/student_projects/Chicago%20Inventions/crackerjack.html> [accessed 18 April 2022].
42 'History of Popcorn', *What's Cooking in America* <https://whatscookingamerica.net/history/popcorn-history.htm> [accessed 7 April 2022].
43 Samantha Chmelik, 'Frederick Rueckheim', *Immigrant Entrepreneurship: German-American Business Biographies, 1720 to the Present* <https://www.immigrantentrepreneurship.org/entries/frederick-rueckheim/#edn9> [accessed 7 April 2022].
44 Alex Jaramillo, *Cracker Jack Prizes* (New York: Abbeville Press: 1989), p. 8; Chmelik.
45 Jaramillo.

46 'Betty Crocker Fudge Brownie Mix', *Betty Crocker,* General Mills <https://www.bettycrocker.com/products/betty-crocker-brownies-and-dessert-bars/fudge-brownie> [accessed 10 April 2022].

47 'Manischewitz Passover Brownie Mix Pan Included', *H-E-B* <https://www.heb.com/product-detail/manischewitz-passover-brownie-mix-pan-included/174712> [accessed 10 April 2022].

48 John D. Riley, *A History of the American Soft Drink Industry: Bottled Carbonated Beverages, 1807–1957* (Washington, DC: American Bottlers of Carbonated Beverages, 1958); Tristan Donovan, *Fizz: How Soda Shook Up the World* (Chicago: Chicago Review Press, 2013).

49 *Frank Leslie's Illustrated Historical Register of the Centennial Exposition 1876*, ed. by Frank H. Norton (New York: Frank Leslie's Publishing House, 1877), p. 123.

50 *Visitors' Guide to the Centennial Exhibition and Philadelphia*, p. 9. Ten cents in 1876 is roughly equal to $2.69 in 2022 [accessed 16 April 2022].

51 *Official Directory*, p. 513.

52 *Soda Parts* <https://www.sodaparts.com/> [accessed 16 April 2022].

53 Rebecca Strassberg, 'The History of Deep-Frying Foods at Fairs', *Thrillist* <https://www.thrillist.com/eat/nation/the-history-of-deep-frying-food-at-fairs> [accessed 12 April 2022].

54 'The Origins of Two American Fried Dough Classics: Funnel Cakes and Elephant Ears', *Gold Medal*, Gold Medal Products Co. <https://www.gmpopcorn.com/resources/blog/the-origins-of-two-american-fried-dough-classics-funnel-cakes-and-elephant-ears> [accessed 12 April 2022]; see 'Zalabia' in Gil Marks, *Encyclopedia of Jewish Food* (New York: Houghton Mifflin Harcourt, 2010).

55 Michaela Liebl, 'Fair Food: Funnel Cake', *YouTube*, 2 October 2018 <https://youtu.be/wGhozkYe6a4> [accessed 18 April 2022].

56 George Geary, *Fair Foods: The Most Popular and Offbeat Recipes from America's State and County Fairs* (Solana Beach, CA: Santa Monica Press, 2017), pp. 64–89.

57 *Gilroy Garlic Festival*, Gilroy Garlic Festival Association <https://gilroygarlicfestivalassociation.com/> [accessed 12 April 2022]. On 25 April 2022, festival officials announced cancellation of the festival due to the cost of insurance (Danielle Echeverria, 'In Gilroy, it's Over for Garlic Festival', *San Francisco Chronicle*, 25 April 2022, pp. C1–C2).

58 'Garlic Festivals 2021', *Garlic Seed Foundation* <https://www.garlicseedfoundation.info/festivals.htm> [accessed 12 April 2022].

59 Kathie Fry, 'United States Onion Festivals', *Do It in the Americas* <http://www.doitintheamericas.com/us/events/onion-festivals.htm> [accessed 12 April 2022].

60 Kathie Fry, 'U.S. Food Festivals by Name'.

61 Mark J. Perry, 'The "Miracle of the Marketplace" Brings Bananas to Your Local Grocery Store from Thousands of Miles Away for Pennies', *American Enterprise Institute* <https://www.aei.org/carpe-diem/the-miracle-of-the-marketplace-brings-bananas-to-your-local-grocery-store-from-thousands-of-miles-away-for-pennies/> [accessed 19 April 2022]; 'The Big Pineapple' (advertisement), *San Francisco Chronicle*, 27 February 1915, p. 4.

62 Chmelik.

63 Scott, 'Vienna Beef History Museum opens in Chicago', *The World's Fair* <https://worldsfairchicago1893.com/2018/05/31/vienna-beef-history-museum-opens-in-chicago/> [accessed 18 April 2022]

64 'Birth of the Hamburger', *Erie County Agricultural Society* <https://www.ecfair.org/p/info/about-the-fair/birth-of-the-hamburger> [accessed 18 April 2022]; Mari Takahashi, personal communication with author, 30 March 2022.

65 Vaccaro, pp. 124–27.

66 Dare.

67 Darran Anderson, 'World's Fairs and the Death of Optimism', *Bloomberg* <https://www.bloomberg.com/news/articles/2018-10-03/the-necessity-of-a-pessimistic-world-s-fair> [accessed 12 April 2022].

68 Dick Preston, 'Beyond the Trivia: State Fairs', *KRCG-TV* <https://krcgtv.com/features/beyond-the-trivia/beyond-the-trivia-state-fairs> [accessed 12 April 2022].

69 Monica Watrous, 'How Deep Fried Twinkies Came to Be', *Food Business News*, 15 August 2016.

Dispatches from the Chicken Sandwich Wars

Michael A. Johnson

Introduction: The US Chicken Sandwich Wars as Necropolitics

Let's start with the name: Chicken Sandwich Wars. There is something comedic about juxtaposing a sacrosanct concept like war with something as apparently low stakes as a chicken sandwich. The juxtaposition invites easy dismissal of the Chicken Sandwich Wars as an entertaining but ultimately harmless rivalry between fast-food industry competitors, analogous to the 1980s-era cola wars, little more than a quick-burn viral phenomenon. Even a reported fatality (a Maryland man stabbed to death in a Popeyes parking lot for cutting in line) might be dismissed as a fringe example of consumerist fervour along the same lines as a Black Friday riot, Beanie Baby mania, or the NFT land grab. The mistake with this line of reason, of course, is that it assumes chicken sandwiches are a politically neutral product with low stakes and zero body count. It's either that or, more cynically, it acknowledges the necropolitical dimension of the fast-food industry and tacitly accepts that the lives ruined and lost to it are somehow necessary sacrifices. The impulse to dismiss the Chicken Sandwich Wars as inconsequential belies an unexamined necropolitical pact at the heart of neoliberal reason.

The Chicken Sandwich Wars, and the fast-food industry generally, participate in what Achilles Mbembe calls the 'necroeconomy' in which contemporary, largely minoritized, populations are managed through the experience of danger and risk. For Mbembe, war is 'the sacrament of our times' and the deployment of necropolitics creates what he calls *deathworlds*, 'new and unique forms of social existence in which vast populations are subjected to living conditions that confer upon them the status of the living dead'.[1] Along similar lines, writing about 'big food' and obesity in the US, Lauren Berlant offers the descriptor 'slow death', explaining that specific populations are 'marked out for wearing out' and the conditions of being worn out and dying are intimately linked with 'the ordinary reproduction of [daily] life'.[2]

Keeping Mbembe's conception of *necropolitics* and Berlant's of *slow death* at the forefront, I would like to examine the Chicken Sandwich Wars as a war in the literal and more sobering sense of the term. Who are the belligerents? Who wins and who loses? What are the body counts?

Chick-fil-A and Popeyes: The Companies

The Chicken Sandwich Wars broke out in 2012 when the president of Chick-fil-A Dan

Cathy's alignment with anti-LGBTQ+ organizations, and his defence of the 'biblical definition of marriage' in interviews, drew criticism and actions from LGBTQIA+ rights groups who called for a boycott of the fast-food chain. Jim Henson Co. pulled its Muppet toys from Chick-fil-A kids' meals, and the mayors of Boston and Chicago made public statements unwelcoming the chain from their cities. Meanwhile, conservative Christians rallied around Chick-fil-A. Former Arkansas governor, Mike Huckabee, inaugurated a 'Chick-fil-A Appreciation Day'; numerous conservative Christian politicians like Rick Santorum and influencers voiced their support of the fast-food chain.

The conflict might have remained a page in the larger chronicle of the culture wars if it weren't for the August 2019 release of Popeyes's answer to Chick-fil-A's chicken sandwich. Public reception of the Popeyes chicken sandwich in the US came just short of messianic fervour. *The Advocate* described it as 'thicker and spicier' than Chick-fil-A's with 'more flavor, less homophobia'.[3] *The New Yorker* food critic Helen Rosner praised it in the highest terms: 'The meat is flavorful and juicy, encased in a spiky, golden sea urchin of batter—surprisingly light, uncommonly crispy'.[4] Popeyes' social media strategists invoked the power of Black Twitter to counter the cultural whiteness of the Chick-fil-A brand, thus making the sandwich a viral phenomenon inviting analysis and commentary from critics who usually steer clear of fast food. *The New Yorker* suggested (with a wink) that the sandwich could 'save America' while the *Los Angeles Times* claimed that the Popeyes chicken sandwich could serve as an 'economic indicator' of the country's fiscal health.[5] The Popeyes chicken sandwich sparked such a frenzy of consumer enthusiasm that numerous violent incidents occurred in the months following its release in response to shortages of the chicken sandwich.

Unable to match Popeyes' social media savvy, Chick-fil-A lost 15% of the market share in chicken sandwiches in the space of three months. After gay-rights protests led to rapid closure of Chick-fil-A's first store in the UK in October of 2019, Chick-fil-A announced a change to its philanthropic and marketing strategies that moved its branding in a more cosmopolitan, diversity-embracing direction. During that same month, in an iconic instance of social media trolling, Popeyes posted a video skit centred on a highway exit food sign on which the logos for Chick-fil-A and Popeyes appear side by side. The camera pans from the Chick-fil-A logo, which boasts 'closed Sunday', to a shot of a worker updating the Popeyes logo to include the words 'open Sunday'. By placing emphasis on Sunday closure, the marketing team presents Popeyes as the consumer-friendly choice while also making oblique reference to Chick-fil-A's anti-LGBTQ stance. Most crucially what the publicity stunt accomplishes is to cement Popeyes' place as an alternative to Chick-fil-A, as the other chicken sandwich.

Should Popeyes be declared victor of the Chicken Sandwich Wars? Certainly, Popeyes wins the social media battle. It also managed to claim a chunk of the market share on chicken sandwiches. But the reality is that both fast-food companies reported tremendous growth between August 2019 and the present. In February 2021, *QRS Magazine* reported that Popeyes had increased per-restaurant earnings by $400k with the addition of the chicken sandwich, which amounts to a 38% jump in same-store sales. The company continues to grow despite, or more likely as the result of, the pandemic with a reported increase of 32% in global system-wide sales in May 2021. By November of 2020 Chick-fil-A's market share of chicken sandwich sales returned to pre-chicken-sandwich-war levels (around 45%) while Popeyes levelled out at around 20% of the total market share. Chick-fil-A's revenue climbed to $5.8 billion last year, well ahead of the $4.3 billion it appreciated in 2020 and $3.8 billion the year prior.[6]

While it may be true that both sides win, it does not necessarily follow that there are no losers in the Chicken Sandwich Wars. Take, for example, the workers.

'Essential' Employees

The next phase of the Chicken Sandwich Wars extends into the first year of the COVID-19 pandemic when Burger King, KFC, Churches, and various other fast-food chains released their own versions of the chicken sandwich during lockdowns and mask mandates. The various supply chain and workforce disruptions caused by the pandemic, combined with the politicization of face coverings and vaccines, put fast-food workers in the spotlight. During the first lockdowns, food service workers in the US were classified as 'essential employees'. Several memes and viral videos highlighting America's hypocritical attitudes towards food service industry workers (deemed 'unskilled' and therefore 'disposable' yet also 'essential') circulated. Cathartic photos and videos of food service workers quitting gave social media users an outlet for their own workplace frustrations. The viral photo of a Burger King marquee in Lincoln, Nebraska that reads 'We all quit. Sorry for the inconvenience' circulated globally and inspired a wave of similar resignations throughout the US. Several videos circulated of restaurant employees caught on phone cameras quitting their jobs in a spectacular fashion. At the end of their rope with the added stress of enforcing mask and vaccine mandates, these food service workers were applauded by social media users as they uttered variations on 'I'm not paid enough for this shit'. Of course, labour conditions had been poor before the pandemic. The franchise model, which predominates in fast food, incentivizes labour violations. A study conducted by researchers at the UCLA Labor Center found that fast food workers in Los Angeles County had systematically faced labour issues related to safety and injury, workplace violence, harassment, retaliation, and wage theft before the pandemic and that these conditions only worsened when the pandemic began. The study found that nearly a quarter of respondents reported having contracted COVID-19 since March 2020, a rate higher than that of the county's population. Worker testimony and complaints also showed egregious failures in communicating COVID-19 outbreaks in restaurants to workers, resulting in the highest increase in mortality of any occupation for limited-service cooks during the pandemic according to the same study.[7]

The visibility of poor working conditions on social media combined with worker shortages led franchise owners to offer higher salaries. Wages for hourly limited-service restaurant workers climbed 10% from 2020 to 2021 according to various industry trackers: a small victory for fast-food restaurant employees – although, of course, restaurant employees are only the final link in a supply chain that depends on labour from farmers, meat packers, food scientists, and truckers, among others. Eric Schlosser's 2001 *Fast Food Nation* exposed the grotesque degree to which farmers and meat packers in the fast-food supply chain are exploited, a situation that worsened considerably during the era of neoliberal deregulation (from Reagan onwards).[8] Despite increased awareness and visibility, work conditions for meat packers remained as egregious as Schlosser's exposé described them until the advent of the pandemic at which point they become precipitously worse. In an article published in *The Guardian* in early May, Nina Lakhani reported on an investigation into the meatpacking industry done by the congressional select subcommittee on the coronavirus. Lakhani writes:

> The meatpacking industry, which includes slaughterhouses and processing plants – is one of the most profitable and dangerous in the US. It is a monopoly business, with just a handful of powerful multinationals dominating the supply chain which, even before Covid, was bad news for farmers, workers, consumers and animal welfare. Yet in April 2020, Trump issued an executive order invoking the Defense Production Act to keep meat plants open following a flurry of communication between the White House chief of staff, Mark Meadows, the vice-president's office, USDA allies and company executives. The order, which was proposed by Smithfield and Tyson (whose legal department also wrote the draft), was an overt attempt to override health departments and force meat plant workers – who are mostly immigrants, refugees and people of color – to keep working without adequate protections while shielding the industry from lawsuits.[9]

In the first four months of the pandemic, the Occupational Safety and Health Administration (OSHA) received a massive spike in complaints from meatpacking workers. During the following few months, nearly 200 meatpacking workers died of Covid; by September 2021, Covid had killed at least 298 meat plant workers. *ProPublica* reports:

> The effect that the meatpacking plant outbreaks had on the early spread of COVID-19 is staggering. […] [A]cademic researchers have found that by July 2020, about 6% to 8% of all coronavirus cases in the U.S. were tied to packing plant outbreaks, and that by October 2020, community spread from the plants had generated 334,000 illnesses and 18,000 COVID-19-related deaths.

It is hard to overstate the necropolitical cynicism of the meat industry leaders who cried wolf about the danger of domestic meat shortages to drive up domestic prices and suspend safety measures while also exporting meat internationally. In a classic instance of disaster capitalism, the meatpacking industry managed to triple its net profit margins since the pandemic started. The necropolitical drive of the industry emerges clearest in its deployment of classist and racist tropes to provide cover for its homicidal neglect of worker safety. Smithfield CEO Kenneth Sullivan wrote, for example, '[s]ocial distancing is a nicety that makes sense only for people with laptops'. A Tyson official blamed outbreaks on the workers' crowded living and commuting arrangements, referring to multi-generational households of Mexican and Central American meatpacking workers by claiming, 'this is a culture issue'. These industry leaders don't even bother to conceal the belief that their workers' lives possess less value than 'people with laptops'.[10]

Although workers at meatpacking plants and fast-food restaurants were both deemed essential workers, and both experienced disproportionately high Covid infection and death rates, restaurant employees were the only ones able to achieve some degree of success in advocating for higher salaries and improved workplace conditions. While fast-food employees were able to harness the power of their smart phones and social media, the threat of deportation for undocumented workers, ag-gag laws, and draconian control over worker time prevent documentation and social media sharing of poor work conditions in meat packing plants. The Chicken Sandwich Wars may have brought limited attention to the plight of restaurant workers. However, the consumer frenzy generated by the Wars may also have occluded the plight of even more precarious agricultural workers and poultry workers. It's a classic false consciousness structure: the spectacle of a righteous struggle between a homophobic company with a white public voice and an anti-homophobic company with a Black public voice serves to occlude both companies' reliance on a massive disposable workforce composed largely of Black and Brown (BIPOC) bodies.

Chicken: The Official Animal Protein of Neoliberal Late-Stage Capitalism

What role do chickens play in the Chicken Sandwich Wars? Chicken, I would like to suggest, is uniquely suited to the cultural and economic demands of neoliberal late-stage capitalism. Chickens are more efficient in converting feed into meat protein, which reduces the amount of land, fertilizer, and energy involved, resulting in a carbon footprint that is roughly one-tenth that of beef. The lower carbon footprint provides a selling point while its measurability makes for more predictable scalability and cost-benefit ratio. Because of their genetic malleability, chickens brought to poultry processing plants are now uniform in size, making it possible to automate previously manual stages of the slaughtering process. Everything aside from the butchering of whole broilers into packaged breast, thighs, tenders, and so on, is now automated in poultry processing plants.

From a market research perspective, the chicken sandwich taps into multiple trends. It appeals to people who are cutting back on red meat consumption. Because of its affordability and ubiquity, chicken is a comfort food for many throughout the globe, and comfort food consumption has increased since the beginning of the pandemic according to various industry measures. Chicken sandwiches are portable; they hold up well in delivery; and the younger demographic buys them more frequently, according to NPD Group data. Technomic found that sales at limited-service chicken chains rose 9% in 2021 compared to 0.3% growth at burger chains and negative growth in sales across all other fast-food categories.[11]

Because chickens lend so well to technological intervention, the twentieth century saw an explosion in the numbers of domesticated chickens across the globe. From 1990 to 2013, global poultry production increased by 165 percent compared to a 23 percent increase in global beef production. The standing population of chickens is 22.7 billion. The world slaughters another 65.8 billion chickens every year, producing a rate of carcass accumulation that is unprecedented in the natural world. Broiler chickens are now so ubiquitous on the planet that their bones will be written into the fossil record as a mark of the Anthropocene. Chicken bones will remain alongside plastics, fertilizers, fossil fuels, and radioactive deposits from nuclear weapons as part of the permanent geological record of humanity's presence on the planet.[12]

Although chicken may have a smaller carbon footprint than beef, the scale of chicken farms and relative lack of regulation in the US produces a worrisome degree of environmental damage. *Vox* journalist Leah Garces writes, 'the poultry industry is essentially given license to use America's public waters as its own unregulated, open sewer system.'[13] Such poor waste management creates oxygen depletion and toxic algae blooms in public streams and rivers. And because the poultry industry consumes most of the world's feed crops (which occupy a third of the world's cropland altogether), the water footprint of chicken winds up being roughly six times that required to grow a grain such as oats per calorie. Much like the choice between Chick-fil-A and Popeyes, the choice of chicken over beef is presented to consumers reassuringly as a virtuous one while ultimately neither of these animal proteins are environmentally sustainable given current practices.

The life of a chicken raised for fast-food consumption occupies a uniquely necroeconomic temporality. The fast-growing breeds preferred by companies like Chick-fil-A and Popeyes take only thirty days to reach slaughter weight. Those thirty days are spent in conditions that barely qualify as a life. The quantification of every aspect of a chicken's life is so thoroughly calculated in advance that its living is predetermined and outweighed by its death. This mirrors uncannily the lives of poultry processing plant workers. Both live in bodies that are commodified in diachronic descension: that is, the choicest bits are extracted first and then, as the factory line progresses, the quantity of value that can be extracted from them declines until they are fully used up, as workers' bodies wear out and they are reassigned to increasingly inhuman(e) and deadly jobs in the plant, as the final extraction of protein paste and bone meal from the chickens' bodies, now unrecognizable, is complete. This is the collapsed temporality of a life reduced to its extractable value. Because being used up is their telos, because their total final value has been quantified in advance, neoliberal reason requires both chickens and poultry plant workers to be already dead in a way, or in Mbembe's terms, to be living dead.

The Consumers: Woke Capitalism and Consumer Fanaticism
The release of a viable alternative to Chick-fil-A's chicken sandwich produced a great sigh of relief from a public disciplined by vote-with-your-dollar consumerism. Before Popeyes released its chicken sandwich, the number of articles encouraging anti-homophobic consumers to take their dollar elsewhere was matched or exceeded by

Chick-fil-A-as-guilty-pleasure memes. Take, for example, the apologetic bear meme ('I support gay marriage but I'm addicted to Chick-fil-A') or the Chick-fil-A rainbow pass ('Made this for my friends. Feel free to use this to eliminate any guilt over eating Chick-fil-A despite their homophobic stuff'). Even after the release of the Popeyes chicken sandwich, Chick-fil-A-as-guilty-pleasure memes continue to circulate on social media feeds. For example, an image of Miss Krabappel from *The Simpsons* entering a movie theatre incognito, with sunglasses and a headscarf; the caption reads: 'Me going to chick-fil-a to get waffle fries to eat with my Popeyes chicken sandwich'. And, of course, during Pride Month each year, tweets such as the following recirculate: 'Ok straights. Consider giving up Chick-fil-A for Pride month. Think of it like lent for allies'.

Much like certain manifestations of sexual shame, consuming a Chick-fil-A chicken sandwich entails the weighing of one's desires over one's qualms, followed by a dopamine/serotonin reward and, soon enough, a lurching sense of guilt, the promise not to do it again, and so on. How profoundly consumerism has become enmeshed with Protestant morality! The 'give up Chick-fil-A for Pride Month' tweet speaks both to the commodification of queer identities and a kind of false consciousness, a flattening of morality in framing allyship in such reductively consumerist terms. It encapsulates the bad faith dynamic of woke capitalism in much the same way as Popeyes's highway exit sign publicity stunt. By inviting consumers to associate Sunday closure with Chick-fil-A's homophobia, Popeyes sells its extended opening hours as the virtuous choice while occluding its own inhumane labour politics. Hard as it might be to imagine in the current political moment, it should not be forgotten that Christian groups and organized labour groups worked together to advocate for maintaining Sunday closure throughout the 1980s as North American blue laws were being overturned. Indeed, Popeyes has one of the worst records among fast-food chains where labour is concerned. Its workers are paid poverty wages while its parent company is infamous for countering labour organization with punitive retaliatory measures. Popeyes is no virtuous choice, but its marketing team grasped the underlying pharmacopornographic principle (making a virtuous choice comes with its own dopamine/serotonin reward) and seized the opportunity presented by the moralized consumerism of the Chicken Sandwich Wars: Eat Popeyes and feel better![14]

Consumer enthusiasm for the Popeyes chicken sandwich was so great that the chain was unable to keep up with demand. Shortages gave way to violence, injury, and death as customers were eager to get their hands on the legendary sandwich. Third-party services emerged selling the sandwich for an upcharge. The rapper Quavo joked on Instagram that he was selling the sandwiches out of the back of his car for $1000 a pop. Press coverage of violent incidents in Popeyes locations led to increased news coverage and public interest in the sandwich and further intensification of consumer fanaticism. Critics began to write thought pieces; chicken sandwich costumes became a Halloween trend; the whole phenomenon took on the dimensions of a defining cultural moment for North American consumerism on par with the Cabbage Patch Kids riots of 1983.

The promise of virtue offered by an alternative to Chick-fil-A is not enough to explain the intensity of consumer response to the Popeyes chicken sandwich. The huge jump in sales which led to shortages took place immediately after Popeyes tweeted, 'Y'all good?' in response to a Chick-fil-A tweet attempting to shift the emphasis away from hatred/homophobia ('Chicken + bun + pickle = love'). Invoking the powerful voices of Black Twitter with an unmistakably Black cultural form ('Y'all good?' said with the right intonation is a classic example of throwing shade), the Popeyes chicken sandwich became immediately associated with the frustration and defiance encapsulated in contestatory hashtags #BlackLivesMatter (2013), #ICantBreathe (2014), #OscarsSoWhite (2015), and #SayHerName (2015). The initial framing of Popeyes as a virtuous alternative to Chick-fil-A sets the stage for this invocation of Black Twitter. But here the emphasis shifts from virtuous choice to joyful self-expression, from vote-with-your-dollar LGBTQIA+ activism to intersectional social justice activism. Buying a Popeyes chicken sandwich offers the promise of bearing witness to and participating in something bigger than oneself, a culturally defining event, a moment of Black defiance and Black joy.

As inspired and brilliant as this social media strategy was, as empowering as it may have been for Black and LGBTQIA+ consumers, it does not negate the company's execrable labour politics. For many, because of its consumerist framing, the spectacle felt disconnected from reality, like a distraction from 'real' (i.e. electoral) politics. In response to images of long lines around Popeyes locations, Janelle Monaé tweeted 'Perhaps we put voting booths at every Popeyes location? While we wait on that sammich you can register and vote @popeyes holla'. Monaé's tweet reflects (albeit more diplomatically) the same sentiment expressed by a Twitter user named #PettyPendergrass who posted: 'BLACK PEOPLE WILL STAND IN LINE FOR HOURS FOR CHICKEN BUT WON'T VOTE OMG OMG'. In response, John Legend tweeted: 'Popeye's would kill for lines like the ones outside of black polling stations. 1. Black people vote 2. We would vote even more if people like the GA Gov didn't intentionally limit poll access in ways that create these ridiculous lines. (pictured: four GA voting booths with unreasonably long lines)' which he followed with 'Andddddd fried chicken is delicious and nobody should be ashamed of enjoying it!' Building on John Legend's intervention, several Twitter users took the opportunity to engage the anti-racist cultural critique of the Chicken Sandwich Wars. For example, Mikki Kendall tweeted, '[t]he reactions to Black joy always prove that racism is woven so deeply into the social fabric of our culture that people can't even imagine not trying to shut down Black bodies publicly experiencing pleasure. It's fascinating how pressed some of y'all are about a sandwich.' Mikki Kendall continued: 'I have now seen the sandwich used to bludgeon Black people about voting, education, marriage & reparations. A $4 fast food sandwich & some jokes has people ready to create a narrative of pathological flaws in Blackness. And they think they're enlightened. It's amazing.' In response to these, and other, critiques, Janelle Monaé tweeted the following apology, 'I think the tweets that I posted about registering and voting were insensitive and wrong – specifically they ignored the very real issues of voter suppression that have impacted my community for years and me directly.'

Dispatches from the Chicken Sandwich Wars

Mikki Kendall's tweet offers crucial insight into the necropolitics of fast food and illuminates one of Berlant's key arguments in 'Slow Death (Sovereignty, Obesity, Lateral Agency)'. The impulse to view Black consumerism through a pathologizing lens as a failure of individual will mirrors the arguments of big food lobbyists who would direct attention away from systemic forces by locating responsibility with the individual. This bad faith misdirection posits a strong notion of sovereignty that contradicts big food's investment in forms of knowledge production, like user research and marketing, that view consumers in the aggregate, behaving according to predictable rules depending on what systemic constraints are in place – bodies of knowledge that assume a weak notion of sovereignty or even non-sovereignty. Likewise, the impulse to discipline Black bodies any time they are (re)presented en masse, as in the long lines around Popeyes, draws on the racist white liberal tradition of hygienics concerned with disciplining minoritized populations with the aim of cultivating a healthy productive workforce. (David Harvey observes in *Spaces of Hope* that sickness in late capitalism is defined as the inability to work.)[15] While outwardly benevolent in aim, this impulse holds little room for Black agency or Black enjoyment.

All consumers are caught in the aporia opened up between a strong (but fictive) notion of individual sovereignty and decidedly unsovereign notions of mass behaviour patterns. Black, working-class, and other minoritized consumers are caught in a second aporia that mirrors the former: caught between the weak notion of sovereignty assumed by sociologically informed public health and nutrition research and the strong notion of sovereignty desired by activism that aims to 'empower consumers to make better choices'. The point here is that both the cynicism of big fast food and the activism/knowledge production around wellness and nutrition participate in necropolitics. Necropolitics, as Mbembe explains, involves the large-scale management of risk and danger in populations. Necropolitics resides in the play between the wearing out of bodies and the maintenance of those bodies.

Where fast food is concerned, the worker and the consumer overlap significantly, largely working class, largely BIPOC. The supply chain involved in producing chicken sandwiches exploits and damages individual and social bodies while on the consumer end, eating unhealthy and oft-contaminated fast-food entails further wear to these same already-worn-down bodies. Everyone is subject to necropolitics. However, as both Berlant and Mbembe point out, minoritized and Black bodies are specially marked for wearing out given that necropolitics emerge in a historical continuum with colonial-era forms of exploitation like slavery at its origin. For these reasons, researchers in food studies, particularly those who work in nutrition and public health, should heed the insights from Black Twitter about the Chicken Sandwich Wars. While it is necessary to continue to shine a light on the fast-food industry's necropolitics, we must also remain vigilant when it comes to how we frame the question of fast-food consumption because necropolitics emerges here too in the impulse to police the enjoyment of working-class and BIPOC folks.

Conclusion

In *Fast Food Nation*, Schlosser takes great pains to demonstrate that the fast-food industry today is not the inevitable result of transparent market developments. It took billions in state infrastructure (highways, etc.), union busting and lobbying, migration precipitated by trade agreements, and many failed experiments in food technologies and consumer and market research to get us where we are today. Schlosser also cites the existence of fast-food chains such as In-N-Out Burger that serve ethically-sourced meat, rely on local supply chains, and pay their employees equitable wages. These alternative fast-food joints cater to a privileged customer base willing to pay more than double the price for a burger or chicken sandwich. These alternative fast-food joints do nothing to address the necropolitics of the fast-food industry as a whole; if anything, they perpetuate the classist myth of the sovereign consumer. However, the simple fact that such alternatives exist demonstrates that the exploitation-cum-efficiency model that currently predominates is not the inevitable telos of the industry's seventy-plus years of knowledge production. In other words, fast food does not only draw on a ruthless science of efficiency and exploitation; it also draws on a joyful science of aesthetics and hedonics. Fast food is a technology that can be hacked for pleasure. The success of Popeyes' marketing campaign and the intensity of people's investment in the Chicken Sandwich Wars proves this.

Notes

1. Achilles Mbembe, *Necropolitics* (Durham, NC: Duke University Press 2019), pp. 2, 92.
2. Lauren Berlant, 'Slow Death (Sovereignty, Obesity, Lateral Agency)', *Critical Inquiry*, 33.4 (Summer 2007), 754–80 (pp. 761 n. 20, 762).
3. Trudy Ring, 'Popeyes' Sandwich v. Chick-fil-A's: More Flavor, Less Homophobia', *The Advocate*, 19 August 2019 <https://www.advocate.com/business/2019/8/19/popeyes-sandwich-v-chick-fil-more-flavor-less-homophobia> [accessed 18 March 2023].
4. Helen Rosner, 'The Popeyes Chicken Sandwich is Here to Save America', *The New Yorker*, 20 August 2019 <https://www.newyorker.com/culture/annals-of-gastronomy/the-popeyes-chicken-sandwich-is-here-to-save-america> [accessed 18 March 2023].
5. Connor Sen, 'Commentary: Popeyes Chicken Sandwich Is an Economic Indicator', *Los Angeles Times*, 21 August 2019 <https://www.latimes.com/business/story/2019-08-21/popeyes-chicken-sandwich-is-an-economic-indicator> [accessed 18 March 2023].
6. Danny Klein, 'Who's Really Winning the Chicken Sandwich Wars?', *QRS Magazine*, 23 February 2021 <https://www.qsrmagazine.com/fast-food/whos-really-winning-chicken-sandwich-wars> [accessed 18 March 2023].
7. Kuochih Huang and others, *The Fast-Food Industry and COVID-19 in Los Angeles* (Los Angeles: UCLA Labor Center and Labor Occupational Safety and Health; Berkeley: UC Berkeley Labor Center and Labor Occupational Health Program), February 2020 <https://www.labor.ucla.edu/wp-content/uploads/2021/03/FastFood_Report_2021_v3_4-28-21.pdf> [accessed 18 March 2023].
8. Eric Schlosser, *Fast Food Nation: The Dark Side of the All-American Meal* (New York: Houghton, Mifflin, Harcourt, 2001).
9. Nina Lakhani, 'Trump Officials and Meat Industry Blocked Life-Saving Covid Controls, Investigation Finds'. *The Guardian*, 12 May 2022 <https://www.theguardian.com/environment/2022/may/12/meatpacking-industry-trump-downplay-covid-threat> [accessed 18 March 2023].
10. Michael Grabell, 'The Plot to Keep Meatpacking Plants Open During COVID-19', *ProPublica*, 13

May 2022 <https://www.propublica.org/article/documents-covid-meatpacking-tyson-smithfield-trump> [accessed 18 March 2023].

11 Jessica Wohl, 'What Marketers Can Learn from the Fast-Food Chicken Sandwich Wars: The Hard-Fought Battle for Poultry Preeminence Offers Valuable Lessons for Brands in All Competitive Categories', *Advertising Age*, 92.5 (19 April 2021), p. 1.

12 Carys E. Bennett and others, 'The Broiler Chicken as a Signal of a Human Reconfigured Biosphere', *Royal Society Open Science*, 5180325180325, 12 December 2018 <https://doi.org/10.1098/rsos.180325>.

13 Leah Garces, 'Replacing Beef with Chicken Isn't as Good for the Planet as You Think', *Vox*, 4 December 2019 <https://www.vox.com/future-perfect/2019/12/4/20993654/chicken-beef-climate-environment-factory-farms> [accessed 18 March 2023].

14 Beatriz Preciado, *Testo Junkie: Sex, Drugs, and Biopolitics in the Pharmacopornographic Era* (New York: Feminist Press at CUNY, 2013).

15 David Harvey, *Spaces of Hope* (Edinburgh: Edinburgh University Press, 2000).

Istanbul's Portable Food Practices: The Case of *Dolma*

Pırıl Kadırgan

Introduction

Dolma or *sarma*, stuffed vine leaves with minced meat, rice, and onion cooked in butter or with rice, onion, and spices cooked in olive oil, is one of the most distinctive flavours in Ottoman-Turkish cuisine. Derived from the Turkish verb '*doldurmak*' (to stuff something), *dolma* is a cooking type involving stuffing a vegetable or fruit with rice or bulgur. While peppers, cabbages, eggplants, and tomatoes are the most common plants used to make *dolma*, it can be made with any fresh or dried vegetable or fruit that can be stuffed. Since *dolma* is an ancient and widely used cooking technique, it can be prepared in a variety of ways and has numerous variations. *Sarma* (stuffed vine leaves) is one of the most common *dolma* variations. The term '*sarma*' is derived from the Turkish verb '*sarmak*' (to wrap something), and it refers to a dish made by wrapping leaves. Although vine and cabbage leaves are commonly used in the preparation of *sarma*, many other types of leaves and herbs are also used throughout Anatolia.[1]

Sarma, a meatless dish cooked in olive oil, was known as counterfeit (*yalancı*) *dolma* in Istanbul. It was primarily a culinary method created by the Christian community in the Ottoman Empire during their fasting periods.[2] The practice of preparing *dolma* in olive oil also began to appear in nineteenth-century cookbooks. Mehmet Kamil's *Aşçıların Sığınağı* (Refuge of Cooks), published in 1844, contains the earliest known mention of the recipe, which calls for rice, onions, spices, pine nuts, currants, olive oil, and vine leaves.[3] It became widespread in Istanbul's cuisine in the nineteenth century.[4]

Dolma is classified as a hot cuisine when it is prepared with meat; nevertheless, *dolma* prepared with olive oil is classified as a cold food. *Dolma* was a staple dish for citizens of Ottoman Istanbul and was consumed in a variety of social settings, including weddings, banquets, and Ramadan dinners. Because of its portability, *dolma* was eaten in streets, hammams, and promenades. *Dolma* prepared with olive oil was popular because it was so convenient to travel and enjoy.

Dolma may be transported anywhere once it is placed on a plate or in a basket. Because a fork or knife is not required, the little cold pieces make an ideal finger meal. This feature of the *dolma* made it appropriate for both a promenade and a palace feast.

About the History of *Dolma*

Dolma is a symbol of rich cultural history. It is among the most sophisticated types of

Istanbul's Portable Food Practices: The Case of *Dolma*

vegetable consumption. '*Dolma* is the only dish to have stood the test of time and retain its status despite its many rivals,' claims Marianna Yerasimos.[5] *Dolma* initially springs to mind even though Ottoman cuisine has a very diverse selection of vegetable dishes. From the old Ottoman territory to Sweden, we find *dolma* and *sarma* of various nations, but the name's etymology comes from Turkish:

> Dolmas are vernacular food in Turkey, the Balkans, the Southern Caucasus, Iran, C. Asia (where the word differs in form according to the local Turkish language: dolâma in Turkmen, tulma in Tatar), and in Egypt, the Fertile Crescent, and Arabia.[6]

It should be noted that in Ottoman cuisine, *dolma* is not only made of vegetables. Its origins as a cooking technique go back to former times. Günay Kut notes that a type of *dolma* called *sımsımrak* (stuffed mutton intestines) was identified in *Divan-ı Lügat-ı Türk*.[7] This shows us that the stuffing technique in Central Asian Turkic cuisine has a long history.

Ottoman texts first record stuffed fruits and vegetables with meat from the beginning of the fifteenth century. For instance, apple *dolma* was included in Muhammed bin Mahmud Shirvani's cookbook.[8] Moreover, *dolma* was one of the most popular dishes at banquet and entertainment tables and was frequently included in such menus. At the feast that was held in 1539 to celebrate the circumcision of Suleiman the Magnificent's sons, stuffed gourd and eggplant were among the foods given to the guests.[9] According to Yerasimos, *dolma* holds a special place among vegetable dishes, including others like *borani* (rice with spinach or a similar vegetable eaten with yoghurt) and *kalye* (made by adding various fruits and vegetables to meat roasted in oil). Quince or apple *dolma* were sometimes replaced by *dolma* with spinach, eggplant, and cabbage.[10] In addition to being a part of palace cuisine, *dolma* also frequently appears in everyday life and has been sold in cookshops since its beginnings. The seventeenth-century traveller Evliya Çelebi lists stuffed gourd, stuffed sheep sausage, stuffed onion, stuffed eggplant, and stuffed cabbage as *dolma* variants available in Ottoman-era Istanbul cookshops. Mustafa Altıntaş also mentions that *dolma* was sold by the piece in Istanbul's cookshops.[11]

Dolma has survived and evolved to this day. Although not all of the classic *dolma* and *sarma* recipes are being used, some are still alive and are passed down from generation to generation.

Female Vendors Selling Stuffed Vine and Cabbage Leaves: Arab Nannies

Street food vendors were an important part of Istanbul's culinary culture, and the richness of the cuisine was also reflected in public spaces from the fifteenth to the twentieth centuries, as evidenced by travellers' accounts and archival materials. Many people and businesses sold food, and street vendors were essential participants in this network.

During the nineteenth century, *dolma* was sold as a portable food on Istanbul's streets by Arab *bacı* (called Arab nannies) who were trying to make a living in the city. Their existence hints at the complex and varied profile of late Ottoman Istanbul's street food vendors. Pertev Naili Boratav, a well-known folklore researcher, has observed that these women were brought to the Ottoman lands by piracy or slave traders from various parts of Africa. As domestic servants, they were responsible for childcare, housekeeping, and other routine tasks. After leaving the mansions, they used to work as vendors to support themselves. These women travelled around Istanbul selling various *dolma*, sesame halva, cookies, and roasted chickpeas (*leblebi*).[12]

Arab nannies used to sell their homemade food all over town, from the baths to weddings, in bazaars and mosques. Looking through the Ottoman archives reveals that the women who sold *dolma* requested support from the state. For example, an archival document dated 1860 shows that an Arab woman named Zülgaffar petitioned the state for help: the document states that this woman sold *dolma*.[13]

In a book that gives detailed information about nineteenth-century Ottoman customs and ceremonial traditions in Istanbul, an Ottoman gentleman named Abdülaziz Bey provides a detailed description of 'women who sell *dolma*':

> These women would leave the house wearing a chador, a sash tied around their waist, a headscarf around their necks, a *kaşbasti*, a sesame basket in their hands, and a stuffed pot on their heads. Arab nannies used to sell the *dolma* they made at market exits and around the mosque to be closer to shopkeepers and bachelors. They would frequently stand at the bazaar gate on the Kaşıkçılar side of the Bayezid Mosque, place the stuffed pot on a small chair they had brought with them, place the lid under the pot, sit next to them, and wait for customers. They had stuffed vine leaves with olive oil and stuffed cabbage for sale.[14]

Dolma as a Portable Food in Promenade

Eating out has a long and notable history. People ate outside their homes for a broad range of reasons in the early modern period, including military expeditions, feasts, and pilgrimages to holy places. According to current literature, eating out became a pleasure and leisure activity in Europe in the second half of the eighteenth century. With nineteenth-century urbanization, escaping the crowds and noise of the city and experiencing nature began to gain importance for urban people.[15] Eating more freely in the countryside, breaking from specific table rules, and spending time in nature turned into important activities. Picnics first appeared during hunting parties in early modern Europe, but they evolved into their current meaning over time.[16]

Compared to Europe, open-air-eating practices during the promenade (*mesire*) were an old tradition in Ottoman culture. In the Ottoman context, although the tradition of having fun by eating in the open air has existed since early modern times, it has been institutionalized since the eighteenth century and has gained more profound meanings.[17] The word *mesire* comes from the Arabic root *syr*, which means 'place to visit'.[18] For the early modern period, Fatma Yaşar points out that:

Picnic, as a term, refers to both a space and an action. On the other hand, *mesire* was defined as a space (*ism-i mekan of seyr*), or *mesiregah*, where people ride for amusement and excursion (*tenezzüh* and *teferrüc*), which are defined as acts related to *mesire*.[19]

Since the early modern period, the promenade had been a prominent part of Ottoman social life at certain times of the year. There were gardens and promenade areas primarily used by dynastic family members, ordinary people, and the upper class. While there were gardens that were only open to the dynasty in the eighteenth century, the public began to use gardens and promenades more widely in the nineteenth century.[20]

People in Istanbul went on promenades for a variety of reasons: to eat some neighbourhood's delicious food, to visit a holy place, or just to have a good time in an area with a nice view and clear air. When the residents of the capital went on a promenade, they usually ate something particular to the place they visited. People used to go to Beykoz (a district on the Anatolian side of the Bosporus) to eat lamb's feet or to Göksu (a neighbourhood in Beykoz) for its famous corn and eggplant.[21]

In her memoirs, Samiha Ayverdi writes, 'It was customary to go to Joshua's Hill together in July and August, especially on Fridays. People used to make plans to visit the promenade and Joshua's grave, and they would accompany them, preparing their halvah and dolma.'[22] When people went to such open spaces, they brought food that was easy to transport. *Dolma* was a must-have for these menus, which feature prominently in novels, memoirs, and travel accounts dating from the early period of the Ottoman Empire to those of the Republican era. According to Balıkhane Nazırı Ali Rıza Bey, the traditional Kağıthane promenade menu consists of cold dishes, *dolma*, *söğüş* (cold lamb cuts), and *sütlü irmik helvası* (semolina halva with milk).[23]

These recreational days allowed Istanbul's residents to socialize and spend time with their family members and friends. Meals on the promenade were vital parts of the entertainment, and almost everyone had a similar menu. Serious preparations were made before going on a promenade; cold and easily transported food was prepared, with *dolma* being the most ubiquitous.

Hammam Entertainment

Hammam culture had an essential place in Ottoman social life. The hammam functions as a place to visit once a week for a bath and as a centre where men and women gather separately and socialize. Especially for women, going to the hammam has a remarkable place in their social lives. In addition to weekly bath days, women also have special days, such as the bridal and the postpartum hammam, where they have fun among themselves. Likewise, the preparations for the promenade included going to the hammam, and suitable, easily transportable food was prepared for the bath. Hammam entertainment was widely featured in many works.

In one of his writings, the well-known author Musahipzade Celal explains that:

The postpartum woman was taken to the hammam on the fortieth day after birth. Midwives, relatives, and neighbors had been invited to the bath. Some postpartum hammams were crowded like a wedding. Dancers were performed in the bath. For the guests, stuffed peppers (*Biber dolması*), pickles (*turşu*), sherbet, and coffee were served.[24]

This description by Musahipzade Celal implies that consuming *dolma* was embedded in many daily practices of Istanbul's people. Besides being a popular and commonly consumed dish, people preferred *dolma* on special occasions like hammam meetings. In another example, Hagop Mintzuri, an Armenian author who lived in Istanbul, mentioned *dolma* consumption at the hammam. According to him, the menu for the bath day was vibrant, and women brought anything that should be eaten cold – for example, beans, *dolma*, halva, sausage, pastrami, olives, and onions. Women prepared several types of *dolma* and halvah for these events by using a variety of materials such as tomatoes, peppers, and vine leaves.[25] Being suitable for consuming cold was essential for portability in every entertainment.

Conclusion: Stuffed Mussels as a Portable Legacy in Today's Istanbul

Stuffed vine leaves and cabbage are no longer seen on vendor tables in modern Istanbul. This culture, like Arab nannies, has vanished from Istanbul's streets. However, *dolma* is still sold by vendors, who sell *midye dolma* (stuffed mussels), made by stuffing black or blue mussels with rice that has been spiced.

Stuffed seafood, such as mussels, has been made in this city since the Ottoman Empire and has recently gained popularity due to newly opened restaurants. Furthermore, vendors selling *midye dolma* can now be found on Istanbul's streets at all times. This traditional flavour, which we still eat, was originally a Lenten food among Istanbul's Orthodox-Christian community. At certain times of the year, the Orthodox Church's fasting rules impose certain restrictions on eating and drinking. During the fasting period, orthodox community members are not permitted to consume meat or foods containing animal fat.

The majority of vendors selling *midye dolma* in modern-day Istanbul are from Mardin, a city in southeastern Turkey. They arrived in Istanbul as a result of internal migration movements that began in the 1950s, and many of them had never seen the sea before.[26] The growing number of *midye dolma* sellers and new shops serving traditional *midye dolma* with new serving options begs the question of how this traditional and local food is adapted to today's needs and trends as a portable food.

The *dolma* sellers from Midyat, who have been the subject of television programmes, documentaries, and interviews, have also given new life to Istanbul's traditional cuisine.[27] Some stores have begun to sell stuffed mussels in plastic buckets for convenience. The mussels are traditionally sold on a bench by street food vendors. Stuffed mussels made with rice and spices in the original recipe began to be made with bulgur as an Anatolian touch, and they are even served with new sauces like curry.

Istanbul's Portable Food Practices: The Case of *Dolma*

In a nutshell, *dolma* was a type of portable finger food that could be taken anywhere, including to jails, during a visit. According to the *Sabah* newspaper, on 9 June 1891, a child delivered a box of *dolma* to a prisoner named Corci, who had been convicted of theft in Izmir (a city located in the Aegean Region of Turkey). The truth was revealed when the prison officer examined the box. It was suspected that there were only a few *dolmas* in the pot, with the rest being marihuana wrappers covered in yoghurt.[28]

Acknowledgments

I am grateful to Associate Professor Özge Samancı for her most helpful feedback on this paper's earlier drafts. I thank Aylin Öney Tan who shared her symposium experiences with me. Also, I would like to thank the Özyeğin University Graduate School of Social Sciences for its financial support.

Notes

1. *Asırlık Tariflerle Türk Mutfağı* (Ankara: Kültür Ve Turizm Bakanlığı Yayınları, 2021), p. 109.
2. Marianna Yerasimos, *Osmanlı Döneminde Rum Mutfakları*, ed. by A. Bilgin and Ö. Samancı (Ankara: Kültür ve Turizm Bakanlığı, 2008), p. 226.
3. *Asırlık Tariflerle Türk Mutfağı*, p. 109.
4. Özge Samancı, 'The Cuisine of Istanbul between East and West during the 19th Century', in *Earthly Delights: Economies and Cultures of Food in Ottoman and Danubian Europe, c. 1500-1900*, ed. by Angela Jianu and Violeta Barbu (Leiden: Brill, 2018), p. 92.
5. Marianna Yerasimos, *500 Years of Ottoman Cuisine,* 4th edn (İstanbul: Boyut Matbaacılık A.Ş, 2010), p. 176.
6. Charles Perry, 'Dolma', in *The Oxford Companion to Food*, ed. by Alan Davidson and Tom Jaine (Oxford: Oxford University Press, 2014), p. 1062.
7. Günay Kut, *Ağız Tadı Türklerde Yemek Kültürü Eski Türk Edebiyatı Araştırmaları IV* (Istanbul: Simurg, 2021), p. 96.
8. Mohammed bin Mahmud Shirvani was a fifteenth-century palace physician. Shirvani's most famous work, despite the fact that the true title of the work is unknown, is referred to as 'a cookbook from the fifteenth century'. Shirvani translated Muhammad ibn-ul Hasan ibn Muhammad al-Katib al Baghdadi's book into Ottoman Turkish and added new recipes. In the work, he also discussed the relationship between food and medicine, particularly which foods are beneficial for treating which diseases.
9. Kut, pp. 229-30.
10. Yerasimos, *500 Years of Ottoman Cuisine*, pp. 174–76.
11. Mustafa Altıntaş, *Osmanlı İstanbul'unda Ta'âm Bişirüb Satanlar: Aşçılar, Başçılar, Büryancılar, Börekçiler, Tatlıcılar (1500–1800)*, İnsan ve Toplum Dizisi (Istanbul: Kitap Yayınevi, 2020), p. 57.
12. Pertev Naili Boratav, *Folklor ve Edebiyat 1* (Istanbul: Adam Yayıncılık, 1982), pp. 124–25.
13. Istanbul, the Ottoman Archives of the Republic of Turkey's Presidential State Archives Türkiye Cumhuriyeti Cumhurbaşkanlığı, Devlet Arşivleri Başkanlığı, Osmanlı Arşivleri, BOA.A.DVN.d.145-98.
14. Abdülaziz Bey, Osmanlı Adet, *Merasim ve Tabirleri Toplum Hayatı,* ed. by Kâzım Arısan and Duygu Ansan Günay, Belgesel 4 (Istanbul: Tarih Vakfı Yurt Yayınları, 1995), p. 318.
15. Marc Jacobs and Peter Scholliers (eds.), *Eating out in Europe: Picnics, Gourmet Dining, and Snacks Since the Late Eighteenth Century* (Oxford: Berg, 2003), p. 3.
16. Jane Davidson, 'Picnic', in *The Oxford Companion to Food*, ed. by Alan Davidson and Tom Jaine (Oxford: Oxford University Press, 2014), p. 2253.
17. Jacobs and Scholliers, pp. 3–4.
18. Sevan Nişanyan, 'Mesire', *Nişanyan Sözlük Çağdaş Türkçenin Etimolojisi* <https://www.nisanyansozluk.com/kelime/mesire> [accessed 29 May 2022].

19 Fatma Tunç Yaşar, 'Women in Early Modern İstanbul: The Use of Space' (unpublished masters thesis, Boğaziçi University, 2004), p. 110.
20 Shirine Hamadeh, *The City's Pleasures: Istanbul in the Eighteenth Century* (Seattle: University of Washington Press, 2004), pp. 129–38.
21 Sermet Muhtar Alus, *İstanbul'un Geçmiş Günlerinde Yeme İçme,* ed. by Tuncay Birkan (Istanbul: Can Yayınları, 2021), pp. 57–58.
22 Samiha Ayverdi, *Boğaziçi'nde Tarih*, 8th edn (Istanbul: Kubbealtı, 2013), p. 381.
23 Balıkhane Nazırı Ali Rıza Bey, *Eski Zamanlarda İstanbul Hayatı,* 5th edn (Istanbul: Kitapevi, 2017), p. 111.
24 Musahipzade Celal, *Eski İstanbul Yaşayışı, Beşyüzüncü Yıl Serisi* (Istanbul: Türkiye Yayınevi, 1946), p. 23.
25 Hagop Mıntzuri, *İstanbul Anıları 1897-1940*, Belgesel-1, 4th edn (Istanbul: Tarih Vakfı Yurt Yayınları, 2002), p. 91.
26 Musa Dağdeviren, '*Mardin, Midyat, Midye Mutfağın Sokaktaki Fedaileri*', *Yemek ve Kültür*, 2012, 68–74.
27 '*Midye Dolmanın Ustaları Mardinliler*', *Yeşil Doğa*, CNN TÜRK, 2019.
28 *Sabah*, 9 June 1891, p. 3.

Provisioning for Adventures: The Meanings of Paddington Bear's Marmalade Sandwiches

Laura Kitchings

Introduction

Michael Bond's 1958 book *A Bear Called Paddington* introduced audiences to Paddington Bear, a bear who migrated from 'darkest Peru' to London's Paddington station with a label stating, 'Please Look after this Bear.' Paddington's adventures have been told in children's books (at various reading levels), television programmes, movies, mobile apps, ice shows, and other media. Paddington's stories are often repeated using both the original visuals created by illustrators known to Michael Bond (1926-2017) and new images controlled by the current license holder. While Paddington's larger community has changed throughout the years, the main cast remains largely consistent. Paddington's London community consists of four members of the Brown family (Mr. and Mrs. Brown, Judy, and Jonathan). There is also the housekeeper, Mrs. Bird; the antagonistic neighbour, Mr. Curry; and the Hungarian immigrant antiques dealer, Mr. Gruber. Paddington's Peruvian family consists of his Aunt Lucy, who lives at the Home for Retired Bears in Lima. Paddington generally wears a blue duffle coat and carries at least one marmalade sandwich under his large hat. Extra sandwiches are often found in his suitcase, and Paddington often eats these sandwiches away from the table, during his adventures. These adventures are often instigated by Paddington's eating a marmalade sandwich in a context where eating them is deemed inappropriate by the food rules of his new world.

This paper considers Paddington as a transmedia property with a single universe created through various media platforms. Transmedia consultant Christine Weitbrecht, who works with film studios, discusses how transmedia projects use a variety of media to enhance the world of a property, not just to repurpose content from an original form of media (Doro 2012). This paper does not consider the version of Paddington in one form of media superior to the Paddington presented in another form of media created under an official Paddington Bear license. The first license for Paddington's image to be reproduced by someone other than Bond and his illustrators was granted in 1972 to Shirley Clarkson who developed the original Paddington stuffed bear ('Plush History' 2022).

This research took a non-random approach to sampling a variety of Paddington books, movies, and television shows produced between 1958 and 2022, with the sampling done between 2018 and 2022 in Massachusetts. The paper uses the familiar

Paddington Bear to refer to the character but notes that this name was given to him by his foster family, the Browns. While Paddington was raised in 'darkest Peru', it is an imagined Peru with limited resemblance to the current country.

Bond's autobiography discusses growing up in a food-insecure household and developing a preference for expensive foods later in life. Addressing Paddington's preference for marmalade, Bond claims that the preference reflects Bond's own preference for marmalade over honey (1997: 157). Bond likely wanted to differentiate Paddington from A.A. Milne's Winnie the Pooh, associated with honey. British marmalade is a condiment generally made of ingredients from the Global South (sugar and oranges) connected to the history of European colonialism. This links Paddington to the colonialist history of the Global South.

Paddington is not the only Western children's text character linked to the Global South. Babar travels to Europe and adopts Western dress and food rules. These food rules include primarily eating food at a proper table. Once Westernized, Jean de Brunhoff's Babar returns to 'the land of Elephants' and becomes the region's new leader, ruling them in a Western fashion (1933).

Philip Nel's *Was the Cat in the Hat Black* (2019) discusses how animal characters in Western literature are stand-ins for non-white characters and may explain why many people of colour often identify with Paddington. While Nel's work shows how children's texts often support systemic racism, Bond placed Paddington as a marginalized Londoner in allyship with other marginalized communities.

In contrast to Babar, Paddington maintains some of his own food rules, eating a marmalade sandwich when and where he chooses, instead of bowing to British food traditions and eating only at the family table. Unlike many other animal characters in British and United States children's texts, Paddington's foster family and community generally support his eccentric consumption of marmalade sandwiches, even when the act leads to societal disruption. For example, during a high-society partnered dance competition in *Paddington Takes the Air*, when Paddington drops a marmalade sandwich down the back of his new dance partner, the pair wins the competition (Bond 1970).

The First Era of Paddington: Understanding Paddington's Food Rules

In the first era of Paddington, Bond had complete control of the Paddington empire. This period started with the publication of his first Paddington book and lasted through the stop-motion animated series of the 1970s and 1980s.

The first two Paddington books feature the bear arriving from Peru, meeting his foster family, the Browns, and acclimating to life in their household. In the first story, how Paddington and the Brown family manage Paddington's post-migration hunger is comically presented to the reader, with Paddington standing on the table while having tea at Paddington Station. While dealing with the immediate hunger issue, it becomes clear that Paddington does not understand the food rules involved in having tea in London. Paddington and Brown eventually end up in a cab, where the driver immediately charges extra for a bear, and then charges extra again for a sticky bear. These initial adventures demonstrate that Paddington will continue to follow his own food rules in London, even if they lead to micro-aggressions by Londoners.

In the third Paddington story, 'Paddington Goes Underground', Paddington begins his adventures in and around London. He initially provisions for his outings with the Brown family by packing leftover bacon in his suitcase, but soon finds it impractical. By the sixth story, Paddington brings a marmalade sandwich with him to the theatre, and while the sandwich disrupts the experience of theatregoers, it all ends well. Paddington's eating sandwiches during adventures away from the Brown's family table becomes standard behaviour (Bond 1958).

In later written adventures of this era, Paddington misses a catch at a cricket match due to his focus on eating his marmalade sandwich. Such adventures always work out in Paddington's favour, and he is rarely criticized for not assimilating to British food rules. Notably, none of the other characters begin carrying marmalade sandwiches under their hats, so Paddington's actions are not seen as prescriptive, just a way for him to mitigate his emotions and physical hunger.

The Paddington stop-motion television series of the 1970s and early 1980s continue the tradition of Paddington's sandwich being accepted as a sign of his otherness as he manages his life in a new city. Similar to the books, the animated series utilizes a third-person limited omniscient point of view, so the audience has some idea of Paddington's thoughts, but also has knowledge of situations without Paddington's presence. In the animated series, Paddington often uses the sandwich as a mental health device, to centre himself in an unfamiliar location. For example, in 'Trouble at the Launderette' (1976), he eats a sandwich after placing the clothes in the washing machine.

Throughout his autobiography and in multiple interviews, Bond states that his inspiration for Paddington's creation was based on children of the Kindertransport, Jewish child refugees to Great Britain at the beginning of World War II. Bond saw these children in England with labels attached to their clothing, and a few children briefly lived with the Bond family. These children dealt with physical hunger and with suddenly being in a different country where they had to cope with food rules likely different from their experience. Bond likely did not view Paddington as a prescriptive character for British children but rather as a reflection of how he perceived these children dealing with major life changes (Bond 1997).

Similarly, sociolinguist Angela Smith views Paddington as a surrogate for members of the Windrush generation, a term generally referring to people arriving in the UK from the Caribbean between 1948 and 1971. The first migrants arrived on the HMT *Empire Windrush*. Smith shows how a reading of 'Paddington Finds a Cure' (1968) demonstrates that the physical characteristics describing Paddington in the story are also those used to describe the members of the Windrush generation. She details how this generation of migrants performed hospital work but also dealt with micro and macro racial aggressions (Smith 2006).

The end of World War II saw an immediate increase in migration from the Caribbean to Great Britain. Bond lived in the Notting Hill area of London in the 1950s, a neighbourhood that became a cultural centre for this community. In *Belly Full* (2017), Riaz Phillips presents a historical summary of British-Caribbean foodways while introducing the reader to several existing restaurants in the United Kingdom. Phillips discusses how new Caribbean migrants not only held on to their food traditions for survival but also used food to fuel nostalgia for their lives in the Caribbean. These Caribbean food traditions evolved in the UK to become their own foodways and now serve as an important part of Britain's diverse food traditions. The late 1960s saw the beginning of the Notting Hill Carnival which became a place to showcase and celebrate Caribbean culture in London. Bond was likely influenced by the new foodways developed in London by his Caribbean neighbours to give Paddington a unique food tradition.

In these early Paddington stories, the migrant Paddington often shares a mid-morning snack with Hungarian immigrant Mr. Gruber. They drink cocoa together in Mr. Gruber's shop. Paddington does not eat marmalade sandwiches while in the shop, signifying possible feelings of comfort with a fellow newcomer to London. Mr. Gruber supports Paddington's choice of eating marmalade sandwiches during their adventures outside of the store. It is likely that Mr. Gruber understands that Paddington's presentation as a bear may lead to aggressions that limit his access to food. In this era of Paddington, the sandwiches serve as consumable comfort objects in addition to providing sustenance and marking Paddington as an 'other' in post-World War II London. In this era, Paddington's immediate contacts do not ask him to change his foodways to eat only at the acceptable table but support him while he faces discrimination in London.

The Second Era of Paddington: Commensality

In 1981 Bond stepped away from day-to-day management of the growing empire with his daughter, Karen Bond Jankel, taking control of the growing empire (*The Resident*, 2014), marking the beginning of Paddington's second era. Paddington's image was licensed for various products and animated television shows. During this period, Bond largely worked on other projects including creating the characters of Olga da Polga and Monsieur Pamplemousse (Bond 1997).

In his autobiography Bond discusses how marmalade sandwiches were used in promotions for Paddington books and stuffed toys (1997). These promotions expanded under Jankel's control. One series of promotions included libraries in the United States holding Paddington Day, obviously meant to encourage children to read, but also providing a reason for libraries to show the stop-motion series. These events also included children eating marmalade sandwiches in the libraries, in the spirit of commensality with Paddington and each other. While children in the United States understood that eating at the library did not follow codified Western food rules, they could participate in Paddington's food rituals without being expected to appropriate them with their families (Fernandez 1985).

Hanna-Barbera Cartoons, Inc. created a new Paddington animated show in 1989, adding a white cousin from the United States to the Brown family. The plots of these cartoon shows featured moments familiar to the audience of United States sitcoms and did not achieve a significant statistical following (Browsh 2020: 413).

In the cartoons, the sandwiches are a moment of connection between Paddington and another character. In 'Paddington Meets the Queen', Paddington and the Queen bond over a shared fandom of marmalade sandwiches. Paddington and the Queen eat the sandwiches on couches. Often the non-Paddington character has a chance to explore Paddington's food rules of eating in unexpected places, with the sandwich used to instigate a commensal experience.

The 1997 animated television series focuses on Paddington's travels with Mr. Gruber. These travels take Paddington to Egypt, Ireland, and other recognizable locations. In 'Paddington in the Ring', Paddington and Mr. Gruber visit a training centre for sumo wrestlers in Japan. When a staff person tries to take Paddington's suitcase (including a marmalade sandwich), Paddington grabs the suitcase back. This encounter leads to Paddington (and the viewers) learning about traditional Japanese customs. The sandwich often serves as an impetus for Paddington to explore the societal rules, including food rules, of other countries, participating in a cultural exchange without the other cultures appropriating Paddington's marmalade sandwich.

In 2007, Paddington's management decided that Paddington should replace the marmalade in his sandwich with Marmite, a commercial yeast extract, in a series of advertisements. The tagline for this campaign was 'You either love it or hate it'. It featured Paddington enjoying adding Squeezable Marmite to sandwiches, while everyone around him dislikes Marmite on sandwiches. The message being that if Paddington tried Marmite, so should the general public, and then they can form their own opinions on the yeast extract. After fans made their displeasure known, Bond publicly stated that he did not write the advertisements and that Paddington was not moving away from marmalade. This showed that fans had claimed ownership of Paddington and had ascribed their own meanings to his marmalade sandwich (*Manchester Evening News* 2007).

The Third Era of Paddington: Paddington Codified

The third era of Paddington is marked by Bond wanting to establish Paddington's legacy and canonize parts of his backstory, knowing that others would eventually take control of the empire. Bond wrote several new Paddington stories, in which he explicitly discusses Paddington's legal status in London. In these new stories, the Brown family worries about Paddington's legal status in a country with a strong nationalistic movement. The citizenship crisis is solved when Paddington's Uncle Pastuzo visits from Peru, where he sells drinks to miners in the Peruvian Andes. After visits to several tourist destinations in London, the uncle places a passport and other needed legal papers in Paddington's suitcase (Bond 2008).

In *Love from Paddington* (2014), Bond re-writes several Paddington stories from Paddington's first-person point of view, using a series of letters (in chronological order).

Bond avoided stating a year in which each letter was written by simply numbering the letters because bears do not use calendars. Bond wrote the introduction from his own point of view, noting that Paddington's view of many of his earlier adventures is slightly different from that of the narrator of the earlier stories. The Preface, told from Paddington's point of view, has Paddington's Aunt Lucy placing him in the lifeboat in Peru with several jars of marmalade and access to drinking water. Throughout the book, Paddington is shown to have been in regular contact with Aunt Lucy. In the third letter, Paddington clarifies that he has always had a marmalade sandwich under his hat as a family tradition. These retellings of the stories connect Paddington's marmalade sandwich directly to his Peruvian heritage. They show Paddington's eating of marmalade sandwiches as an unapologetic solitary activity. For example, in the retelling of a story of eating a sandwich at the theatre, the fact that it fell on a patron does not bother him. He also shows annoyance that while participating in a cricket match the other players did not wait for him to finish his sandwich before expecting him to join the game.

The book ends with a letter from Aunt Lucy explaining why she put Paddington in the lifeboat and her reasons for sending Paddington to London. She explains that before Paddington's birth she met an explorer from London, explaining her connection with London, who developed the Home for Retired Bears, her current residence. She states that she made sure Paddington had food and water during his migration. In this new telling, Paddington experienced little food insecurity during his migration to London, which conflicts with the version told in the 1958 stories.

This book also provides more information about Paddington's connection with Mr. Gruber and their ritual. Mr. Gruber is now a Hungarian immigrant who spent time in South America before coming to England as a refugee. Bond explicitly codifies Mr. Gruber's refugee status. Paddington regularly shares 'elevenses' with cocoa with Mr. Gruber, and tellingly while in Mr. Gruber's shop Paddington does not rely on his marmalade sandwiches for mental health purposes (Bond 2014).

These new stories were written during a time when activists were discussing the food needs of refugees. For example, in *Food Ethnographic Encounters* (2011), Elizabeth Cullen Dunn considers the lack of cultural significance behind the bland macaroni provided to refugees, instead of the spiced vegetables the community was used to. While the macaroni met bare nutritional needs, it provided nothing in terms of connecting the refugees with their past or encouraging the development of new culturally based food traditions.

The 2014 film *Paddington* begins with a new backstory for Paddington, specifically defining him as a refugee. The early scenes show a British explorer introducing marmalade to Paddington's Peruvian family along with stories of London. Paddington and his family adopt the introduced marmalade into their own family food traditions. This evolves into a family tradition of making marmalade on the day the oranges have ripened. After an earthquake takes the life of Uncle Pastuzo and destroys their home, Aunt Lucy places Paddington in a boat headed for London telling him to find the explorer. She surrounds him with jars of marmalade for the journey and gifts him his deceased uncle's hat. While placing him in the boat, she tells Paddington the story of the Kindertransport, assuring him that Londoners will accept him. She provides for his health and mental well-being on the journey. Once in London, Paddington uses the sandwich to remain connected to his past, while finding his place in London. In the movie, the sandwich also helps defeat the villain, but otherwise it serves as a connection between Paddington and his Peruvian family. This film also includes possibly the first mention that one marmalade sandwich contains all the daily vitamins and minerals a bear needs, assuring the audience of Paddington's nutritional security. Uncle Pastuzo explicitly tells Paddington to always keep a marmalade sandwich under his hat in case of emergency, codifying the sandwich as a family tradition.

The movie makes non-food connections between Paddington's experience and the 1950s experiences of Caribbean migrants to London, showing his connections to marginalized communities. While in Peru, Paddington's aunt and uncle are seen listening to records advising the bears how to behave in London which invokes pamphlets given to Caribbean migrants on how to behave in London. A calypso band is seen during many of Paddington's adventures in London. Near the film's beginning, they sing 'London is the Place for Me' made famous by calypso singer Lord Kitchener on the *Windrush*'s arrival in London (Spencer 2011).

In 2017 Paddington's support of refugees became codified when Vivendi, owner of the Paddington brand as of 2016, partnered with UNICEF to have Paddington champion UNICEF's goal of protecting children's rights (Vivendi and UNICEF 2017). The marmalade sandwich of this era serves as a signal that, like Paddington, all refugees need culturally appropriate foods that meet their nutritional needs.

In this era, Bond incorporated the fans' reading of Paddington as a refugee and the idea of the marmalade sandwich helping Paddington deal with the trauma of migrating to London. Explicitly defining Paddington's status as a refugee placed Paddington as a member of a marginalized group living in London.

The Fourth Era of Paddington: Paddington and British Traditions

The 2017 film *Paddington 2* expands Paddington's backstory in Peru. The film depicts Aunt Lucy and Uncle Pastuzo eating marmalade sandwiches on a bridge overlooking a waterfall in Peru while discussing plans to migrate to London within a month. They see a drowning bear cub and decide to raise the cub (Paddington) as their own child and put off their migration. During the rescue, Pastuzo's marmalade sandwich falls into Paddington's mouth. It becomes the first food he eats with his new family. It can be assumed that Paddington faced food insecurity before meeting his Peruvian family.

The film revolves around Paddington's mistaken incarceration in a London jail. Paddington teaches the prison's head cook (Mr. Knuckles) to make marmalade sandwiches for the prisoners. This commensality over marmalade sandwiches leads to prison reform, with the prisoners beautifying the prison and the guards reading the prisoners bedtime stories. Many of the prisoners are shown starting their own bakeries after being released from prison.

The sandwich allows for commensality among the prisoners and guards while connecting Paddington to his past. Still, the film also clarifies that Paddington regularly gets his vitamins and minerals. The marketing features Paddington and sandwiches, showing the association among the general public of Paddington with marmalade sandwiches. The film ends with Aunt Lucy arriving in London and celebrating her birthday with Paddington's London community. This means that Paddington is no longer displaced from his family of origin, but part of a global community. The sandwich places Paddington as an ally of other marginalized communities in the prison.

Conclusion: Tensions in the New Era of Paddington

Current Paddington projects are now developed without the leadership of Michael Bond. In the new Paddington television series, shown on Nick Jr. in the United States, the adult human characters remind the human child character that marmalade only contains the vitamins and minerals needed for bears, not for humans. The food rules in the television series are meant to be prescriptive to childhood viewers. This places marmalade sandwiches in the older tradition of children's texts, reminding children of the importance of maintaining normalized Western food traditions.

In an episode of the series featuring the Lunar New Year, Paddington bonds with Asian migrants to London while they maintain their own traditions such as flying dragon kites. The episode ends with a moment of commensality with Paddington eating dumplings at the table with a Chinese immigrant neighbour. It is noteworthy that the antagonist Mr. Curry is shown eating a sandwich alone away from the table in a park. The viewer is led to understand that this is improper behaviour and serves as a punishment, unlike earlier versions of Paddington where Paddington celebrated eating his sandwich alone (Hayday Films and StudioCanal 2019).

The newer texts show Paddington moving away from making marmalade as a family tradition to making pancakes, which is more familiar to a United States audience. This event occurs in a kitchen with the family eating pancakes at a table. Paddington can have marmalade with his pancakes, but only at the table. Now Paddington can have his culturally significant marmalade, but only as a small part of a more-normalized family tradition (Capucilli 2022). While Paddington does continue having a marmalade sandwich under his hat, it is more a simple way of maintaining family tradition, and not as serving a mental health need.

The brand is also trying to tie the sandwich to the British monarchy, having Paddington share a sandwich with the Queen during her jubilee, and thanking her for her service to the British people after her death. In contrast, many of the fans post social media photographs of themselves sharing a sandwich with Paddington at his statues and adding captions that they are eating with Paddington as a member of their community that was oppressed in some way by Britain's history of colonialism. For South Asian immigrants, members of the Queer community, and others, sharing a marmalade sandwich with Paddington is a moment of solidarity between members of marginalized communities. Future Paddington texts will need to manage these tensions, which seem similar to those at the end of the second era of Paddington, when Paddington's management tried to change the marmalade sandwich, and as a result Bond needed to repair the relationship between Paddington's management and his fans. There are hopeful signs: the official Paddington Instagram account regularly reposts images of people, often representing marginalized communities, sitting with Paddington's statue in London's Leicester Square while trying to bite his sandwich.

References

Bond, Michael, *A Bear Called Paddington ... With Drawings by Peggy Fortnum* (London: Collins, 1958)
——, *Bears & Forebears: A Life so Far* (London: HarperCollins, 1997)
——, *Love from Paddington* (London: HarperCollins Children's Books, 2014)
——, *Paddington Here and Now* (New York: HarperCollins, 2008)
——, *Paddington Takes the Air* (London: Collins, 1970)
Bond, Michael, Lesley Young, and R.W. Alley, *Paddington's Cookery Book* (London: HarperCollins Children's Books, 2011)
Browsh, Jared Bahir, *Hanna-Barbera: A History* (Jefferson, NC: McFarland & Company, 2022)
Capucilli, Alyssa Satin, *The Adventures of Paddington: Pancake Day!* (New York: HarperCollins, 2022)
De Brunhoff, Jean, *The Story of Babar*, trans. by Merle S. Haas (New York: Random House, 1933)
Doro, 'Transmedia Storytelling around the World: Christine Weitbrecht', *StoryFusion*, 1 February 2012, <https://storyfusion.de/tmsb-archive/personen/interviews/transmedia-storytelling-world-christine-weitbrecht-20120201/> [accessed 23 May 2022]
Dunn, Elizabeth Cullen, 'The Food of Sorrow: Humanitarian Aid to Displaced People', *Food: Ethnographic Encounters*, ed. by Leo Coleman (London: Bloomsbury Academic, 2011), pp. 139-49
Fernandez, Lourdes, '"L" is for Library, Learning', *Florida Today: Indian River*, 2 December 1985, p. 3A.
Manchester Evening News, 'Paddington "Not Switched to Marmite"', *Manchester Evening News*, 19 September 2007 <https://www.manchestereveningnews.co.uk/news/greater-manchester-news/paddington-not-switched-to-marmite-1004663> [accessed 23 May 2022]
Nel, Philip, *Was the Cat in the Hat Black?: The Hidden Racism of Children's Literature, and the Need for Diverse Books* (New York: Oxford University Press, 2019)
Paddington, dir. by Paul King (Films StudioCanal, 2014)
Paddington 2, dir. by Paul King (Films StudioCanal, 2017)
Paddington Goes to the Movies, dir. by Barry Leith (FilmFair, 1980)
'Paddington in the Ring', *The Adventures of Paddington Bear*, Cinar Corporation and Protecrea 1997
'Paddington Meets the Queen', *The Funtastic World of Hanna-Barbera*, Hanna-Barbera, 1989
'Paddington's First Chinese New Year', developed by Jon Foster and James Lamont, *The Adventures of Paddington* (Hayday Films and StudioCanal, 2019)
Phillips, Riaz, *Belly Full: Caribbean Food in the UK* (London: Tezeta Press, 2017)
'Plush History', *Paddington*, 2022 <https://www.paddington.com/us/back-in-1958/plush-history/> [accessed 23 May 2022]
Spencer, Neil, 'Lord Kitchener Steps off the Empire Windrush', *The Guardian*, 15 June 2011.
Smith, Angela, 'Paddington Bear: A Case Study of Immigration and Otherness', *Children's Literature in Education*, 37 (2006), 35-50
The Resident, 'Karen Jankel on Life with Paddington Bear', *The Resident*, 27 November 2014 <https://www.theresident.co.uk/news/21477279.karen-jankel-life-paddington-bear/> [accessed 23 May 2022]
'Trouble at the Launderette', *Paddington* (FilmFair, 1976)
Vivendi and UNICEF, *Paddington to Become a Champion for Children in Support of UNICEF*, 2017 <https://www.vivendi.com/wp-content/uploads/2017/09/PR170920-UNICEF-Paddington-partnership-launch.pdf/> [accessed 23 May 2022]

Serving the Global Nation: Menus from the Royal Dutch Airlines

Charlotte Kleyn and Aimée Plukker

'KLM Royal Dutch Airlines is delighted to say "Welcome", and to offer you this foretaste [preview] of the menu you will enjoy,' opens a KLM Royal Class menu from an early 1970s Amsterdam-Chicago flight. It continues: 'For many decades, KLM has been linking the great continents of the world, and you'll find this reflected in the cuisine we offer. Familiar foods or the adventure of something new we invite you to choose whichever you prefer.' The menu offered standard continental cuisine, but also 'typically Dutch cuisine', consisting of *boerenkool* with three different kinds of Dutch meats: *rookworst, speklapje,* and *varkenscoteletje*. The dishes were not the only Dutch features on the KLM menu, which also presented information about Amsterdam canal houses, including a note that 'with the growth of overseas empire and trade, a playful ornamentation appeared'.[1] These examples illustrate how this KLM menu highlights both the national (the Dutch dishes, the canal houses) as well as the international and the global (the airline connecting continents, the continental menu, the reference to Dutch colonial history). This paper will look at the politics of what is served and consumed in the airplane's cosmopolitan space, specifically aboard the Royal Dutch Airlines (Koninklijke Luchtvaart Maatschappij, KLM). The paper zooms in on airplane menus from the immediate post-World War II period (1945-1970s) covering both the peak of the 'jet age', with its international, glamourous class known for frequent travel and global tastes, and the shift towards the beginning of mass air travel.[2]

In the years just after the establishment of KLM in 1919, no food was served in the air. Travelling by airplane was uncomfortable, noisy, and cold since no heating was provided, and pilots flew low to the ground in order to orient themselves, which meant constant turbulence and sick passengers.[3] In the 1930s, travelling by plane became more enjoyable for passengers, with comfortable chairs, heating, and some food and drinks: cold sandwiches and coffee, tea, and broth – 'beef tea' – from thermos flasks.[4] When travelling far, planes had to stop for fuel regularly, which enabled passengers to eat and sleep at hotels overnight. Dinners and accommodations were included in the ticket price, which was very high.[5] As early as the 1920s and 1930s, different airlines experimented with serving hot meals, trying out inflight ovens using charcoal, denatured alcohol, or engine heat to warm premade food.[6] Technological innovations for heating and cooling food in the air were stimulated during World War II, finding their way aboard

Serving the Global Nation: Menus from the Royal Dutch Airlines

commercial aircrafts soon after the war ended.⁷ Efficient electric convection ovens proved the best solution to serve meals while operating in a small space and dealing with weight restrictions.⁸ In 1947 KLM decided to include frozen meals, to be reheated in the air, a practice which is still being used up to this day.⁹

In the 1940s and 1950s these frozen KLM meals were predominantly French, since French cuisine still evoked the cosmopolitanism of high society. Only from the 1970s onwards did the menus include dishes from the destination, such as a Chinese stir-fry and a Dutch cheese platter. This paper will look at changes in dishes on KLM airplane menus, but also at how the dishes and the destinations were displayed on the menus. How did food and how it was presented create national and cosmopolitan identities? Through an examination of post-World War II KLM menus, this paper will investigate the interplay between the national and the global with a particular focus on types of food and visual representations.

Serving French International Cuisine

After World War II, KLM shifted its focus to transatlantic flights, due to declining demand for flights to Indonesia caused by the Indonesian War of Independence (1945-1949). KLM introduced two flights a week to New York, and from 1950 onwards flights took off daily. Initially, a transatlantic flight took 25 hours and 30 minutes, including two stops. In the early post-war period flying was still reserved for the elite, as tickets were expensive, and services on board these airplanes were adapted to these customers, including the most luxurious food and drink. In 1952 KLM introduced the Tourist Class, an initiative of the International Air Transport Association (IATA). Tourist class was meant to attract new groups of potential customers and reduce the competition between airlines to offer the most luxurious and abundant flights, a non-profitable practice. Tourist Class tickets were 30 percent cheaper, and the food and drink service offered to its passengers was simpler: the IATA dictated that passengers could not receive more than one starter, soup, or fruit juice; one main course, including two side dishes; and one piece of fruit or dessert. Tourist Class proved a great success: 340,000 passengers travelling the North-Atlantic route in 1951 increased to one million in 1957, of whom 770,000 travelled in Tourist Class.¹⁰

Since air travel was new, travellers received detailed instructions on how to travel and what to expect. The Fodor's *Women's Guide to Europe* (1954) had a section titled 'Everything you need to know' in which travelling from the United States by air was explained. The guidebook underscored that air travel was 'by no means limited to moguls and movie stars. Tourist air rates of today are designed for stenographers and art students, for the average housewife and the busy businesswoman'. The section on air travel included what to expect when boarding an airplane. It also provided the female traveller with tips, including the following remark related to food: 'She [the hostess] will tell you when to set your watch ahead, the approximate landing time and what to do for a queezy tummy (this is not likely to result from air travel, but we warn you to forego lobster thermidor).'¹¹ The quote illustrates the predominance of French cuisine in this time period.

KLM was no exception. The menus from the 1950s and 1960s were usually written in both French and English, regardless of the destination: English, the new lingua franca of the world, and French, the language of the culinary world's dominating cuisine.[12] For example, First Class (Royal Class from 1966 onwards) KLM menus from the 1960s consisted of lavish, mostly French dishes, using luxurious ingredients. People travelling from Amsterdam to New York on a flight in 1961 could start their dinner with fresh caviar on ice with vodka or Champagne, or fresh lobster with Champagne or Riesling. After the soup, one had a choice between 'tournedos à la catalane,' stalk celery, mixed salad, and Parisienne potatoes, or roast spring chicken, green peas in butter, mixed salad, and diced potatoes, accompanied by Bordeaux.[13] The menu shows how luxury in first class menus and service was defined, and to what extent it was intertwined with French cuisine. As anthropologist Christine R. Yano has pointed out in her account of Pan American World Airways (Pan Am) 'Nisei' (second generation Japanese American) stewardesses: 'practices, knowledge, and foods are embedded within a global hierarchy of cultural capital and prestige.'[14] She concludes that 'things European carried their own unmistakable stamp of worldliness'; to Americans, Europe epitomized 'class'.[15] KLM menus from the 1950s and 1960s confirm Yano's points on cultural capital, prestige, and class by showing similar food practices at a European airline, with a particular focus on anything French.[16]

The prestige of 'Old World' food has also been described by historian Bryce Evans in his book about food on Pan Am airplanes. He describes how Pan Am created a President's Special menu (1948) and even cooperated with the famous Parisian restaurant Maxim's for meals in 1952 to compete with other airlines like KLM, Air France, and the British Overseas Airways Corporation. Serving French dishes meant 'accumulating greater cultural capital by appropriating Old World authenticity,' writes Evans. Serving Pan Am customers 'European – and particularly French – culinary culture was part of the company's contribution to the American post-war ideal of "glamour"'.[17] However, KLM menus show that this creation of 'glamour' through serving 'Old World' food was not typically American, but a transatlantic phenomenon.

Serving Imperial Legacies

Beginning in the late 1940s, hot food served on KLM flights did not refer to the destination at all. Nor was food included from KLM's home base, the Netherlands, except for Dutch Gouda cheese, which was for instance served in 1955 and has long been a famous Dutch export.[18] Another exception from December 1953 is a special Saint Nicolas-themed menu, a popular children's feast celebrated yearly in the Netherlands on December fifth. The menu, written in Dutch, is full of references to Sinterklaas (Saint Nicolas), a white bishop who is said to live in Spain and travel to the Netherlands by steamship to give presents to well-behaved children. The racist depiction of Zwarte Piet (Black Pete), Sinterklaas's Moorish servant, is also included on the menu.[19] The dinner started with 'hors-d'oeuvre St. Nicolas' and champagne, continued with a 'Spanish soup', '*Hoofdschotel à la Stoomboot*' (Main Course *à la* Steamship) with red wine, and ended with typical cookies from this feast: *pepernoten*.[20]

KLM menus from the 1950s and 1960s contain frequent depictions of Dutch culture. A menu of tulips was followed by a whole series of flower-themed menus, including ones depicting a narcissus and a hyacinth. The text on these menus presented the Netherlands as a 'Land of Flowers', promoting bulb-growing as a major national industry.[21] Flowers were also often promoted in guide- and photobooks of that period, such as in the photobook *Holland* (1957) and the depiction of the Amsterdam Flower Market in the book *Amsterdam* (1968).[22] The flowers on menus displayed flowers as an industry, tourist attraction, and souvenir to be brought home (as bulbs and flower seeds). Other stereotypical depictions of Dutch culture included windmills, Delftware (also known as Delft Blue), canal houses (Figure 1) and traditional costumes, various Dutch towns, and works of Dutch artists, such as a series of menus with etchings by Rembrandt to celebrate the Rembrandt year in 1969.[23] Just like the flowers, the Rembrandt house and his paintings in the Rijksmuseum were major tourist attractions promoted in many guidebooks. It is also worth thinking about the menus as souvenirs themselves; with colourful illustrations, the airplane menus could have been kept by KLM customers as a reminder of travels abroad.[24]

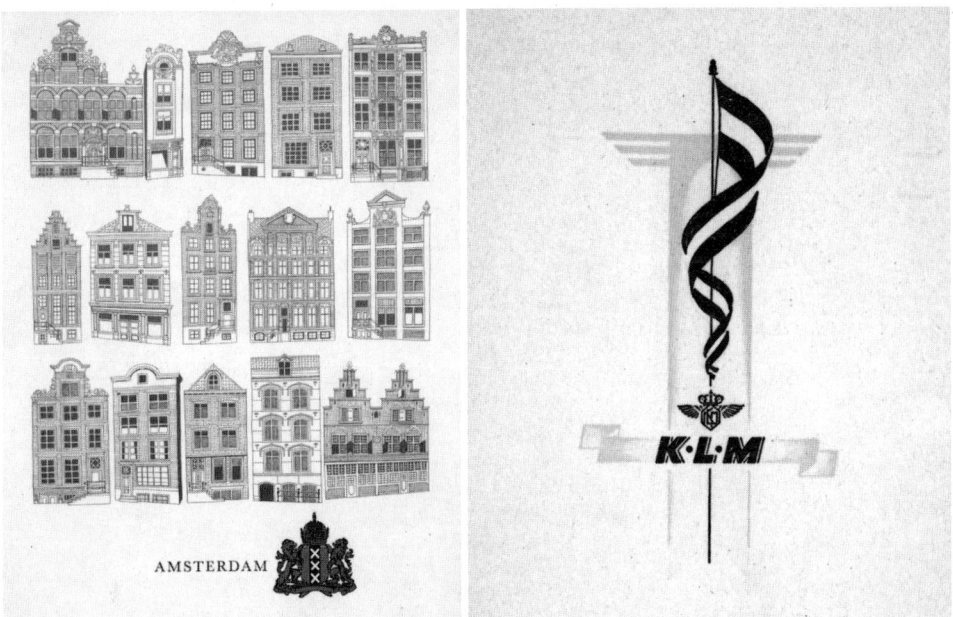

Figure 1 (left). Canal houses in Amsterdam, Allard Pierson, Collection Menus of the Culinary Museum Amersfoort, fol. 2 subfile 27 item 417. Figure 2 (right). Dutch national flag, Allard Pierson, Collection Menus of the Culinary Museum Amersfoort, fol. 2 subfile 27 item.
N.B. *The authors aimed to trace the copyright of the images used in this paper. Whoever believes they own the right, please contact the publisher.*

The covers of a series of menus from 1959 and 1960 show colourful illustrations, one of the city of Alkmaar and its famous cheese market, an attraction also featured in tourism promotion. The other three covers are illustrations of ships, a busy carnival, and the Noordeinde Palace in The Hague and the Golden Coach, that carried the Royal family to deliver the Speech from the Throne at the Ridderzaal from 1901 until 2015. The last three covers all depict Dutch flags (Figure 2), emphasizing Dutch nationalism. The depiction of the Golden Coach is another example of the 'hidden' Dutch empire in KLM menus: the gold came from Surinam, and its left panel portrays the Tribute from the Colonies, showing people from the Dutch colonies presenting gifts to the queen.[25] The cover illustrations celebrate the nation: cheerfulness at the carnival and around the palace, international cooperation at the harbour, and peaceful trade at the cheese market. Inside, the menus emphasize that 'several of KLM's chefs are members of la confrerie de la chaîne des rôtisseurs, the seven hundred years old honorary international society of culinary masters,' and in some menus the name of the chef is specified.[26] The celebration of 'national and Old World glory' continues in the design of the menu by the inclusion of the words 'Voortgezette Compagnie van Verre' (Continued Company of the Far Away) in a gothic font, an explicit reference to the Compagnie van Verre, the predecessor of the Dutch East and West India Company (Vereenigde Oostindische Compagnie and Westindische Compagnie).[27]

An undated KLM menu portrays Dutch imperial history even more directly.[28] On the cover we see two Wayang puppets, used in traditional Javanese shadow puppet theatre (Figure 3). One Wayang, depicted in green, wears the typical attire of Prince Arjuna, a main protagonist in most Wayang stories from the *Ramayana*, while the other puppet, in red, is dressed in traditional Dutch clothing.[29] The message of the cover could be interpreted as Indonesian culture still being part of the Netherlands, or as a celebration of the Dutch colonial legacy. By appropriating traditional Indonesian culture and combining it with Dutch traditional culture, KLM used Dutch colonial history to create its own self-representation. In the 1950s and 1960s colonial 'Old World' notions were combined with French cuisine, an especially interesting phenomenon in the context of post-Wolrd War II decolonization. Another example of this celebration of the colonial past is the menu cover depicting a statue of Peter Stuyvesant, the Dutch colonial officer and last director-general of New Netherland, with as background the old and new New York (Figure 4).[30] KLM menus not only served food: they also served a specific view of the past, which emphasized global connections.

Serving the Cosmopolitan

During the 1970s, menu options on board of KLM planes were more diverse and international. This development happened around the same time as the increase in lower-cost air travel, which became more widely available to the middle and upper-middle class.[31] As sociologist Marco D'Eramo writes, the rise of low-cost airlines globalized tourism.[32] We can also trace this process of globalization in KLM menus.

Serving the Global Nation: Menus from the Royal Dutch Airlines

For instance, the use of French on menus was sometimes replaced with the destination language, such as Arabic, Korean, Spanish, and Dutch. Travellers flying from Amsterdam to Tokyo in 1970 could now choose between a prawn cocktail followed by a *consommé à la parisienne*, or they could pick a Chinese menu, consisting of entremets 'Fujiyama,' 'chong-sik ngau phah: beefsteak Chinese style' and 'kway-far-chee: shark's fins with scrambled eggs'.[33] A menu from Amsterdam to Manila, on the stretch between Bangkok and Delhi, shows five European-style dishes, like tournedos and civet of hare with red wine sauce, but also 'a special oriental meal', which consisted of lamb and radish, duck with almonds, stewed abalone, crab claws, and steamed white rice.[34]

These 'oriental' options were not exclusively served on flights to Asia. On several flights, for example one from Amsterdam to Paramaribo, Surinam, travellers could order a 'special oriental meal' which consisted of Dutch-Indonesian dishes such as *sate babi, udang* Singapore, *ayam blado, acar babi, sambal bajak, krupuk,* and *nasi*.[35] This cuisine was presented as part of Dutch culinary identity by KLM, as can be seen by the accompanying text on the same menu: 'Three hundred years of close ties with the Far East have placed oriental food so frequently on Dutch tables that the Dutch no longer refer to it as "foreign"!'[36] The phrasing 'close ties' does not quite encapsulate the Dutch

Figure 3 (left), Wayang Puppets, Amsterdam, Allard Pierson, Collection Menus of the Culinary Museum Amersfoort, fol. 2 subfile 27 item 409. Figure 4 (right). Peter Stuyvesant and New York City, Amsterdam, Allard Pierson, Collection Menus of the Culinary Museum Amersfoort, fol. 2 subfile 27 item 417.

colonial history in Indonesia, but even more striking is the fact that 'oriental' food is presented as being served in the Netherlands for hundreds of years. Dutch-Indonesian cuisine originated in the colonial households of the Dutch Indies, with Indonesian cooks working for Dutch families. After the independence of Indonesia, many Dutch-Indonesian (Indo) people moved to the Netherlands between 1949 and 1967, adapting dishes from their home country with the limited ingredients they could find in the post-war Netherlands.[37] The menu also explains how the 'Oriental Adventure' must be eaten ('with spoon and fork [...] start by taking a little rice onto your spoon, surrounding it with food from the side dishes').[38] Here we see how food in the air also served to educate KLM travellers on how to eat different kinds of food; the cultural capital that could be attained by travelling accrued from the moment that travellers entered the plane.

The following text was placed next to the menus from the 1970s that also presented 'oriental' options:

> KLM introduces Dutch Culinary Art on flights out of Amsterdam. A tasteful combination of all that is excellent in Dutch cooking and produce. The finest home-reared meats. The freshest home-grown vegetables. The most delicious home-caught fish. Blended, after more than a year of research, experimentation and sheer creativity, by KLM's Award-winning chefs into something special in in-flight cuisine. A series of dishes which, we hope, will charm and surprise our on-board guests.[39]

The 'oriental' meal was probably used to surprise travellers. The non-'oriental' meal included, just as for decades beforehand, caviar, lobster, and foie gras. Only in the main course is some Dutch influence visible: Medallion of Dutch veal, sauce 'Boerenmeisjes' (dried apricots in liquor) or fillets of sole with shrimps and mustard from Zaandam.[40] The text and menu also emphasize to what extent promotion of the Netherlands was intertwined with meals being served in the air, turning the non-place of the airplane into an extension of the nation.[41]

Other 1970s KLM menus also include typical Dutch food, such as the menu of the flight to Chicago with which we started that offered travellers 'continental or typically Dutch cuisine'. Besides Dutch food, the menu included a special selection of Dutch drinks (*oude jenever* and *jonge jenever*, spirits flavoured with juniper, and *advocaat*, a sort of eggnog) and Dutch cheeses. The menu, which by now consisted of several pages, instead of the one-page menus of the 1950s and 1960s, contains a description of *boerenkool* as the typical food eaten by farmers. It is also stated: 'Whether your preference is for Dutch or "International" food, we offer you the traditional Dutch wish: "Smakelijk eten!....Eat well!"'[42] Interestingly, these menus were not written in Dutch, but still in English and French: the target audience was therefore international. Dutch food was not only served on flights to and from the United States. On a flight from Manila to Amsterdam, breakfast included Dutch options, such as 'hagelslag (chocolate sprinkles), krentenbrood (currant bread), beschuit (Dutch rusk) and ontbijtkoek (honey cake)'.[43]

Serving the Global Nation: Menus from the Royal Dutch Airlines

In an increasingly globalized world, KLM made sure that the nation was well presented, including its colonial past referenced through hybrid Indo-Dutch dishes.

Serving the Cold War

Besides glorifications of the past, the KLM menus also provide insight into contemporaneous politics and events. Many KLM menus served a particular occasion or purpose, such as the menu made for the Rembrandt year. Other menus for special occasions include the yearly Christmas menus, a menu for the addition of Moscow to KLM's network in 1958, and a menu for the Tokyo Olympics in 1964.[44] Between 1955 and 1962 a series of menus with famous composers, such as Wagner, Bach, and Mozart, was designed for flights that transported the symphony orchestras of Minneapolis, Philadelphia, and Boston.[45] These menus emphasize the role of airplane travel in transatlantic cultural exchange, an important element of Cold War cultural politics.[46] Of particular interest in this context is the menu designed for the flight of the Buffalo Schwaben Verein (a German heritage organization) from Stuttgart via Amsterdam to Buffalo in September 1961. The menu includes a photograph of a group of people standing in front of a KLM aircraft holding a sign stating: 'KLM Royal Dutch Airlines Welcomes the Buffalo Schwaben Verein on the German American Friendship Tour.'[47] The rhetoric of German American Friendship and friendship tours was common in the early Cold War period, when such transatlantic exchanges were used to strengthen the ties between the US and West Germany.

With the word 'royal' in its company name, it is hardly surprising that KLM celebrated the birthday of Queen Juliana (1909-2004) with themed menus throughout the years. The designs illuminate the special occasion: a little crown, the colours of the Dutch flag (red, white, blue), and a lot of orange, the national colour of the Netherlands that represents the last name of the royal family (which literally translates as 'from Orange', a principality in France between 1163 and 1713). In 1954 the meal itself included small references to the queen's birthday: a 'Queen soup' (*consommé Reine Hortense*) and an 'orange bavarois'.[48] The prominent place of the royal family in KLM menus also relates to the depiction of the Golden Coach discussed above. Besides strong ties to the royal family it is worth noting that KLM was a government-subsidized airline, part of what historian Jenifer van Vleck called the 'empire of the air': airlines linking metropolitan capitals to far-flung colonial outposts, a legacy also clearly reflected on KLM menus.[49]

Other examples of politics being served on board KLM are the United Nations Day menus from 1960 and 1961 for flights from Amsterdam to New York. The menus were in English, and some dishes were presented with themed names such as 'mixed salad UNESCO'. and 'mayonnaise all nations style'. The menu from 1961 included even more themed dishes, such as 'stalk celery unicef', 'universal coffee', 'cold supper of fraternization', 'cold chicken all nations', 'waldorf salad Geneva', and 'breakfast of understanding.'[50] With these specially curated names, the menu promoted the UN's many goals and organizations. These menus are typical examples of 'gastrodiplomacy', the use of food as soft power to achieve political goals, which Paul Rockower has defined as 'a broad cultural diplomacy venture to communicate culinary culture to foreign publics in a manner that is more diffuse and tries to influence broader audiences rather than high-level elites'.[51] A particular aspect of gastrodiplomacy is its focus on taste and emotional connection. Airplane menus were a perfect venue, combining taste with ways of seeing and understanding. By presenting the dishes in a fun way, the propagandistic goals of the United Nations Day are easily consumed.

Conclusion

This paper has explored the politics of post-war KLM menus. What was being served in the air and how dishes and destinations were displayed provides insight into the self-representation of a major airline, but it also reflects the development of international tourism and airline travel. The 'Old World', cosmopolitan style of French food in the 1950s and 1960s slowly evolved to more international and Dutch-national fare as airline travel grew globally. KLM menus promoted what the Netherlands had to offer in terms of industry and tourism, but it also presented a one-sided view of its colonial past. This celebratory view of the Netherlands happened at the same time as decolonization took place, and the Cold War context influenced views of the past and present in friendly terms like 'close ties', 'friendship', and connections, instead of Iron Curtains and exploitation. Although innocent and mundane at first sight, the KLM menus highlight the politics of portable food.

Notes

1 Amsterdam, Allard Pierson (AP), Collection Menus of the Culinary Museum Amersfoort, KLM Menus 1949–1971 (KLM), fol. 2 subfile 27 item 423. *Boerenkool* is kale (on the menu mistranslated as curly cabbage), probably mixed with mashed potatoes for the typical dish stamppot. *Rookworst* is smoked sausage, *speklapje* fried bacon and *varkenscoteletje* is a small pork chop.
2 Vanessa R. Schwartz, *Jet Age Aesthetic. The Glamour of Media in Motion* (New Haven, CT: Yale University Press, 2020).
3 Bram Bouwens and Frido Ogier, *Welcome aboard! Een eeuw KLM. Het officiële jubileumboek* (Zwolle: WBooks, 2019), p. 161; Richard Foss, *Food in the Air and Space: The Surprising History of Food and Drink in the Skies* (New York: Rowman & Littlefield, 2015), p. 31.
4 Bouwens and Ogier, p. 204; Foss, p. 24.
5 A one-way ticket from Amsterdam to Batavia (nowadays Jakarta) in the Dutch Colony Indonesia cost 1000-1500 guilders, which is nowadays about €15.000–20.000.
6 Foss, p. 31.
7 Foss, p. 105.
8 Marc Dierikx, *Luchtspiegelingen. Cultuurgeschiedenis van de luchtvaart* (Amsterdam: Boom, 2008), p. 114; Foss, pp. 105–06.
9 René de Leeuw, *KLM: de geschiedenis van de KLM vanaf 1919* (Weesp: Unieboek, 1984), p. 78.
10 Bouwens and Ogier, pp. 79, 80, 166, 80. By 1958, the larger and quicker jet plane had arrived, and IATA launched Economy Class, with even lower prices. Service had to be simple: passengers could only receive a sandwich, and no food and drink menus were used.
11 Eugene Fodor (ed.), *Woman's Guide to Europe*, rev. ed. (New York: David McKay, 1954), pp. 10–11.

12 By 'French cuisine', we refer to the 'international haute cuisine before there was "haute cuisine" (coined in the twentieth century)' as defined by Katie Rawson and Elliott Shore in *Dining Out: A Global History of Restaurants* (London: Reaktion, 2019): a style of cooking and serving food found in hotels and restaurants around the world from the nineteenth century onwards.
13 AP, KLM, fol. 2 subfile 27 item 427. According to Auguste Escoffier's *Le Guide Culinaire* (London: Mayflower, 1979), Catalan sauce consists of tomato, onions, grilled bell peppers, parsley, and poivrade sauce, whereas *Larousse Gastronomique* (London: Hachette UK, 2018) refers to artichoke hearts and grilled tomatoes.
14 Christine R. Yano, *Airborne Dreams: "Nisei" Stewardesses and Pan American World Airways* (Durham, NC: Duke University Press, 2011), p. 119.
15 Yano, pp. 119–20.
16 In 1962 KLM employed Japanese and later (since 1968) also Thai stewardesses on flights to Tokyo and Bangkok, in order to serve their Japanese and Thai customers in First/Royal Class. Several Dutch newspapers covered the story: '*Japanse lijn bij KLM*,' *De Volkskrant*, 22 December 1962 <https://resolver.kb.nl/resolve?urn=ABCDDD:010876483:mpeg21:a0324> [accessed 27 May 2022]; *Trouw*, 12 June 12 1968 <https://resolver.kb.nl/resolve?urn=ABCDDD:010814095:mpeg21:a0160> [accessed 27 May 2022].
17 Bryce Evans, *Food and Aviation in the Twentieth Century: The Pan-American Deal* (London: Bloomsbury, 2021), pp. 60–61, 72.
18 AP, KLM, fol. 2 subfile 27 item 407. According to *The Oxford Companion to Cheese*, ed. by Catherine W. Donnelly (Oxford: Oxford University Press, 2016), p. 323, cheese from Gouda, Alkmaar, and Edam has been exported since the twelfth century.
19 The blackened face of Black Pete is supposedly caused by coming down the chimney to distribute candy and presents to children, but the figure of Black Pete is clearly a stereotypical representation: 'thick red lips, golden earrings, an Afro wig, clad in a colorful Moor's costume, and, until recently, wielding a quite deplorable grammar, "dumbspeak"' (Gloria Wekker, *White Innocence: Paradoxes of Colonialism and Race* (Durham, NC: Duke University Press, 2016), p. 139).
20 AP, KLM, fol. 2 subfile 27 item 430.
21 The text states: 'To-day, Holland is exporting bulbs to the value of 150,000,000 guilders a year […] the official catalogue of Dutch tulips lists 6,600 different varieties.' Ithaca, Kroch Library Cornell University, Division of Rare and Manuscript Collections, Special Collections Archives 8020 K-122 A 1 C, Box 1 Airline Menus, KLM Menu and AP, KLM, fol. 2 subfile 27 items 403-405. Typical Dutch symbols were also used in the design of dishes and cutlery, such as plates and cups decorated with orange tulips for First Class passengers in the late 1970s.
22 E. Elias and Ed van Wijk, *Holland* (The Hague: W. van Hoeve, 1957), pp. 104, 108–09; Ton Koot, *Amsterdam*, trans. by Yda Ovink (Munich: Knorr & Hirth Verlag GMBH, 1968), p. 79.
23 For instance: AP, KLM, fol. 2 subfile 27 items 421 and 427.
24 The same goes for collecting menus of cruise ships: Sandra van Berkum and Tal Maes, *Captain's Dinner: Koken met de Holland-Amerika Lijn* (Schiedam: Scriptum, 2011).
25 The Golden Coach was given to queen Wilhelmina in 1898. The panel was painted by Nicolaas van der Waay.
26 AP, KLM, fol. 2 subfile 27 item 426.
27 This could also be a reference to Nyenrode Business University, a private school in the Netherlands. The student association of the university is called Voortgezette Compagnie van Verre. Relations between Nyenrode and KLM have been strong since its foundation in 1946, since KLM director Albert Plesman initiated the foundation of the university: <www.nyenrode.nl/alumni/alumni-verbondenheid/nyenrode-alumni-vcv> [accessed 32 May 2022].
28 AP, KLM, fol. 2 subfile 27 items 408-409. Undated, but probably from the early 1960s.
29 We would like to thank Eric Tagliacozzo for helping to identify the Wayang figure.
30 AP, KLM, fol. 2 subfile 27 items 417–418.

31 Schwartz, p. 5.
32 Marco d'Eramo, *Il selfie del mondo. Indagine sull'età del turismo* (Milan: Feltrinelli, 2019; 2017), pp. 24–26.
33 Amstelveen, Archive of the KLM, no item number, KLM Menu Amsterdam-Tokyo, 1970.
34 AP, KLM, not yet described. The menu is undated, but comparing the design with other menus, probably from the early 1970s.
35 AP, KLM, fol. 2 subfile 27 item 431.
36 AP, KLM, 431.
37 *Post-Colonial Immigrants and Identity Formations in the Netherlands*, ed. by Ulbe Bosma (Amsterdam: Amsterdam University Press, 2012); Anneke H. van Otterloo, 'Chinese and Indonesian Restaurants and the Taste for Exotic Foods in the Netherlands: A Global-Local Trend', in *Asian Food: The Global and the Local*, ed. by Katarzyna J. Cwiertka and Boudewijn Walraven (Honolulu: University of Hawaii Press, 2001), pp. 153–66; Ann Laura Stoler, *Carnal Knowledge and Imperial Power: Race and the Intimate in Colonial Rule* (Berkeley: University of California Press, 2002).
38 AP, KLM, 431.
39 AP, KLM, fol. 2 subfile 27 item 423. The depiction of a Rembrandt painting on the menu also underscored the words 'Culinary Art'.
40 AP, KLM, 423.
41 Marc Augé, *Non-places: Introduction to Supermodernity* (London: Verso, 2009).
42 AP, KLM, 423.
43 AP, Menu Collection (no number).
44 AP, KLM, fol. 2 subfile 27 items 411, 433, 435.
45 AP, KLM, fol. 2 subfile 27 item 407.
46 See for instance Penny M. Von Eschen, *Satchmo Blows Up the World: Jazz Ambassadors Play the Cold War* (Cambridge, MA: Harvard University Press, 2004).
47 AP, KLM, fol. 2 subfile 27 item 425.
48 AP, KLM, fol. 2 subfile 27 item 437.
49 Jenifer van Vleck, *Empire of the Air: Aviation and the American Ascendancy* (Cambridge, MA: Harvard University Press, 2013), p. 15.
50 AP, KLM, fol. 2 subfile 27 item 440.
51 Paul Rockower, 'A Guide To Gastrodiplomacy', in: *Routledge Handbook of Public Diplomacy*, ed. by Nancy Snow and Nicholas J. Cull (New York and London: Routledge, 2020), pp. 205–212, 207.

'No kind of food? Just rubbish?': Food Options and Family Meals of Visitors to Canadian Prisons

Else Mare Knudsen

I spent a good part of 2013 riding buses to prisons and eating snack foods. I was conducting fieldwork for a qualitative study of the experiences of children of prisoners in Canada, and one way I recruited participants was to volunteer with charitable prison transportation services for families of prisoners. This involved spending long days on a bus, handing out juice boxes, chatting with mums, and creating elaborate play-dough pizzas with children. I learned about their experiences, built rapport, and distributed my study flyer to the parents. Those who trusted me enough (usually after sizing me up for several trips) gave me a call, and we arranged interviews with them and their children.[1]

When I eventually interviewed these children and youth with a parent in prison, at their homes or a fast-food restaurant, I asked them about their perceptions and opinions about parental incarceration. They told me about their anger, fear, sadness, and frustration at having a parent 'inside'. Unexpectedly, they also spoke a great deal about crisps. They shared their excitement about getting crisps from the prison vending machines in the visiting room, their frustration at the high price of Cokes, and their disgust at the days-old, dried out hamburgers in the 'hot' machine. Though I had not set out to ask about food, my participants inevitably raised the topic, and it quickly became clear that food (and specifically the options, prices, quality, limitations, and sharing of food) was a key element of these families' experiences of visiting a loved one in prison. As families are prohibited from bringing any food whatsoever into prison visits, they are constrained to the snack food offerings of the prison, offerings that have the worst aspects of 'portability' and which were roundly criticized by my participants.

This article explores the ways that food and family meals play a role in families' experiences of the incarceration of a loved one, the subjugating power of the restrictions on food in prison visits, and how the poverty faced by most families of prisoners affected their experiences of meals 'inside'. The foods available to most visitors to prisons are not home cooked and are so limited, restricted, expensive, and nutritionally poor that it is difficult to turn the sharing of food into a family meal during the visit. Given that family meals are important not just to family life but also to health and well-being, these limits are, I argue, a type of harm that reflects the dismissiveness and systemic invisibility that define social policy approaches to families of prisoners.

Family Meals

Food is more than caloric intake: sociologists and historians of food have developed a rich literature on its meanings, power and impact. Food practices can be social, caregiving, coercive, or even therapeutic, and food can be a site of socialization, regulation, learning, negotiation, intergenerational connections, and culture.[2] Food within families is of specific interest; food is argued to be central to the constitution of family identity, and eating family meals found to be associated with a variety of benefits to health and well-being.[3] Ochs and Shohet write that meals are 'central to defining and sustaining the family as a social unit'.[4]

Higher frequency of family meals in higher income countries is associated with both healthier and higher quality dietary intake, as well as with weight benefits and reductions in disordered eating.[5] Parents consistently perceive family meals to benefit family connection and communication.[6] However family meals are also deeply stratified by class, with children more likely to eat more family meals if they have high socio-economic status and if their mother is not employed outside the home.[7]

While most research into family meals defines them simply as meals occurring in the home and with family members present, a meta-analysis by Dallacker, Hertwig, and Mata found that it is these six elements that are associated with the benefits of family meals: parent modelling of healthy eating behaviour, high food quality, a positive atmosphere, involvement of children in meal preparation, not having the TV on, and longer meal duration.[8]

Food in Prisons

Food in prison tends to have none of these elements. While the 'bread and water' diet of some nineteenth-century gaols is no longer used as punishment, the food that prisoners around the world consume is often poor quality and implicitly part of the punitive systems of the prison. The meals at most Canadian prisons are industrially prepared in central 'cook chill' facilities and reheated in large bags at each prison, at a typical total cost per prisoner of around $6.00-$9.00 per day (under £4-£6). A typical meal for a prisoner is as follows:

> Whole wheat toast/bread 2, margarine 7 g, chicken bologna 90 g, baked beef steakette 140 g, mashed potatoes 175 ml, canned pears 125 ml, enriched fruit drink crystals 1 each.[9]

A 2019 audit by the federal Correctional Investigator in Canada found a variety of food 'deficiencies' in federal prisons, including: failure to meet national Food Guide requirements regarding nutritional content of meals 21% of the time, inadequate inventory and inspection controls that directly affect the quality and quantity of food service delivery, and inconsistent or substandard meal portion sizes.[10] The Investigator found that prisoners were purchasing food from the canteen to 'supplement or substitute for meals or portion sizes that are unappetizing, inadequate, poor or inconsistent quality'.[11]

Food Options and Family Meals of Visitors to Canadian Prisons

Food is, unsurprisingly, also constitutive of and defining to the experience of the prison for incarcerated people. Scholars have shown the ways in which prisoners use licit and illicit food preparation as a practice of taking back agency and power, building solidarity, undertaking coercion, and asserting race and gender.[12] So central is food to prisoners' agency and experience of the institution that prisoners, in Canada and elsewhere, have rioted and held hunger strikes to protest poor food quality.[13] Ugelvik argues that food in prison allows prisoners to transcend the prison and connect to their 'outside' lives because of the way it serves as a reminder of home, culture, and connection with friends and family.[14]

Combined with the context and the importance of the family meal, it is little wonder that prisoners and their visitors value the opportunity to eat meals together. Prisoners describe family visits as a 'reprieve' from prison and an opportunity to transcend the institution and be included in the family, in part through visiting activities like sharing food.[15] The topic of food experiences of prison visitors has been little studied by scholars, with the exception of Johnna Christian and Megan Comfort, the latter of whom describes women visitors as seeking to create a 'domestic satellite' in the visiting room.[16]

Findings from a Study of Families of Prisoners in Canada

In my study of the self-reported experiences of Canadian children and their mothers on the 'outside', most held deep ambivalence about prison visits, both relishing the opportunity to see their loved one, but also criticizing the many barriers to doing so. It was clear that achieving that home-like 'domestic satellite' was the result of a long, difficult journey.

Before their loved one is ever incarcerated, these families are more likely to face a variety of challenges. Like prisoners, they are more likely to be living in poverty and to experience racism due to being racialized or Indigenous.[17] Almost all of my study participants were living on general assistance benefits. Children of prisoners are more likely to have parents who are unemployed or experiencing substance use disorder.[18]

Once the person is incarcerated, families face a number of additional challenges as a result. Parental incarceration itself appears to have negative impacts on children, including mental health concerns, and unsurprisingly increases the chance of caregiving instability, changes to family structure, and entry into the foster care system.[19] Having a parent in prison has been added to the list of Adverse Childhood Experiences (ACEs) which predict health concerns in adulthood.[20] For families, having a loved one in prison can have a variety of direct and indirect costs, from the emotional impacts, time spent navigating the court and prison systems, stigma from their communities, as well as the financial costs of lost income, expensive phone systems, sending money in for canteen, and travelling to visits.[21]

Maintaining contact with a loved one in prison is rife with the challenges of regulation, precarity, expense, and frustration. Visitors to prisons must meet a long list of detailed requirements that they may find confusing, unclear, or arbitrary, such as rules about their clothing and items that can be brought in. For example, straps on tank tops must be over a certain width. Up to $40 may be brought inside each visit, but only in coins and contained in large clear plastic bags (not provided).[22] Brien, a Canadian scholar who herself experienced familial incarceration, writes about the perseverance and resourcefulness that is needed for families to 'navigate the criminal (in)justice system'.[23]

Prison visits are deeply precarious, and visitors must brace for uncertainty: they might be turned away, after hours of travel, for issues that are difficult to predict or control, such as having booked improperly, the prison being on lockdown, or the notoriously imperfect ion-scanning technology detecting traces of drugs.[24] On the buses I rode, an intricate cleansing ritual would begin as we approached the prison: hands, glasses, watches, jewellery, identification cards, and shoes would be wiped in sanitizer to keep the scanner from detecting microscopic particles of drugs that might have brushed on them in transit. Items needed for the long journey but prohibited from the visit (such as water bottles, lip balm, snacks, money, phones, and cigarettes) would be stored on the bus to avoid using the waiting room storage lockers, which cost money and were sometimes unreliable. All of the participants in my study, including children, described fear and frustration about being turned away from a visit. Here, Sue described the time invested in:

> … getting money together, cleaning your ID, making sure you wipe down your, washing all your clothes because yeah, ion-scan sets off and does whatever. Then you come three hours and you get turned away. First visit I ever went to, the dogs sat on me. And then I started crying, I'd come all the way from [several hours away], I had her and she was just a little baby and then I said 'well, what am I gonna do?'[25]

Once inside the visit, families face another challenge in the equally precarious task of creating home-like family meals out of the offerings in the visiting room vending machines, which is the only food available to them. One part of this problem is the selection, usually limited to tea and coffee, fizzy drinks, traditional snack foods, milk, and occasionally sandwiches in refrigerated vending machines; sometimes warm vending machines offer foods like hot dogs or hamburgers.

The vending machines are run by private, for-profit companies that the prison service contracts (but does not pay) to provide the machine, supply the food therein, and to pay that prison's 'Inmate Committee' a commission of their earnings.[26] The vending machine prices are high because, though they are paid by disproportionately poor people who have no choice but to buy this food, they must cover not only the cost of the food, but also that of the machines, the labour for their upkeep, the profit for the providers, and the commission. The prison service places no upper limits on pricing.

In all, only one of the 'elements of the family meal' identified by Dallacker, Hertwig, and Mata can be met with the food options in the visiting room: longer meal duration. While the visits were cherished family connections to participants in my study, most meals during visits consisted of nutritionally poor snack foods, with no fresh or home-cooked foods, eaten in an atmosphere that none considered positive.

Meals in this context – of prisoners and visitors craving 'family' and 'home' – thus contain the seeming worst elements of portability. Their ability to be eaten without utensils or plates underscored the non-'home-ness' of the prison. Their non-perishability and durability reinforced that they are neither freshly made nor home cooked, and contribute to their poor quality.

Four themes about food and prison visits emerged from my interviews.

Food as a Key Element of the Visiting Experience

Food (and in particular cost, options, and quality) was the first and key element that participants mentioned about their visits to the prison, and food appears to define many of their experiences. For example, seven-year-old Rob reports:

What do you like about your visits [to the prison]?
I gets lots of treats from my daddy..
You get lots of what from your daddy?
I get lots of treats from my daddy.
You get lots of treats from your daddy! what kind of treats?
Um I got, I get pop, [//you get pop] chips, and popcorns.
Okay. And what do you like best about visits with daddy?
The stuff he gives me.

Likewise, Grace, also seven, identified food when asked both about what she liked about visiting her father in prison, and even about having an incarcerated parent itself:

Anything you like about the day visits?
Uhm that we get to have food and we get to play and I get to talk to my daddy and get to talk to my mommy.
[…]
Okay is there anything good about having dad in jail?
Uhm that he gives me candies.

Similarly, when asked about the difference between the two different prisons her father had lived in, ten-year-old Darcy used the food offerings to assess her experience:

And is there one [of the prisons] you liked better or were they about the same?
Uhm … the one that he's in now, there's like TV and stuff. [Okay] And the vending machines, they have better stuff.
Oh. What was it like in the first place?
They had like chips and sandwiches and drinks that I don't really like …
Okay gotcha. But how is the food in this one he is in now?
They taste good.

Portable Food

Food as Prohibitively Expensive

The poverty that families of prisoners are more likely to face was often evident in participants' relationship with food in the prison. For example, ten-year-old Phoebe was so keenly aware of the financial burden of buying food at visits that when I asked her what changes she would make at the prison if she were in charge, she suggested:

> Change the prices on the food. ... Because everything is almost like $3.50, they should change about, to like $2.00.
> *Okay. How does it affect families?*
> With their money? Because they need money for groceries ... and bills.

The high cost of food was particularly raised in relation to the Private Family Visit (PFV), a system by which families can occasionally visit prisoners in a private bungalow inside the prison grounds for a weekend, after everyone meets a variety of conditions. While visitors bring clothes and personal care products in, they are not allowed to bring food; the prisoner must purchase the food with money from their account (which family members can contribute to) and families told me prices were inflated. Further, the PFV trailers are empty before families arrive, so basic staples like salt, ketchup, or oil must be purchased in this food order. Egregiously, to several of my participants, food may not be taken home after the visit by the prisoner or their families, meaning that a whole bottle of ketchup is purchased, used for two days and then discarded.

Sue explains that she spent up to $300 (£185) on food for a two-day visit (none of which could be used after that visit) while her entire monthly income from benefits was under $2000 (£1200). In this way, food was one medium through which poverty defined the prison family experience:

> Sometimes you can spend $40 on condiments going in on the [PFV] trailer. So sometimes we're guessing it's gonna be like a $300 bill because you need to make sure you get enough milk, to make the food. 'Cause you can't go out again once they have cereal or whatever, you're done. So, that's cause, we're spending almost 200 when me and [daughter] and [prisoner husband] were there. And just me and him last time and we did $124 and we made it just by the skin of our teeth.

Poor Quality Food

Another key issue raised by families was the poor and inconsistent quality of the food in the vending machines. For example, twelve-year-old Summer and her mother Cathy explained:

> *What about the food in the visits?*
> Summer: it's usually gone bad.
> *It's usually gone bad? All this $40 you spend?*
> Cathy: We spent almost $20 in milk trying to get cold chocolate milk. And it was all gone bad. And my husband's sitting there tasting each one of them, like

they were disgusting.
And you paid for it and they didn't ...?
Cathy: And the food, my husband almost puked in the garbage can from the food, it was ...
Summer: and they force me to eat it.
They did?
Cathy: we have to eat it in there.
Yeah, there's nothing else, right? Some of the institutions, like the, I know they have the traditional vending machines with the chips and pop and whatever, but they also have the ones with the sandwiches and stuff like that.
Cathy: That's the one that's always bad.
And is it ever empty, like do you ever not have any food except ...
Cathy: There's days we've gone in on the Saturday and it's been empty, and people have had to go out and say 'look, you have a full' [room full of kids] room full, get them in here' So they did end up coming in. But there's days we go in it's very ...
[...]
Summer: They got rid of the Whistle Dogs, the only good thing in there
Cathy: I think we should be allowed to bring food in. I honestly, that's the way I feel.

Another mother did not mind the food but had inconsistent experiences, with fruit sometimes being available in the vending machines but more often not. When asked what she thought the prison should do to support families, Bree argued:

There should have somewhere you can get actual food, not these, just vending machines. ... [The food in the prison visiting areas is] chips and chocolate and pop. People are going to be in there with kids from 9:30 to 3:30, no food? No kind of food? Just rubbish? Garbage?

Limited, Non-Home Food Choices

Many families visiting prisons sought to (re)create a type of 'home life' at the visiting table, but the highly regulated and limited choices of the food available prevented this. Being prohibited from bringing food in meant that many important family-constituting meals were missed; families are prevented from sharing meals cooked specifically for the loved one, special foods to mark family celebrations or holidays, favourite family items, or culturally-specific items. The provision of only processed, non-perishable, and portable foods can thus be seen as prohibiting partners and children from creating or 'doing' family and culture at the visit.

The limited food choice was a particular hardship to families with any less normative food needs. One parent and advocate, Nathalie, noted that her vegetarian family had difficulty finding anything at all to eat:

> I don't think you realise it but I think so much of our emotions are tied to, like, hunger or, like, the quality food we need and stuff, never mind children, right? [...] We're vegetarian so they have vending machines with, like, the salami sandwiches and whatever. But there's nothing vegetarian except for chips. [...] You know you're hungry, it's all day long. Like, an all-day visit, like, you need to eat.

Stringent rules around packaging further ensured that the food that prisoners and their families share during visits were less family meal-like. For example, Cathy raised her frustration at security rules around food that she deemed unreasonable:

> [My husband] tried to bring down an open thing of cookies [from his cell, to a visit], all there was, was like 3 cookies out of the package, but because it was opened, they threw the whole container in the garbage. He was just bringing them down for the kids, this was his treat to give for the kids and they threw it in the trash.

While ostensibly based on concerns about drugs and other contraband being passed, Ugelvik argues that such strict regulations on food in prisons are the 'continuation of the more general attacks on [prisoners'] identity' by the carceral space through their inability to enact agency over their meals.[27] Extending this analysis to families of prisoners helps make sense of the frustration that many participants reported around food options during visits.

Discussion

That visitors to Canadian prisons found the food to be expensive, limited, poor quality, and unlike home-cooked food would surprise no criminologist. While the conditions of the prison should not be in themselves punitive, it is common to find explicitly punitive staff attitudes and conditions in prisons.[28] The challenges that the food systems in the visiting rooms pose to families are arguably punitive in their effect; they are certainly perceived as punishing and unfair by some families.

These findings support the theoretical argument that families of prisoners become de facto 'subjects' of the prison themselves when their loved one is sentenced. For example, several child participants showed a detailed orientation to the prison environment and its practices, becoming attuned to the regulations and functioning of the institution. In this way, children and families are also affected by 'secondary prisonization', a term that describes the ways in which partners become quasi-inmates, adapting to the practices and cultural life of the prison.[29] Through processes that include the food systems in the visiting room, they become governed and regulated by the carceral power, constituted as its subjects by its 'mentalities of rule', after Foucault.[30]

Insofar as the food on offer at visits is so roundly criticized by families and fails to meet their desire to create a family meal at the visit, these findings also support the argument that families of prisoners are ignored at all aspects of criminal justice policy, as part of a process I have termed 'systemic invisibility'.³¹ As subjects of the policies of the prison visiting room, one might expect the prison system to consider the needs, barriers, and best interests of prisoners' families when designing these policies. However, there is virtually no evidence of this in the design of the food systems in visiting rooms. For example, in the tender process for a vending machine contractor at one federal prison, minute details about the requirements for the machine are included, but there is no mention of the food preferences of visitors, the need for child-friendly foods, or any requirements about the cost of items.³²

Recommendations

Human-rights-based solutions to these problems would begin by asking visitors themselves what they need. When asked how she would improve the experience of prison visitors, Casey pointed to the role of true family meals:

> If [families] could even bring a picnic lunch or something. Okay, no utensils or whatever [which might raise a safety concern]. Sandwiches or whatever, and even if it means we get the pop from [the prison staff] so it's not glasses coming out, you know. Whatever the issues.
>
> So you're there for that length of time, the person can enjoy eating a cooked meal just for that one time that they see you. So they know what they're looking forward to.

As Casey notes, there is much that could be done to allow families to have true family meals during visits, with all their associated benefits. For example, family visitors could be offered meals freshly cooked in the prison kitchens from high quality ingredients, at low or no cost. Indeed, international models exist for programmes in which prisoners cook for or with their visitors and then share the meal.³³

Another promising model is seen in prisons in the US and UK that host a visiting centre, run by charities and located just outside the gates, to support families before and after their visits. These centres have friendly staff to receive travellers after long journeys, restrooms, places for children to play while rules are explained and paperwork is filled out, and free or low-cost coffee, healthy snacks, and sometimes meals.³⁴

While such programmes would come at a cost to the state, the social policy, legal, and therapeutic justifications are manifold. Legal scholars Donson and Parke note that the UN Convention on the Rights of the Child requires signatory states to develop and implement child-sensitive approach to prison visitation, and they argue that this entitles children to specialized family visits during which they can play and eat a meal together with their incarcerated parent.³⁵ In line with elements of the positive family meal, prisoners and visitors would be able to model healthy eating behaviour by having fresh, cooked foods available, and they could connect in a more positive, family-focused environment.³⁶ Eating together during visits to the prison could shift from consuming portable foods to engaging in family meals.

Notes

1. Full description of the study methodology, including extensive discussion of ethical research practice with vulnerable participants, can be found in Else Marie Knudsen, 'The Experiences of Canadian Children of Prisoners' (unpublished doctoral thesis, London School of Economics and Political Science, 2016).
2. Samantha Punch and Ian McIntosh, '"Food Is a Funny Thing within Residential Child Care": Intergenerational Relationships and Food Practices in Residential Care', *Childhood*, 21.1 (2014), 72–86; Elinor Ochs and Merav Shohet, 'The Cultural Structuring of Mealtime Socialization', *New Directions for Child and Adolescent Development*, 2006.111 (2006), 35–49.
3. Punch and McIntosh.
4. Ochs and Shohet.
5. Dianne Neumark-Sztainer and others, 'Family Meal Patterns: Associations with Sociodemographic Characteristics and Improved Dietary Intake among Adolescents', *Journal of the American Dietetic Association*, 103.3 (2003), 317–22.; Georgia Middleton and others, 'What Can Families Gain from the Family Meal? A Mixed-Papers Systematic Review', *Appetite*, 153 (2020).
6. Middleton and others.
7. Neumark-Sztainer and others.
8. Mattea Dallacker, Ralph Hertwig, and Jutta Mata, 'Quality Matters: A Meta-Analysis on Components of Healthy Family Meals', *Health Psychology*, 38.12 (2019), 1137.
9. Corey Mintz, 'Food in Hospitals and Prisons Is Terrible – but It Doesn't Have to be that Way', *The Globe and Mail*, 10 May 2016 <https://www.theglobeandmail.com/life/food-and-wine/food-trends/food-in-hospitals-prisons-is-notoriously-bad-but-it-doesnt-have-to-be-that-way/article29951216/> [accessed 28 February 2023].
10. Office of the Correctional Investigator Annual Report 2018-2019 (June 2019) <https://www.oci-bec.gc.ca/cnt/rpt/annrpt/annrpt20182019-eng.aspx#fn52-rf> [accessed 28 February 2023].
11. Office of the Correctional Investigator Report.
12. Amy B. Smoyer and Linda Kjær Minke, *Food Systems in Correctional Settings: A Literature Review and Case Study* (Copenhagen: World Health Organization, 2015) <https://apps.who.int/iris/handle/10665/326323> [accessed 28 February 2023]; Rod Earle and Coretta Phillips, 'Digesting Men? Ethnicity, Gender and Food: Perspectives from a Prison Ethnography', *Theoretical Criminology*, 16.2 (2012), 141–56.
13. Earle and Phillips; Mintz; Melissa Gouge and Jennifer Hostetter, 'Starving for Rights: Hunger Strikes as Weapons of Resistance inside Farms and Prisons in the United States', *Food and Power: Proceedings of the Oxford Symposium on Food and Cookery 2019*, ed. by Mark McWilliams (London: Prospect Books, 2019), pp. 115–25; 'Review Finds Food, New Prison Management Contributed to 2016 Saskatchewan Penitentiary Riot', *CBC News*, 27 May 2018 <https://www.cbc.ca/news/canada/saskatoon/corrections-riot-report-1.4595637> [accessed 28 February 2023].
14. Thomas Ugelvik, 'The Hidden Food: Mealtime Resistance and Identity Work in a Norwegian Prison', *Punishment & Society*, 13.1 (2011), 47–63.
15. Shenique S. Thomas and Johnna Christian, 'Betwixt and Between: Incarcerated Men, Familial Ties and Social Visibility', in *Prisons, Punishment, and the Family: Towards a New Sociology of Punishment?*, ed. by Rachel Condry and Peter Scharff Smith (Oxford: Oxford University Press, 2018).
16. Johnna Christian, 'Riding the Bus: Barriers to Prison Visitation and Family Management Strategies', *Journal of Contemporary Criminal Justice*, 21.1 (2005), 31–48.; Megan Comfort, *Doing Time Together* (Chicago: University of Chicago Press, 2009).
17. Statistics Canada, 'Adult Correctional Services Survey 2013/201', *Juristat* (Ottowa: Canadian Centre for Justice Statistics, 2015); Lauren E. Glaze and Laura M. Maruschak, *Parents in Prison and Their Minor Children* (Washington, DC: US Department of Justice, Bureau of Justice Statistics, 2008), pp.1–25 <https://www.ojp.gov/ncjrs/virtual-library/abstracts/parents-prison-and-their-minor-children> [accessed 28 February 2023].

18 Joseph Murray and David P. Farrington, 'Parental Imprisonment: Effects on Boys' Antisocial Behaviour and Delinquency through the Life-course', *Journal of Child Psychology and Psychiatry*, 46.12 (2005), 1269–78.
19 Joseph Murray and others, 'Effects of Parental Imprisonment on Child Antisocial Behaviour and Mental Health: A Systematic Review', *Campbell Systematic Reviews*, 5.1 (2009), 1–105; Susan D. Phillips and others, 'Disentangling the Risks: Parent Criminal Justice Involvement and Children's Exposure to Family Risks', *Criminology & Public Policy*, 5.4 (2006), 677–702.
20 Vincent J. Felitti, 'Future Applications of the Adverse Childhood Experiences Research', *Journal of Child & Adolescent Trauma,* 10.3 (2017), 205–6.
21 Joyce A. Arditti, Jennifer Lambert-Shute, and Karen Joest, 'Saturday Morning at the Jail: Implications of Incarceration for Families and Children', *Family Relations*, 52.3 (2003), 195–204.
22 Knudsen, 'The Experiences of Canadian Children of Prisoners'.
23 Natasha Brien, 'Self Reflexivity: A Narrative Analysis of a Poem Titled "Crying"', *Journal of Prisoners on Prisons*, 24.2 (2015), 88–100.
24 Stacey Hannem, 'The Ion Mobility Spectrometry Device and Risk Management in Canadian Federal Correctional Institutions', in *Security and Risk Technologies in Criminal Justice: Critical Perspectives*, ed. by Stacey Hannem and others (Toronto: Canadian Scholars Press, 2019), pp. 87–109; Knudsen, 'The Experiences of Canadian Children of Prisoners'; Christian.
25 All participant names are pseudonyms.
26 Public Services and Procurement Canada. Website: 'Vending Machines for the Two Separate Visitors Areas at Collins Bay Institution (21441/21442)' <https://buyandsell.gc.ca/procurement-data/tender-notice/PW-17-00792153> [accessed 28 February 2023].
27 Ugelvik.
28 A key international human rights standard is that people are to be sent to prison 'as punishment, not for punishment' per *United Nations Standard Minimum Rules for the Treatment of Prisoners* (the 'Mandela Rules') (Geneva: General Assembly of the United Nations 2015) <https://documents-dds-ny.un.org/doc/UNDOC/GEN/N15/443/41/PDF/N1544341.pdf?OpenElement> [accessed 28 February 2023]; Michel Larivière and David Robinson, *Attitudes of Federal Correctional Officers towards Offenders* (Ottowa: Correctional Research and Development, Correctional Service of Canada, 1996); David Garland, *The Culture of Control: Crime and Social Order in Contemporary Society* (Chicago: University of Chicago Press, 2001).
29 Comfort.
30 Kim McKee, 'Post-Foucauldian Governmentality: What Does It Offer Critical Social Policy Analysis?', *Critical Social Policy*, 29.3 (2009), 465–86.
31 Else Marie Knudsen, 'The Curious Invisibility of the Children of Prisoners in Canadian Criminal Justice Policy', *Criminologie*, 52.1 (2019), 177–202.
32 Public Services and Procurement Canada.
33 Smoyer and Minke.
34 Christian; Prison Advice and Care Trust (PACT) Website: 'Visitors' Centres' <https://www.prisonadvice.org.uk/visitors-centre?> [accessed 28 February 2023].
35 Aisling Parkes and Fiona Donson, 'Developing a Child's Right to Effective Contact with a Father in Prison – An Irish Perspective', *Child Care in Practice*, 24.2 (2018), 148–963.
36 Dallacker, Hertwig, and Mata.

Eating in Public in Antiquity: Honourable or Obscene?

Joshua Lovinger

Introduction

In Book II of *Histories* (written before 425 BCE), Herodotus gives an account of the campaign of the Achaemenid Persian king, Cambyses II, son of Cyrus the Great, to conquer Egypt in the sixth century BCE. He then digresses into a lengthy ethnography of the ancient Egyptians – details warranted by 'the number of remarkable things which the country contains'. Notable to Herodotus was their consumption of raw fish and poultry and loaves of bread made of spelt – the dough kneaded with bare feet.[1] Herodotus also finds their manner of eating peculiar:

> the Egyptians themselves in their manners and customs seem to have reversed the ordinary practices of mankind [...]. To ease themselves they go indoors, but eat outside in the streets, on the theory that what is unseemly but necessary should be done in private, and what is not unseemly should be done openly [...].[2]

From his astonishment at the Egyptian practice, one gathers that, in the Greece of Herodotus, etiquette forbade eating outdoors.

Herodotus's *Histories* provoked much criticism. How much, if any, of his account of the Egyptians was based upon first-hand knowledge as opposed to imagination and assumptions? The first century CE Roman Jewish historian Flavius Josephus quoted the Hellenistic priest Manetho – himself an Egyptian and author of his own history of Egypt in Greek, the *Aegyptiaca*, – who 'on many points [...] convicts Herodotus of having given a false account of Egyptian matters out of ignorance'.[3] On our passage specifically, a modern commentator on Herodotus accuses him of over-generalizing. Depictions of Egyptian homes, including pictures in a tomb in Thebes from the Eighteenth Dynasty – a millennium prior to the *Histories* – do show dining rooms, indicating indoor eating, at least for the wealthy; poorer Egyptians would tend to eat in the open.[4]

What shocked Herodotus about Egyptians eating outdoors would be unlikely to bother modern readers, conditioned by a deluge of publications and television programmes extolling street food and authenticity. And yet the ancient Greek avoidance of eating outside in public has been found among multiple geographically and temporally disparate cultures, and has been well-described already by *fin-de-siècle* anthropologists (who might count Herodotus as their forebear). While much of their information is dated and biased, the breadth of their comparisons across cultures demonstrates the scope of our topic.

Eating in Public in Antiquity: Honourable or Obscene?

James Frazer (1854-1941) attributed apotropaic and magical rationales to avoiding eating in public. Some would not just eat alone and indoors – they would even make sure to lock the doors of the home while they ate, to bar the devil or evil eye from entering:

> In the opinion of savages the acts of eating and drinking are attended with special danger; for at these times the soul may escape from the mouth, or be extracted by the magic arts of an enemy present [...]. Precautions are therefore taken to guard against these dangers [...].[5]

In the following section of his work, 'Seclusion of kings at their meals', Frazer notes even more excessive precautions taken by kings to avoid being seen eating and drinking by their subjects, documented, mostly with examples from Africa, for reasons other than countering demonic and magical attacks. In some locales, for a subject to see the king eating was a capital offense:

> Among the Ewe speaking-people [...] the person of the king is sacred, and if he drinks in public every one must turn away the head so as not to see him, while some of the women of the court hold up a cloth before him as a screen. He never eats in public, and the people pretend to believe that he neither eats nor sleeps. It is criminal to say the contrary.[6]

To share a meal is democratizing; it levels its participants, which would damage the hierarchy required to rule one's subjects. If kingship was believed to be associated with a divine right or if the king was believed to be divine himself, public demonstration of the bodily needs of the ruler would shatter this myth.[7]

Ernest Crawley (1867-1924), in a book dedicated to Frazer, copies much of Frazer's discussion verbatim, but adds possible rationales like 'egoistic caution and fear of interruption', and concludes by noting that a similar impulse can be seen in 'the modern small boy who eats his cake in a corner'.[8]

In accounts of the Trobriand Islanders, Bronislaw Malinowski (1884-1942) notes that one of their 'leading characteristics' is 'pride in possessing abundant food'. However, there are no large communal meals:

> Meals are never taken in public, and eating is altogether regarded as a rather dangerous and delicate act. Not only will people never eat in a strange village, but even within the same community the custom of eating in common is limited. After a big distribution, the people retire to their own fireplaces with their portions, each group turning its back on the rest. There is no actual conviviality on a large scale. Even when the big communal cooking of taro takes place, small groups of related people assemble round the pot which has been allotted to them, and which they have carried away to a secluded spot. There they eat rapidly, no one else witnessing the performance [...]. *In fact, eating is rather a means of social division and discrimination than a way of bringing people together.* To begin with,

distinctions of rank are marked by food taboos [...].⁹

Their whole conduct, in the matter of eating in public, is guided by the rule that no suspicion of scarcity of food can possibly be attached to the eater. For example, to eat publicly in a strange village would be considered humiliating and is never done.¹⁰

The inmates of each house eat their meal in common, sitting on the verandah. Relatives and friends from other houses are often present, and are invariably invited to partake of the meal [...]. Although friends from the same village are not ashamed, or afraid, to eat with their hosts, strangers from another tribe would not partake of the food in the presence of others [...] no native would ever eat publicly in a strange village; if, however, there were several natives of one tribe in a strange place they would not be shamed to eat in public [...].¹¹

the climax – as we understand the climax of a feast, i.e., *the eating – is never reached communally, but only in the family circle*. But the festive element lies in the preparations, in the collection of the prepared food, in making it all a common property [...] and finally in the public distribution [...].¹²

For the Trobriand Islanders, in Malinowski's understanding, eating was a means of expressing kinship and distinction. Eating publicly, in the presence of outsiders, would degrade the exclusivity of bonds amongst close neighbours and families.

Similar observations regarding the Balinese were made by Margaret Mead (1901-1978) and Clifford Geertz (1926-2006); however, they give rationales using terms like 'disgust' and 'repugnant' behind an avoidance of eating in public. For Geertz, in Bali, 'Not only defecation but eating is regarded as a disgusting, almost obscene activity, to be conducted hurriedly and privately, because of its association with animality'.¹³ Shelly Errington describes Margaret Mead telling 'of the horror that the Balinese felt when she told them about public restaurants; the thought of eating in public was repugnant to them. Now Bali is filled with little roadside stands selling food, and snack vendors'.¹⁴ In his studies of sex and modesty, Havelock Ellis (1859-1939) cited additional examples from cultures outside of Bali (e.g. Tahitians, Malays, the Warrua of Central Africa, the Bakairi of Central Brazil) where eating in public was viewed as indecent, emphasizing components of shame and disgust in the avoidance. He contrasted this with the indifference with which some of these same cultures viewed the exposed naked human body.¹⁵ Identical terms related to disgust are used by a modern opponent of public eating.¹⁶

Lest one think that such beliefs about eating in public were limited to Africa, the islands of Southeast Asia, or preindustrial societies, we must note that in nineteenth and twentieth century Japan, as well, young people were educated that such behaviour was improper.¹⁷ The seven rules of behaviour adopted by the Aizu clan's school (*Nisshinkan*) includes 'We must not eat in public' sandwiched between 'We must not pick on those who are weaker' and 'We must not talk to girls'.¹⁸ A modern English guide to etiquette in Japan attributes the recent disregard for this rule to the arrival of fast-food chains with limited seating.[19]

That anthropologists gave a prominent place to this widespread taboo in their accounts may indicate much about the observers' home cultures in the West where the taboo was seldom found; all of these scholars were born in the United States or Europe.

The discussions of the anthropologists provide several potential motivations for avoidance of eating in public. Their descriptions also allow us to distinguish between different meal settings, which may indicate the possible driving rationale for the taboo in each context. A meal may be indoors, but either in a group (private eating) or alone (solitary eating). Similarly, a meal may be outdoors, on the street, but in a group or alone. Other distinctions are possible as well.[20]

Ancient Greece and Rome

Picking the story up from Herodotus, other Greeks had an opposite view. The legendary king of Sparta, Lycurgus, is described by ancient historians as having instituted drastic communal reforms, including public meals. Eating in common was intended as a remedy to extreme luxury. Transparency was expected to lead to moderation, and to curb excess. As the Greek historian Xenophon (fourth century BCE) described:

> Before Lycurgus' day the Spartans used to mess at home like the other Greeks. Realizing that under these conditions they became extremely negligent, he brought the messes into the public domain in the belief that in this way the laws given [sc. by him] would be infringed the least. And he specified the rations for the messmates, so they should neither have too much food nor too little [...]. In short, during the mess the table is never without food nor is the fare extravagant [...] *since they mess together as I have described, how could anyone ruin either themselves or their house by gluttony or drunkenness?*[21]

The Greek Plutarch (first century CE) referred to this Spartan institution with admiration, calling it Lycurgus's 'finest reform'. Notable is his description of Sparta prior to these reforms; Plutarch was, in a not so subtle fashion, alluding to (and safely critiquing) the extravagant tables of the wealthy and powerful Romans of his own day:

> This [institution of common messes by Lycurgus] stopped them spending time at home reclining at table on expensive couches, *fattening themselves up in the dark* like insatiable animals on the produce of craftsmen and cooks, and ruining themselves morally as well as physically by indulging every whim and gorging themselves until they needed long sleeps, hot baths, a great deal of quiet, and, so to speak, daily nursing [...]. For when rich and poor went to the same meal, the rich could not even use or enjoy, let alone gaze upon or display, all their paraphernalia. The upshot was – and this was how everyone put it – that Sparta was the only city in the world where Wealth could be seen truly blind [...].[22]

Other classical authors pointed as a model not to Sparta, but to the Roman Republic. With nostalgia for a Rome that he never experienced, Valerius Maximus (first century CE) commended 'the men of old' for their self-restraint: 'The greatest among them were not ashamed to take luncheon and dinner in the open. To be sure, they had no feasts that they blushed to expose to public gaze.'[23] While Valerius presents this mode of counteracting gluttony as self-motivated, the fifth century author Macrobius gives an opposite account in his *Saturnalia*: the Republic was required to legislate to prevent gluttony. Many details of Roman sumptuary law (like the second century BCE *lex Orchia*, named for Gaius Orchius, tribune of the plebs) are known exclusively from the *Saturnalia*, 'an encyclopedic compilation quarried from mostly unnamed sources'.[24] For him, the ancient Romans were no saints:

> It would be a long job, should I wish to catalog all the implements of gluttony that those people applied their wits to dreaming up or their zeal to devising. Of course, that's why so many laws were brought before the people concerning dinners and expenditures, and *why it began to be the rule that people had to eat lunch and dinner with their doors wide open, so that the scrutiny of their fellow citizens would set a limit on their luxury* [...].[25]

While James Frazer quoted example after example of cultures where it was necessary for the king to eat in isolation, Pliny the Younger – a contemporary of Plutarch and Valerius Maximus – praised the Roman emperor Trajan for never eating alone. In contrast with his predecessor, Domitian, Trajan avoided luxury – notwithstanding the risk for excess associated with his office. Nor did Trajan's modest public eating, which was open to all, mask 'secret gluttony and private excesses' in an unseen meal following the public one.[26] This praise of Trajan can be linked to a separate history of criticism of solitary eating (*monophagia*) in Hellenistic literature, spanning from Epicurus (a contemporary of the students of Plato in the third century BCE), who considered it animalistic – 'a dinner of meats without the company of a friend is like the life of a lion or a wolf' – to the author of IV Maccabees, at least four hundred years later, who twice compares the solitary gormandizer with the glutton and drunkard.[27]

An exception to this view among the Romans can be found in the *Res Gestae* of the soldier Ammianus Marcellinus. Ammianus participated in the mid-fourth-century-CE campaigns against the Sassanian Persian Empire led by Emperor Julian and others, but reflected with admiration upon the modesty of his military opponents at the table and in the privy. Similar observations about the Persians were made around the same time by contemporary Jews, which the Babylonian Talmud (composed in Sassanian Persia) records (bBerakot 8b). As in Herotodus, the opposing biological functions of eating and elimination are juxtaposed:

> They avoid as they would the plague splendid and luxurious banquets, and especially, excessive drinking [.... E]very man's belly is, as it were, his sundial;

when this gives the call, they eat whatever is at hand, and no one, after he is satisfied, loads himself with superfluous food. They are immensely moderate and cautious…one seldom sees a Persian stop to pass water or to step aside in response to a call of nature; so scrupulous do they avoid these and other unseemly actions.[28]

Like other Latin authors, there is an implicit criticism of Roman luxury in Ammianus's writings about the Persians.[29] He laments that laws of Ampelius (prefect of Rome, 372-371 BCE) restricting undue eating and drinking, did not last long. Unlike other Latin authors who promoted public eating to curb excess,[30] Ammianus reports that Ampelius's legislation forbade public eating, along with restrictions designed to curtail the activity at public taverns where food and drink were served:

> I wish he [Ampelius] had been steadfast of purpose; for he could have corrected in part, even though to a small extent, the incitements of appetite and gross gluttony, if he had not let himself be turned to laxity and thus lost enduring fame. For he gave orders that no wine-shop should be opened before the fourth hour [...] and *that no respectable man should be seen chewing anything in public*. These *shameful acts*, and others worse than these, had, by being constantly overlooked, blazed up to such unbridled heights [...].[31]

A modern commentary to Ammianus here assumes, without sources, that this view was commonplace, simply remarking that '[e]ating in public was obviously in bad taste'.[32] Aside from the question of a utility to banning (or promoting) public eating for curbing excesses, Ammianus calls the act of chewing in public itself shameful, a perspective that the Balinese might share but not found among other Latin authors.

Diogenes the Cynic

The comments of most classical Greek and Roman authors on eating in public had little influence on subsequent discussions. Not so with Diogenes of Sinope. Diogenes was forced to leave his hometown on the southern shore of the Black Sea, and came to Athens and Corinth where he encountered the students of Socrates – Plato among them.[33] He became known for his attacks on the conventional (including philosophical) wisdom of the day and for exemplifying the Cynic value of *parrhesia*, speaking and demonstrating through his actions the undistorted truth to those around him, despite their status in society. Anecdotes from his life depicting this characteristic are among the most well-known from the lives of ancient philosophers, and have been depicted in Western art for centuries.[34] Most famous among these are probably his encounters with Alexander the Great:

> Philosophers gained audiences with him [Alexander] to tender their congratulations, and he hoped that Diogenes of Sinope, who was living in Corinth at the time, would follow their example. Diogenes, however, continued to live an untroubled life in Craneium, without paying the slightest attention

to Alexander, so Alexander paid him a visit and found him relaxing in the sun. Diogenes raised himself up a bit when the huge crowd of people appeared and looked at Alexander, who greeted him and asked him if there was anything he wanted. 'Yes', replied Diogenes, 'move aside a little, out of my sunlight'. The story goes that Alexander was so struck at being held in such contempt, and so impressed with the man's haughty detachment, that while the members of his retinue were ridiculing and mocking Diogenes as they left, he said, 'But as for me, if I were not Alexander, I would be Diogenes'.[35]

This account exists in multiple forms – in some, the concluding declaration of Alexander and the story of Diogenes' dismissal of the young Macedonian conqueror appear separately.[36] The story's purpose is not only to reflect Diogenes' characteristic of fearlessly speaking truth to power; it also exists in biographies of Alexander (as with stories of him and Socrates), to buttress his legend as the model philosopher-king – who listened when wise scholars spoke, even when critical.[37] Another tale strengthening their link had both men dying on the same day.[38]

Perhaps more famously, Diogenes shocked his contemporaries for his lack of shame, urinating, defecating, farting, masturbating, and having sexual intercourse in public view – behaviours which Western artists thankfully did not choose as subjects.[39] Along with these, and presented as equally disturbing, Diogenes ate in the marketplace for all to see. His third-century biographer, Diogenes Laertius, repeated multiple anecdotes about Diogenes of Sinope's predilection for public eating, comparing sating his hunger to satisfying sexual urges:

> He regularly performed in public the acts associated with Demeter and Aphrodite. He used to make the following sort of argument: 'If to take breakfast is not absurd; then in the marketplace it's not absurd; and it is *not* absurd to take breakfast; so to do so in the marketplace is *not* absurd.' Frequently masturbating in public, he said, 'If only one could relieve hunger by rubbing one's belly.'[40]

This sort of behaviour was among several explanations for Diogenes's nickname, 'the Dog' – as he was called by Plato, bystanders, and even himself.[41] A marble statue of a dog was placed over his grave.[42]

Diogenes' conduct was not random. His public acts reflected a belief that natural urges, like hunger and thirst, were not base and should not be considered shameful. One should follow a life in accord with nature. Rather, it was acts such as theft and falsehood which were unnatural and should provoke men's disgust. Therefore, '[r]eproached one day for eating in the marketplace, he said, "It was in the marketplace that I got hungry"'.[43] In 362 Emperor Julian explained Diogenes' violation of social conventions thusly:

> Let [Diogenes] trample on conceit; let him ridicule those who although they conceal in darkness the necessary functions of our nature – I am speaking of

the expulsion of excrement – yet in the center of the marketplace and of our cities carry out most violent [deeds] which are not proper to our nature: robbery of money, false accusations, unjust indictments, and the pursuit of other such vulgar business. When Diogenes farted or went to the bathroom or did other things like this in the marketplace, which they say he did, he did these things to trample on the delusion of those men and to teach them that they carried out [deeds] far more sordid and dangerous than his. For what he did was according to our common nature, while what they did was not, so to speak, in accord with everyone's nature, but were all carried out because of perversion.[44]

Authentic writings of Diogenes himself have not survived, if they ever existed, and close to 500 years elapsed between the time of his death in 323 BCE and the works of Plutarch and Dio Chrysostom, and about a century more until the lengthiest account by Diogenes Laertius.[45] But despite the time gap, later anecdotes about Diogenes give the same impression as Herodotus (in the fifth century BCE) – that in Greece eating in public was a violation of the norm. Otherwise, the critique of Diogenes would lose its pointedness.

The comment of Diogenes, that he ate in the marketplace because 'It was in the marketplace that I got hungry', was far more influential on subsequent thought about the appropriateness of public eating than the comments of other classical authors. With other authors, remarks on eating were typically made tangentially. With Diogenes, this witticism was central to his programme of critiquing conventional manners, which he exemplified. The repetition of such anecdotes in book 6 of Diogenes Laertius's life of Diogenes of Sinope is not happenstance. That the teaching was packed into a compact, sharp saying, easily recalled and shared, amplified its diffusion. Such comments, known as *chreiai*, were memorized as classroom exercises in Late Antiquity and were an important part of rhetorical education to serve as the building blocks for speeches. Diogenes Laertius's chapter on Diogenes the Cynic itself is composed of these short *chreiai*.[46]

Judaism and Islam

Jewish legal works from the period of the tanna'im (Second Temple period to ~220 CE) specifically list eating in the marketplace as potential grounds for divorce (Sifre Zuta to Deuteronomy and a baraita in bGittin 89a) due to immodesty. The Jerusalem Talmud (yMa'aserot 3:5) cites a tanna'itic legal statement that 'it is not praiseworthy for a scholar to eat in the marketplace'. While that statement seems limited to the scholarly elite, a similar passage in the Babylonian Talmud (fifth/sixth c. CE) has a broader application, to all persons, asserting that:

> One who eats in the marketplace is similar to a dog.
> Some say that he is invalid as a witness [to testify in court].
> R. Idi b. Abin said: The law is in accord with those [that he is invalid as a witness].[47]

The Israeli classicists E.E. Halevy and Joseph Geiger noted that the reference here, in the Talmudic passage to dogs in relation to eating in public, refers to the Cynics, the followers of Diogenes (some of whom were active in the third and fourth century in the Land of Israel and in Syria).[48]

These Talmudic passages expressing concern with eating in the marketplace were copied into post-Talmudic ethical and etiquette compilations (Derek Erez Rabbah, Kallah Rabbati, as well as Baraita de-Niddah, an interesting work of unclear provenance) and in post-Talmudic legal works produced in Baghdad and North Africa (the Geonic period in Judaism, ~600-1100 CE).[49]

Interestingly, similar concerns (stated in nearly identical fashion – with reference to the marketplace and with discussion of the viability of violators as witnesses) can be found roughly contemporaneously in early Islamic literature, including in the classics of the Golden Age of the Abbasid Caliphate – in works by Ibn Qutayba (d.889) and Al-Jāḥiz (776-868/9).[50] While there were debates about their reliability, statements about the permissibility of eating in public were also transmitted in *hadith* from each of the major Islamic legal schools. These discussions – in both Jewish and Islamic legal traditions – and the popularity of this anecdote about Diogenes persisted well into the high Middle Ages.[51]

Notes

1. Herodotus, *The Histories*, trans. by Aubrey de Sélincourt, rev. edn (London: Penguin, 1996), pp. 99, 113.
2. Herodotus, pp. 98–99 (II, 35).
3. *Flavius Josephus: Translation and Commentary, Vol. 10: Against Apion*, trans. by John M.G. Barclay (Leiden: Brill, 2007), p. 52 (I, 73). Barclay comments on Josephus's motive: 'This reminder of Greek unreliability [...] makes Egyptians, like Judeans, victims of ignorant Greek historiography'.
4. Alan B. Lloyd, *Herodotus, Book II. Introduction and Commentary* (Leiden: E.J. Brill, 1976; repr. 1994), II, pp. 150–151 and 'Commentary to Book II', in David Asheri, Alan Lloyd, and Aldo Corcella, *A Commentary on Herodotus Books I-IV*, ed. by Oswyn Murray and Alfonso Moreno, trans. by Barbara Graziosi and others (Oxford University Press, 2007), p. 263 (notes to Book II 32–36).
5. James G. Frazer, *The Golden Bough*, 3rd edn (1911; repr. Cambridge University Press, 2012), III, pp. 116–17.
6. Frazier, pp. 117–19.
7. I have intentionally focused on a description of the phenomenon of public and private eating in this paper, while avoiding attention to the specific subjects and objects of gaze, and both the asymmetry and power dynamic between the two (including the inversion that might occur were the king to be observed eating by those he rules), though this topic could be productively subjected to such an analysis, as has been the focus of much late twentieth century philosophy, from Jean-Paul Sartre's 'Le regard' to Michel Foucault's *Discipline and Punish*.
8. Ernest Crawley, *The Mystic Rose: A Study of Primitive Marriage* (London: Macmillan, 1902), pp. 150–52.
9. Bronislaw Malinowski, *The Sexual Life of Savages in North-Western Melanesia; An Ethnographic Account of Courtship, Marriage and Family Life among the Natives of the Trobriand Islands, British New Guinea* (New York: Readers League of America & Eugenics Publishing Company, 1929), pp. 441–42; emphasis added.
10. Bronislaw Malinowski, 'The Primitive Economics of the Trobriand Islanders', in *Cultures of the Pacific: Selected Readings*, ed. by Thomas G. Harding and Ben J. Wallace (New York: Free Press, 1970), p. 57.
11. Bronislaw Malinowski, 'The Natives of Mailu: Preliminary Results of the Robert Mond Research Work in British New Guinea', *Transactions of the Royal Society of South Australia*, 39 (1915), 545.
12. Bronislaw Malinowski, 'Baloma: The Spirits of the Dead in the Trobriand Islands', *Journal of the Royal Anthropological Institute of Great Britain and Ireland*, 46 (July-December 1916), 373; emphasis added.

The ceremonial distribution of food is known as *sagali*.

13. Clifford Geertz, 'Deep Play: Notes on the Balinese Cockfight', in *The Interpretation of Cultures: Selected Essays* (New York: Basic Books, 1973), pp. 412–453 (p. 420). However, elsewhere Hildred Geertz and Clifford Geertz comment on the uses of the meal to demonstrate kinship and exclude outsiders, as described by Malinowski: 'To eat with someone else from the same dish signifies precise equality, and only a very narrow circle of kinsmen are eligible to do this' (*Kinship in Bali* (Chicago: University of Chicago Press, 1975), p. 129).

14. Shelly Errington, *Meaning and Power in a Southeast Asian Realm* (Princeton: Princeton University Press, 1989), p. 89. See Gregory Bateson and Margaret Mead, *Balinese Character: A Photographic Analysis* (New York: New York Academy of Sciences Publications, 1942), pp. 20–21: 'So uncomfortable are the Balinese when eating meals that many observers have come away from Bali insisting that the Balinese will never eat in public if they can avoid it, and that if others are present they turn their backs on each other, and toss the food into their mouths as quickly as possible [...]. Eat together they must, at feasts, at ceremonies, and as members of work groups who are fed by the host, but always there is the turning away, the search for privacy reminiscent of the search for privacy in defecation.' However, compare 'the institution of the food vendor', pp. 22–23.

15. Havelock Ellis, *Studies in the Psychology of Sex*, 3rd edn (Philadelphia: F.A. Davis Company, 1910), I, pp. 48–49.

16. See Leon R. Kass, *The Hungry Soul: Eating and the Perfecting of Our Nature*, 2nd edn (Chicago: University of Chicago Press, 1999), pp. 129–60, and the criticisms of Steven Pinker, 'The Stupidity of Dignity', *The New Republic*, 238.9 (28 May 2008), 28–31.

17. See for example 'Mrs. Hugh Fraser', *A Diplomatist's Wife in Japan: Letters from Home to Home*, 3rd edn (London: Hutchinson & Co., 1900), pp. 495–96.

18. Quoted in *Remembering Aizu: The Testament of Shiba Gorō*, ed. by Ishimitsu Mahito, trans. by Teruko Craig (Honolulu: University of Hawai'i Press, 1999), p. 6.

19. Boyé Lafayette De Mente, *Etiquette Guide to Japan: Know the Rules that Make the Difference!*, 3rd rev. edn (Rutland, VT: Tuttle Publishing, 2011), p. 62.

20. We might also distinguish, within manners of eating, between a more formal meal and something more temporary or limited, like a snack (as in the difference in English between 'to eat' and 'to dine'). Such a distinction with respect to the permissibility of eating in public is made in medieval Jewish sources. See Kallah Rabbati ch. 10 and R. Jacob b. Meir (= Rabbenu Tam) of Ramerupt, France (d.1171) in Tosafot to bQiddushin 40b.

21. Michael Lipka, *Xenophon's Spartan Constitution: Introduction, Text, Commentary* (Berlin: Walter de Gruyter, 2002), pp. 74–75 (V, 1–4).

22. Plutarch, *Greek Lives: A Selection of Nine Greek Lives*, trans. by Robin Waterfield (Oxford University Press, 1998), pp. 17–18.

23. Valerius Maximus, *Memorable Doings and Sayings*, trans. by D.R. Shackleton Bailey (LCL 492) (Harvard University Press, 2000), I, pp. 162–63.

24. Macrobius, *Saturnalia*, trans. by Robert A. Kaster (LCL 510) (Harvard University Press, 2011), I, p. xii.

25. Macrobius, *Saturnalia*, trans. by Robert A. Kaster (LCL 511) (Harvard University Press, 2011), II, pp. 118–19 (III, 17.1–3).

26. Pliny the Younger, *Letters, Vol. II: Books 8-10. Panegyricus*, trans. by Betty Radice (LCL 59) (Harvard University Press, 1969), pp. 430–33. On luxury under Domitian, see Martial, *Epigrams*, trans. by D.R. Shackleton Bailey (LCL 480) (Harvard University Press, 1993), III, pp. 100–03 (XII, 15).

27. H. Anderson, '4 Maccabees: A New Translation and Introduction', in *The Old Testament Pseudepigrapha*, ed. by James H. Charlesworth (New York: Doubleday, 1985), II, pp. 545-46 (1:27 and 2:7). See discussions in John Wilkins, *The Boastful Chef: The Discourse of Food in Ancient Greek Comedy* (Oxford University Press, 2000), pp. 67–70; *Thirteen Satires of Juvenal*, 2nd edition (London: Macmillan and Co., 1872), I, p. 134 (notes to I:95); Seneca (*Seneva IV: Epistles 1–65*, trans. by Richard M. Gummere (LCL 75) (Harvard University Press, 1917), pp. 130-31 (Epistle 19.10)) quotes Epicurus by name.

28 Ammianus Marcellinus, *History*, trans. by John C. Rolfe (LCL 315) (Harvard University Press, 1940), II, pp. 392–93 (27.6.76–79). For the Talmudic parallel, see Geoffrey Herman, 'Table Etiquette and Persian Culture in the Babylonian Talmud', *Zion*, 77.2 (2012): 149-188. 'Eating and Elimination' – is the name of the seventh chapter in Carl D. Schneider, *Shame, Exposure, and Privacy*, which touches upon eating in public (Boston: Beacon Press, 1977), p. 68.

29 Athenaeus notes that the Persian kings dined alone. Here it is likely a criticism, among several anecdotes he brings comparing *monophagi* (= those who dine alone) to gluttons. See Athenaeus, *The Learned Banqueters*, trans. by S. Douglas Olson (Harvard University Press, 2006), Vol. I: Books I-III.106e (LCL 204), pp. 192–195 (IV.145b) and Vol. II: Books III.106e-V (LCL 208), pp. 28–29 (I.5e) and 44–45 (I.8e).

30 One possible exception: Quintilian relates a quip made by a Roman knight at the expense of Emperor Augustus. 'A Roman *eques* was drinking in the theatre, and Augustus sent him a message to say "If I want lunch, I go home." "Of course, said the *eques*, "you are not afraid of losing your place."' (Quintilian, *The Orator's Education*, Vol. III: Books 6–8 (LCL 126), trans. by Donald A. Russell (Harvard University Press, 2002), pp. 94–95 [6.3.63].) But contrast the comment of Suetonius that Augustus would not eat specifically at home, but rather wherever he felt like it, just as Diogenes did. Suetonius, Vol. I: *Lives of the Caesars*, trans. by J.C. Rolfe (LCL 31) (Harvard University Press, 1913; rev. 1998), pp. 264–265.

31 Ammianus Marcellinus, trans. by John C. Rolfe (LCL 331) (Harvard University Press, 1939), III, pp. 138–39 (28.4.3–5).

32 J. Den Boeft, J.W. Drijvers, D. Den Hengst, H.C. Teitler, *Philological and Historical Commentary on Ammianus Marcellinus XXVIII* (Leiden | Boston: Brill, 2011), pp. 175-176.

33 Diogenes Laertius, *Lives of the Eminent Philosophers*, trans. by Pamela Mensch (Oxford: Oxford University Press, 2018), pp. 269–70 (VI, 20), 282 (VI, 49). For discussion of the circumstances of his exile and its role in the thought of Diogenes and the Cynics, see Robert Bracht Branham, 'Exile on Main Street: Citizen Diogenes', in *Writing Exile: The Discourse of Displacement in Greco-Roman Antiquity and Beyond*, ed. by Jan Felix Gaertner (Leiden: Brill. 2007), pp. 71–85.

34 Diskin Clay, 'Picturing Diogenes', in *The Cynics: The Cynic Movement in Antiquity and Its Legacy*, ed. by R. Bracht Branham and Marie-Odile Goulet-Cazé (Berkeley: University of California Press, 1996), pp. 366–87.

35 Plutarch, p. 323.

36 For example, see Diogenes Laertius, p. 275 (VI, 32) citing Hecaton (Alexander's declaration, without the story about Alexander blocking Diogenes' sunlight) and Arrian, *Alexander the Great: The Anabasis and the Indica*, trans. by Martin Hammond (Oxford University Press, 2013), p. 196 (the story about Alexander blocking Diogenes' sunlight, without Alexander's concluding declaration).

37 See Diogenes Laertius, p. 291 (VI, 68).

38 Diogenes Laertius, p. 296 (VI, 79). Cf. Dio Chrysostom, Vol. I: *Discourses 1–11*, trans. by J.W. Cohoon (LCL 257) (Harvard University Press, 1932), pp. 257-260.

39 For convenient references, see Derek Krueger, 'Diogenes the Cynic Among the Fourth Century Fathers', *Vigiliae Christianae*, 47.1 (March 1993), 29-49 (p. 45 n. 7).

40 Diogenes Laertius, p. 291 (VI, 69). As the editor here notes: 'Demeter was the goddess of grain (and hence of food generally), Aphrodite of sexual love' (n. 41). A nearly identical comment about masturbation and hunger appears earlier (pp. 281-82 (VI, 46)).

41 Diogenes Laertius., pp. 279 (VI, 40), 287 (VI, 60–61), 295 (VI, 78). His predecessor Antisthenes was called this as well: Haplocyon, translated variously as 'downright dog', 'absolute dog', or 'simple / pure / natural dog' (Diogenes Laertius, p. 264 (VI, 13)); references in Aristotle to 'the Dog' likely refer to him rather than to Diogenes (Marie-Odile Goulet-Cazé, 'Appendix B: Who Was the First Dog?' in *The Cynics*, pp. 414–15). Antisthenes taught at a gymnasium outside the walls of Athens called the Cynosarges (= white dog) and the Latinized form of the Greek word for a canine (*cyōn / cynicos*) gave the movement its name (the Cynics).

42 Diogenes Laertius, p. 295 (VI, 78).

43 Diogenes Laertius, p. 286 (VI, 58).
44 As trans. by Derek Krueger, 'The Bawdy and Society: The Shamelessness of Diogenes in Roman Imperial Culture', in *The Cynics*, pp. 233–34.
45 Some attributed to Diogenes over a dozen works, but Sosicrates and Satyrus 'say that Diogenes left nothing in writing' (Diogenes Laertius, p. 296 (VI, 80)). See the cautioning of Bracht Branham ('Cynicism', in *The Classical Tradition*, ed. by Anthony Grafton and others (Cambridge, MA: Belknap Press, 2010), p. 247).
46 Jan Fredrik Kindstrand, 'Diogenes Laertius and the Chreia Tradition', *Elenchos*, 7 (1986), 217–43; Derek Krueger, 'Diogenes the Cynic'. Perhaps surprisingly, the contributions to prior Oxford Food Symposia, in 1991 on the topic of Public Eating, and in 2014 on the topic of Food & Markets, make no mention of Diogenes's subversive eating in the marketplace.
47 R. = Rabbi; b. = ben, son of. In Mishnaic Hebrew and later, in Talmudic Aramaic, the noun used here, שוק, can refer to a street / open area, district, or market / bazaar. See Michael Sokoloff, *A Dictionary of Jewish Babylonian Aramaic of the Talmudic and Geonic Periods* (Ramat Gan: Bar Ilan University Press & Baltimore, MD: Johns Hopkins University Press, 2002), pp. 1123–24.
48 E.E. Halevy, *The Values of the Aggadah and Halakah in Light of Greek and Latin Sources*, Vol. IV (Tel Aviv: Devir, 1982), p. 109 (§6) [in Hebrew]; Joseph Geiger, 'Greek in the Talmud: Three Notes Pertaining to Dogs', *Tarbiz* 51.2 (Tevet-Adar 1982): 303–305 [in Hebrew, with English abstract]. Menahem Luz has described interactions between cynics and rabbis in the Land of Israel in Late Antiquity in a series of articles, including 'A Description of the Greek Cynic in the Jerusalem Talmud', *Journal for the Study of Judaism in the Persian, Hellenistic, and Roman Period* 20.1 (June 1989): 49–60.
49 Other Talmudic and Midrashic passages about public eating and drinking seem to be less germane to this discussion. See bBekorot 44b (Derek Erez Zuta 7:3), bPesahim 86b, and *Midrash Kohelet Rabbah 1–6: Critical edition based on manuscripts and Genizah fragments*, edited by Marc Hirshman with Shaul Baruchi (Jerusalem: Schechter Institute of Jewish Studies, 2016), pp. 150–151 (2:17) = *Midrash Rabbah, Vol. 8: Ecclesiastes*, trans. by Abraham Cohen (London: Soncino Press, 1939), p. 67. The latter case, for example, deals specifically with judges drinking wine in public. (Cf. the prohibition of intoxication for those rendering judgment in bEruvin 64a-b.) The case of bBekorot mentions drinking water and urination only, not eating; bPesahim 86b deals with the unique circumstance of a bride – its applicability to the general public may be limited.
50 For a preliminary account, see: Paulina B. Lewicka, *Food and Foodways of Medieval Cairenes: Aspects of Life in an Islamic Metropolis of the Eastern Mediterranean* (Leiden: Brill, 2011), pp. 351–86. The prominence of Diogenes the Cynic in early Arabic literature is not surprising – many of his teachings (in gnomological form) have survived only in Arabic: see Dimitri Gutas, 'Sayings by Diogenes Preserved in Arabic', in: *Le cynisme ancien et ses prolongements. Actes du colloque international du CNRS* (Paris, 22–25 July 1991), ed. by M.-O. Goulet-Cazé and R. Goulet (Paris 1993), pp. 475–518; for an Arabic version of Diogenes on public eating, see p. 490 (151.1).
51 I hope to treat the Jewish and Islamic discussions of the propriety of public eating and the laws of witnesses more fully in a future study. See, for now, Gideon Libson, *Jewish and Islamic Law: A Comparative Study of Custom During the Geonic Period* (Cambridge, MA: Islamic Legal Studies Program, Harvard Law School, 2003) and the earlier important study of Zevi Taubes (which comes to similar conclusions, but seems to have been unknown to Libson), הגאונים ז״ל בענין פסולי עדות" "הפסקים הפירושים וההוספות של, in *Essays presented to Chief Rabbi Israel Brodie on the occasion of his seventieth birthday*, ed. by H.J. Zimmels and others, Vol. II [= Hebrew section] (London: Soncino Press, 1966), pp. 179–93 (e.g. compare Libson, pp. 105–07 to Taubes, pp. 182–83).

Food for the Final Journey: Feeding the Soul in Hindu Death Rituals

Priya Mani

Prologue

It was a still midsummer morning when my mother's mobile rang. Her face grew pale, and her eyes glassed up as she listened to her brother. My mother's mother had died quite suddenly, as such circumstances usually occur. In that instant, our world sank into despair. When we came to her apartment, family and friends sat in solemn silence around her stiff body dressed in a red sari. Her face looked calm, punctuated with vermillion and silver. My mother was allowed an hour to soak in the new reality of her mother. Shock and a surreal pain enveloped my being as I fathomed the loss.

My uncles and cousins brought a long bamboo bier, and we women watched as they slowly placed grandma on it. Priests arrived to initiate the funerary rituals, and my uncle, her only son, was tasked with the final rites. Shouldering the bier, the men set out for the crematorium. Here she would be cremated, as many Hindus have been for millennia. Five hours later, my uncle returned and revealed two stones wrapped in a white cloth, a fragment from her shroud. These, he explained, would embody grandma's spirit for the next ten days. My aunt and uncle then set up an imaginary home in a corner, drawn with rice flour, conferring a vitality on the stone. They embarked on a ritual of care for the stone like they cared for grandma in her life. Her spirit was on a journey to the afterworld of our ancestors – of her parents and grandparents, they said. In the next ten days, her spirit would prepare itself for life beyond our realm. They offered the stone water and food as if enlivened by life's basic needs. My aunt prepared the choicest sweets every day. As grandma's spirit travelled through the year, we followed her astral journey, slowly uncoupling our dependence on her presence. A rich, storied temporality was bestowed on the permanence of absence, as a routine of care, love, and feeding ensued, all mirroring life itself.

Introduction

The Hindu macrocosm consists of hundreds of castes, communities, and denominations, and the core ideologies of the faith ripple through its tribes and indigenous peoples. Hindu mortuary and funerary rites follow a general blueprint, but practices vary vastly within communities. On death, a body's mortal existence is generally concluded by cremation or burial – the latter seen more in South India. The subtle variations in mortuary preparation and funerary rites are explained as *sampradaya* or *paddathi*, loosely meaning praxis. *Sampradaya* is the ritual culture of a collective, like members of a community or caste in village, clan, or kindred.

Hindus collectively believe in the afterlife. On cremation or burial, the spirit, *preta* (often described as the hungry spirit), is disembodied from its mortal body and must eat to gain a new form. The messengers of death then escort it, transcending realms, on a long, uncertain journey to join his ancestors, *pitṛ* in *pitṛlok,* or *yamalok,* the realm of Yama, the God of Death.

This paper discusses the edible offerings made to the dead on their journey to afterlife and comestible transactions in Hindu death rites and rituals known as *antyeṣṭi*.[1] Food – feeding the spirit, ancestors, bereaved, and the community – is central to mourning in Hindu death rituals. In feeding the dead, albeit metaphorically, we continue to care for them in a way within our limits. Food in the context of *antyeṣṭi* is not a gastronomical endeavour. It has a higher purpose, focused on nourishing the spirit and creating a safe passage for its journey. In the phylogeny of the obsequy, we see that such a response assumes a ritual character.

The Grammar of a Funeral: *Kartā, Karm,* and *Kriyā*

Hindu liturgical practices use the allegory of Sanskrit grammar to explain funeral acts. The person performing funerary rites, or chief mourner, is called *kartā* (subject); the act of the funeral, *karm* (action); while rites and rituals are (*apara-*)*kriya* (passive voice).[2]

Death and Departing for the Journey
Edible Articles in Mortuary Rites

The very moment of death affects the departed, the bereaved, and the larger community differently. Death concerns the mortal body and the bereaved, not the dead itself. On death, or a moment before, many Hindus pour the holy water of the Ganges into the dying person's mouth as the last drink.[3] Many others pour ghee and honey into the mouth of the deceased to give the spirit strength for the journey ahead. Sugar, a tulsi leaf (*Ocimum sanctum*), and a piece of silver or gold are placed in the mouth of the corpse. Among the Kodavas of Coorg, tender coconut water is dripped into the mouth using a basil leaf or cloth.[4]

In Hindu homes, the moment of death is marked by the immediate suspension of the kitchen. The hearth or, as appropriate now, the kitchen is closed until cremation is complete. Only one fire is permitted to burn: the home kitchen is revived when the pyre's fire is extinguished. Food and feeding the soul, the bereaved, and the mourners takes centre stage at ephemeral venues beyond the walls of the home. Feeding and eating in the following days will be from contributed or catered dishes. Castes within communities ordained as funerary priests or traditional brahmin priests are called to initiate funeral rites.

Following customary mortuary rituals, the corpse is dressed, laid for a short wake, and the chief mourner starts funerary preparations. The bereaved sprinkle grains of raw rice, smear honey on the face, or place pieces of turmeric on the eyes according to their *sampradaya*. A bundle of rice and a coin tied to the shroud is the first of many provisions for the deceased. Many communities in south India make a hole on the shroud by the mouth. They fill it with raw rice grains, symbolically feeding the dead person the 'last meal'. The Badagas of Nilgiris place funeral baskets of puffed grains (amaranth, barley, rice) under the funeral cot to provision for the dead.[5] A small flame is kindled in the deceased's home and carried in a terracotta pot to start the funeral pyre.

In the landscape of this world live various mythical creatures, mythological beings, and other psychopomps of human imagination for whom hunger seems universal. Thus, the family offer foods to appease these spirits as the hearse proceeds to the crematorium. *Piṇḍa*, a tightly gathered rice ball with specks of black sesame seeds, honey, and ghee, is the essential edible artifact prepared and offered for funerary rites. These are offered as *pathipiṇḍa* or *pathi bali*, food to appease wayside spirits. In this manner, the first rice ball, *shavapiṇḍa*, is placed at the spot of death. The second *piṇḍa, paanthapiṇḍa,* appeases the spirits residing on the home's threshold, the spirit's metaphorical departure. The third and fourth *piṇḍa, khecharapiṇḍa* and *bhootapiṇḍa,* are for supernatural beings encountered en route to the crematorium. The fifth, *sadhakapiṇḍa,* is offered to the demigods guarding the crematorium.[6] Sometimes popped rice may be used instead of *piṇḍa*.

Cooking the Body

Cremation or burial is a crucial episode in the journey of the dead.[7] The liturgical verses describe the offering of the corpse as '*havis*' (lit. cooked grain, to the fire), exemplifying the meaning of *antyeṣṭi*: the final sacrifice. For the dead, it functions as a vehicle to the sublime plane of the afterlife. Some keep barley or rice flour *piṇḍa* on the forehead, face, shoulders, and chest of the dead body as food for the spirit leaving its mortal body.[8] The corpse's shroud embodies the deceased, metaphysically connecting the spirit to its *kartā* and the rites. Thus, a square fragment is torn and brought back for future rituals or, in some instances, the fragment is worn by the *kartā* himself, briefly personifying the dead. The *kartā* lights the funeral pyre, and the body is ceremonially cooked as an offering to the Gods.

The *kartā* circumambulates the body thrice with a clay pot filled with water. The pot, punctured with a stone, drips water around the pyre, symbolically the 'last drink before departure' in some communities, and interpreted as 'wearing of life' in others. The *kartā* throws the pot backward, shattering it as the skull will on cremation, and walks away, never looking back at it. Two stones gathered at the crematorium, or the stone used to puncture the pot, are brought home and these represent the deceased in the ensuing days. They are called *ashma* or *pashanam*.

When the crematorial fire cools, the *kartā* gathers bones in an urn using a branch of *Solanum virginianum* tied to his index finger with a black woollen rope. Various

Food for the Final Journey: Feeding the Soul in Hindu Death Rituals

ceremonies metaphorically offer food to the disembodied spirit to mark the end of cremation. Among the Khatri caste silk weavers of Kanchipuram, the family marks the completion of cremation on the third day by offering three wheat-flour *piṇḍa* mixed with honey and milk at the spot where the death occurred, where the bier rested, and where the corpse was burnt. Milk is poured over as if to feed at each pit stop.[9]

The *kartā* consecrates the urn in the sea or hangs it from a tree with a rope until mourning ends. Water and milk are given as drink offerings to this urn daily, the rope as its metaphorical ladder. Food, usually *piṇḍa*, or a simple meal of foods dear to the dead person are placed on a leaf under the pot.

Feeding the Spirit

To serve the disembodied spirit on its journey, Hindus first invoke it in artifacts: a small clay lamp that stays lit during mourning, a pot of water with a coconut placed on top, or *ashma*, the stone. In Odisha, the Kalinji caste picks a piece of bone after the cremation and buries it under a pipal tree (*Ficus meligiosa*) feeding the spot daily until the tenth day.[10] The bereaved offer food and sweets (or only sweets) every day in this corner. The foods offered are hyper-regional, and *sampradaya* dictates practices. In some communities, sattvic foods are prepared; in others, foods once loved by the departed are offered.

The bereaved mark a corner in the house with an image of the deceased. The *ashma* stone, as the true entity of the spirit, is installed by the home's entrance, *grihadwarakunda* (lit. a pit by the home's door). The geolocation of its departure, where ash is dissolved, is established with the other *ashma* stone as *naditheerakunda* (lit. a pit by the riverbank). In a unique dramatization of these landscapes, the *kartā* creates these vistas at home by

Figure 1. The twin ritual sites, grihadwarakunda *and* naditheerakunda, *established at the deceased's home representing the home and the point where ashes were immersed.*

constructing both pits with a raised clay wall. The first pit represents the *grihadwarakunda*, the other, *naditheerakunda* and *ashma*, wrapped with thread or darbha grass, the dead. It is placed in the pits, and food and drink are offered to them until the tenth day.

Drinking to Quench Thirst

Cremation is believed to create a great thirst, and hence offering water is the essence of all funerary rites. *Tilodakam*, a drink of water with sesame seeds, is poured over the *ashma* thrice daily, with an extra portion for each passing day, until the tenth day after death. *Vasodakam*, a drink of water, wrung through the shroud fragment folded in three, is provided to the *ashma* in the *grihadwarakunda*. The shroud, by virtue of its contact with the dead body, transcends the feeding to the spirit reliably.

Eating to Gain a Body

On death, a strange predicament presents – death causes the loss of the mortal body. So how might the spirit undertake the journey? The bereaved dedicate almost all the post-death rituals to the pursuit of feeding the spirit to gain a new body for afterlife.[11] Metaphysical intervention helps for such food to be offered by the bereaved and received by the departed. Edible objects are the singular material artifact carried forward in life after death. Of all our life-giving needs, hunger and thirst remain the only human needs bothering the living, the departed, and the ancestors.

Piṇḍa offers the perfect food packed for a journey. There are many regional variations of making *piṇḍa*, but grain (rice, barley, wheat, millet) is its most important constituent. The ingredients are put on a brass plate and mixed to make a homogenous mass, gathered by hand, and compacted to make a sphere, usually the size of a lemon. Recipes, essential in nature, reflect the connection to the land (grain) and the vernacular landscape. In non-brahmanical practices, the favourite foods that the departed person relished are mixed in cooked rice and then rolled into *piṇḍa*.[12] *Piṇḍa* is never eaten by the living.[13]

Figure 2. A piṇḍa *speckled with black sesame seeds.*

Food for the Final Journey: Feeding the Soul in Hindu Death Rituals

Various methods to make a *piṇḍa* are in practice. Some examples of *piṇḍa* recipes include: rice grits, ghee honey, and black sesame seeds; rice flour, water, ripe banana, jaggery, ghee, or black sesame seeds; cooked rice, yogurt, honey, ghee, and black sesame seeds; powdered barley and rice, black sesame seeds, ghee, and honey; and cooked rice, black sesame seeds, and occasionally fish.

The funerary *piṇḍa* is metaphorically divided into three parts. The spirit uses half a *piṇḍa* to recreate a portion of its new body, a quarter to pay its escorts to the other world, and another quarter to nourish its growing body. In *antyeṣṭi*, one *piṇḍa* is offered every day for ten days after death (two *piṇḍa*, a large one for a sumptuous lunch and a smaller one for supper, are provided in some *sampradaya*). The practice of providing *piṇḍa* to the deceased to attain a new body on the tenth day after death and the imagined development of the *preta*'s new body is strikingly similar to all Hindu *sampradaya*, in no way restricted to the brahminical castes.

The disembodied spirit resides in the *ashma*, close to its family. Feeding on *piṇḍa* daily and quenching its thirst, the spirit builds back a new body. The *piṇḍa* offered on the first day creates the head; while eyes, ears, and nose are formed from the *piṇḍa* provided on the second day; the neck, shoulders, arms, and chest on the third; the navel and private parts on the fourth day; while the thighs, calves, legs, and feet are formed from the *piṇḍa* fed on the fifth. On the sixth day, the *piṇḍa* nourishes the vital organs – the heart, liver, and kidneys; on the seventh, the nerves, veins, and nervous system are created; while the *piṇḍa* on the eighth day after death makes teeth and hair. On the ninth day, blood, fluid, and semen/ovum form, and with a *piṇḍa* on the tenth day, the body is complete, and thirst and hunger are born. Like a metonymy for the ten lunar months of gestation when a foetus is nourished by its mother, each *piṇḍa* serves to recreate a specific body part, mimicking the developments in-vitro. Feeding the spirit with *piṇḍa* gives it a new body called *piṇḍasharira* (*piṇḍa*-body), assumed to be ethereal and air-like.

The spirit still desires the living family with this new body, unable to transcend into the afterlife. It experiences a ravenous hunger, and the bereaved feed the spirit. Thus, on the tenth day, a special ceremony called *prabootha bali* is performed, facilitated by *antyeṣṭi* priests. The female relatives prepare foods dear to the spirit to appease its edacious hunger but without salt. They spread the shroud fragment and offer these foods to the spirit, arranging it in a human form. A copious volume of rice is cooked and provided in this cavern. A *piṇḍa* aptly called *prabhutabalipiṇḍa* is offered by the *kartā*. On being served its favourite foods without salt, the spirit perceives that lack as an act of neglect. Devasted, and in a fit of frustration and fury, the spirit decides to eventually leave his family for his last journey. In enacting this, the *ashma* stones from the two pits are also added to this cloth laden with food, gathered, and cast into the sea. This ritual marks the complete evolution of the spirit, ready for its journey into the afterlife.

Many communities perform a version of this feast at the end of the mourning period. Among the Kammas of Andhra and Tamil Nadu, who are agriculturists, the *kartā* makes an effigy representing the dead and invokes the spirit in three small stones tied by darbha. He pours water, and cooked rice with vegetables is offered as food. The chief mourner then submerges the effigy in water, marking the spirit's last meal and the start of his journey into the afterlife.[14]

Today, for the bereaved's convenience, cost, and practicality – or with little understanding of its ritualistic relevance – all ten *pinḍa* are offered on a single day, like the tenth or eleventh day.

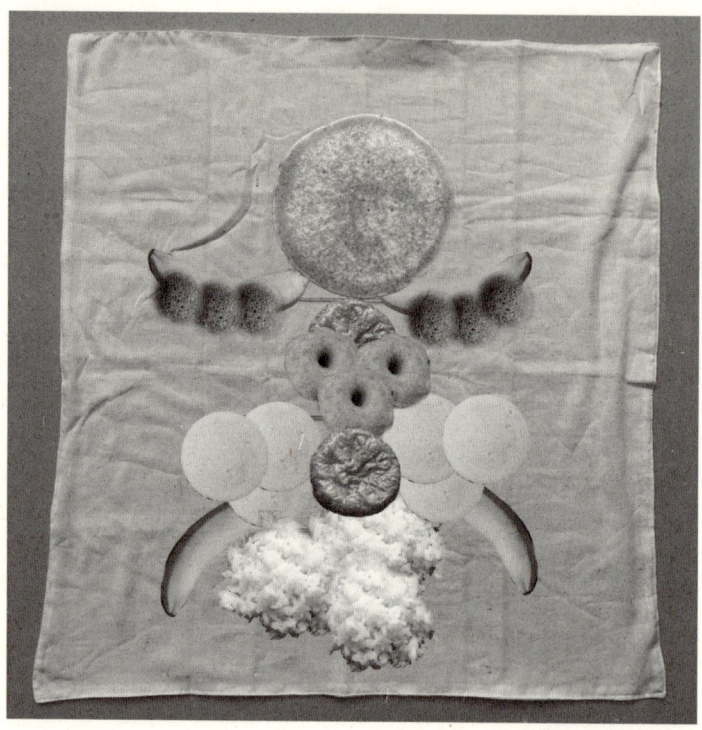

Figure 3. Prabhuta Bali *performed on the tenth day is a meal designed using the human body as a metaphor. Some of the essential preparations for this ritual include cooked rice,* adirasam *(anarsa),* appam, sugiyan, *and* vadai, *all archaic preparations with origins that can be traced back to the Vedic age.*

Eating for the Spirit, Eating for the God, and Eating Sin Itself: *Eka-uddishta śrāddha*

A widely observed *antyeṣṭi* ritual is the return of the full-bodied spirit to its family home on the following day, the eleventh day after death. The family serves *pinḍa* through a metaphysical medium (like water, fire, or crow), and essential articles for the spirit's last journey are donated to priests. The spirit's goal is to be transformed into a soul and be united with his ancestors, the *pitr*.

But among brahmanical castes, a death ritual priest is sought to embody the spirit.[15] On this day, this unassuming ball of cooked rice and sesame seeds becomes the spirit itself. The spirit's new body, *piṇḍaśarira*, is after all formed of the food (*piṇḍa*) he has eaten. The day's rituals begin with the making of thirty-two *piṇḍa*. The liturgical texts describe the spirit's sins and reflect on the need for repentance. The *piṇḍa* thus become the sins, and the priest is ordained to eat all thirty-two *piṇḍa*, consuming the sins of the deceased in a meal. He is given a grand donation (*dāna*) of grains on behalf of the spirit, benefacting it on its journey. Given the impractical gluttony of this meal, most *sampradayas* on this day offer *piṇḍa* into a sacrificial fire and/or to crows, which Hindus consider to be winged messengers of their ancestors. Donations (*dāna*) of articles denotative of the spirit's journey are made to the *mahabrahman* today. Footwear, a fan, pillows, a stick, a bed, grass mats, umbrella, new clothes, blankets, vessels and containers of brass, an iron box, containers, and foods like salt, black sesame, sesame oil, honey, paddy, and black lentils are among the gifts. The *kartā* gives the *mahabrahman* all ingredients for a meal. He cooks and consumes ritualistic foods in an alternate kitchen that mimics the 'real' and 'other' world. Impersonating the spirit, he eats the meal (cooked by himself) and leaves the house quietly, with no one seeing him. The *mahabrahman* thus feeds himself, eating for the spirit. The funeral priests accept food and gifts for their services, establishing a unique reverse transactional value for the act of eating – for one is paid to eat.

The twelfth day after death marks the beginning of the journey of the spirit. A feast is prepared for four brahmins, personifying the spirit (*preta*), ancestor (*pitr*), deities (*viswadevas*), and God, respectively. The brahmins briefly assuming the role of the deceased and his ancestors eat in complete silence, placing their left thumb pressing towards the ground during the meal. This way, they transmit the passage of food to their spiritual counterpart. When the meal is over, donations (*dāna*) are made to sixteen brahmins, each receiving one item. The brahmins once again become a transcendental medium accepting the gifts to benefit the spirit traveling in the afterworld. A stick of sugarcane and an oil lamp provided to the brahmin symbolize the boat needed to cross the Vaitarini river, the river of death.

The Spirit's Travel Itinerary

Once the funerary feast is served, the messengers of death guide the soul to Yama's abode, *yamalok*, crossing Vaitarini. The bereaved have nourished the spirit with food, drink, and care, and will continue to do so on his year-long journey. There are sixteen cities along this way – Yamya, Sauripura, Nagendrabhavana, Gandharva, Shailaagama, Kraunca, Kroorapura, Vicitrabhavana, Bahvapada, Duhkhada, Naanaakrandapura, Sutaptabhavana, Raudra, Payovarshana, Shitaadhya, and Bahubheeti are the various pit stops during the journey. Water and sixteen meals of *piṇḍa* are served through this journey, once every lunar month and some more on ordained days.

We Are What We Eat: Joining the Ancestors

In a more complex staging, not limited to upper castes, many Hindus perform a special ritual, *sapindikarana* (Sa.pindi.karana, lit. putting the *piṇḍa* together) that transforms the spirit into a soul, the *pitr*. The *kartā* unifies the spirit to his ancestors in a theatrical performance using *piṇḍa*. He arranges three *piṇḍa* to represent the deceased's parents, grandparents, and great-grandparents (ancestors), and a larger *piṇḍa* enacting the spirit. Now mashing the spirit's *piṇḍa* he merges it with those of the ancestors. The three *piṇḍa* now represent the deceased and his two immediate ancestors – father and grandfather (if they are dead), thus liberating the fourth ancestor.

Piṇḍa thus first serves as a visual metaphor for food for the dead person. *Piṇḍa* made on the eleventh and twelfth day are equivocal, representing the spirit, the ancestors, or the sins themselves. But *piṇḍa* served at ancestral rituals, *śrāddha*, are seen as mere food and offered to birds, typically crows.

Commensality in Eating Away from the Table

Traditionally, the role of the community is integral to mourning, evident in shouldering the corpse and sharing funerary activities. Family and neighbours feed the bereaved as death momentarily disrupts the order of the home kitchen and the lay of the family table. Fasting is forbidden, and commensalism is key to coping with loss and loneliness for the living.

The day of death is the first day of mourning. In some communities, the affinal family of the dead brings food on this day; close relatives and neighbours do so in others. Among the Izhava of Kerala, a rice *kanji* is served to the bereaved and mourners.[16] A sweet preparation, usually *payesh* (sweetened milk and rice porridge) is prepared and served to the bereaved and the guests, except in the case of untimely death. Among the Kalinji of Orissa, on the day after death, food, made bitter by the addition of *margosa* leaves (*Melia azadirachta*), is offered.

Multi-generational migration to urban areas and the plurality of denominations among Hindus have posed new challenges to the meaning of community. This has led to the emergence of various contractible services in the funerary rite ecosystem. Off-premise caterers feed the community by putting feasts on their plates to feed their mortal bellies. They are familiar with some of the dietary restrictions but not all ritualistic nuances. A good example is an act of frying in oil. Orthodox families refrain from frying in oil or blooming spices in hot oil (*tadka, talippu*) during mourning, but caterers are ignorant of this restriction. Feeding at the funeral can be expensive for the bereaved, and, in a jocular juxtaposition, feeding the community competes for resources with feeding the dead. The female relatives of the deceased perform a vital role in feeding the spirit, reflecting long gendered activities of cooking and caring. They make and offer *piṇḍa* and the saltless feast needed for *prabhuta bali* on the tenth day.

Other Edible Articles in Funerary Feasts

However, the core ritualistic meals for priests and priests personifying spirits have been maintained uncannily from antiquity, perhaps safeguarded by fear of the dead, their wrath, and the dietary code of the priests themselves. Traditional funeral meals thoughtfully exclude many ingredients and include many preparations to mark mourning, even in a moment of gustatory indulgence. The use of alliums and vegetables introduced in India after colonization, like chilli peppers, cabbage, cauliflower, carrots, and peas, are strictly avoided.

Cultivated and cooked foods include the wide varieties of gourds native to India like the cucumber, bitter gourd, and ash gourd. Grains of paddy; tender, furled banana leaves; young coconut, and its water; green bananas; and black sesame seeds are offered. Families typically include white flowers like *Agave amica, Jasminum multiflorum, Tabernaemontana divaricate*, and herbs like the holy basil (*Ocimum sanctum*), bermuda grass (*Cynodon dactylon*), and leaves of the Indian Bael (*Aegle marmelos*) in the offerings.

Plantain and green banana are the most important and, like *piṇḍa*, are deeply symbolic.[17] Turmeric, a symbol of the auspicious, is generally avoided for funerary occasions though *sampradaya* differs. However, turmeric is used for all the rituals if the deceased is a married woman whose husband is still alive. The use of multi-seeded fruits and vegetables, typically a symbol of life, is avoided. Ginger has great importance, and so do milk products.[18] In eastern India, unfired terracotta pots, tender banana leaves, or dried sal leaves (*Shorea robusta*) are used as receptacles for food and drink. The ephemerality of leaves and clay make them ideal foil for funeral foods.

Communal feasting marks the end of mourning, and in this grand culinary spread the bereaved rediscover conviviality. Thus, the expression of *antyeṣṭi* is as a mosaic of practices and not a set of monolithic rules, where foods can heal the living and the dead, and nourish their relationship across time and space.

Conclusion

Food for the travelling spirit is truly food away from the table of mortals. It is meaningful to the participants as a symbolic marking of death, making the social milieu of funerals both a moment of austerity and a feast. Using the metaphors of food, ingredients, and cooking techniques, familiar to people of all castes, communities, and socio-economic backgrounds, rituals carry meanings immanent to life and its portability in the afterlife. Through their funerary foods, Hindu communities have established an identity of their collective and self; with its ingredients they institute a connection to their natural world, an authenticity to their being.

Today *antyeṣṭi* rituals are on the brink of massive change only spurred further by the recent pandemic with its stringent rules around hospitalization and social distancing. Like other professions, many priests took funerals online or relayed on WhatsApp for a family's passive participation, all for a simple bank transfer.

For a faith that has survived millennia, the last few decades have seen rapid change, a loss of oral tradition and practice. Rituals are reduced to mere traditions, strictly followed for fear of the unknown, less often for love. For brevity, the fortnight-long ritual is done on the eleventh and twelfth day, or, pragmatically, all on the last day of mourning. But people have forgotten the spiritual idea of a journey, so bulk feeding is incoherent. Life's last journey is unlike any other. The start of the odyssey, like the destination, is hardly the point. It is how you get there that matters.

Tradition is not the worship of ashes but the preservation of fire.[19]

Notes

1. This paper focuses only on the culinary aspects of *antyesti* practices and hence by no means is a treatise on the ritualistic practices themselves.
2. Both *karm* and *kriyā* stem from the Sanskrit root *kri*, meaning 'to do', the latter specifically 'to offer a libation of water to the dead' (Monier Monier-Williams, *A Dictionary: English and Sanscrit* (London: W.H. Allen and Co, 1851), p. 301).
3. The Ganges has mythical purifying powers, and its waters salvage the spirit from sins. A small copper pot filled with water from the Ganges is packaged and sold in many towns along the river's course as a portable drink for the dying.
4. 'Chaavu-padithi', *Kodava Clan* <https://kodavaclan.com/kodaguheritage/chaavu-padathi/> [accessed 2 May 2022].
5. William A. Noble and Louisa B. Noble, 'Badaga Funeral Customs', *Anthropos* 60.1/6 (1965), 262–72.
6. Priya Mani, interview with Narasimha *ganapadigal* (Mumbai, May 2022).
7. Typically, the son of the dead person claims the right to perform the funeral and its associated rites. In the absence of a son, another male like a brother, father, son-in-law, nephew, or an adopted son may also be allowed to perform the rites. One may also ordain a relative or person as his or her *kartā*. In the absence of family, the dead can be offered a decent farewell by simply uttering '*Govinda*' thrice, generally called *Govinda kolli* in parlance. Women and daughters have largely remained absent in the actual act of cremation, although this is changing dramatically in recent years with many communities allowing women to perform final rites.
8. Among Kashmiri Pandits, three *pindas* of barley flour – the *bodha pinda*, the *makardhwaja pinda*, and the *Yamaduta pinda* – are offered to the deceased before cremation (S.S. Toshkhani, 'Kashmiri Pandits – Funerary and Post Funerary Rites', *Kashmiri Pandit Network*, 2023 <https://ikashmir.net/sstoshkhani/funerary.html> [accessed on 15 April 2022]).
9. Edgar Thurston and K. Rangachari, *Castes and Tribes of South India* (Madras: Government Press, 1909), III, p. 287.
10. Thurston and Rangachari, III, p. 52.
11. The mourning period varies for communities generally agreed to be based on the *varna* system: brahmins, ten days; traders and craftsmen, *vaishyas*, sixteen days; warrior castes, *kshatriyas*, twelve days; and other castes, up to one month, although this may be different in praxis. This period is followed by two days of concluding rituals and a final day of feasting to mark the end of mourning.
12. Thallapally Manohar and Pagidipalli Krishna, 'Role of Dependent Castes in the Conduct of Funeral Rites of Sudra and Ati-Sudra Communities in Telangana: A Cultural Study', *Kakatiya Journal of Historical Study*, 8.1 (2018), 61–80.
13. Except by the *mahabrahmanas* invited for the eleventh day rituals: see *Eka-uddishta śrāddha*, p. 9.
14. Kammakul Makkal Seyyum Eemachadangugal.
15. These priests are known as *mahabrahman* (Uttar Pradesh), *mahapatra / mohapatra* (Odisha), *agradani* (in Bengal), *sapindibrahmana, savandi kottan, ottan* (Tamil Nadu, Kerala); they are invited to eat a special feast on behalf of the spirit on the eleventh day.
16. Thurston and Rangachari, II, p. 418.
17. They form the leg in *prabhuta bali*, offered in lieu of meat and fish.
18. 'Serving ginger is akin to a thousand vegetables', goes the saying.
19. Attributed to Gustav Mahler.

Picnicking in Iran: From Cemeteries to Ski Slopes, Caves, Traffic Circles, and Beaches

Nader Mehravari

In every Persian Garden there is a takht, or mud platform, in the shade of the trees, and if possible, near the running water [...]. A pleasant trait about Persians is that all are free to come and picnic in these retreats, the owner apparently being flattered when parties of merrymakers invade his solitude.[1]

In addition to being great food lovers, people of Iran are admirers of the outdoors. They use any excuse – cultural events, certain religious occasions, holidays, social occasions – to get away from their homes and indoor spaces to enjoy food in an outdoor setting with friends, family, and even invite strangers in the vicinity of their picnic setting to join the festivities. References to Iranian picnic rituals fill Persian- and English-language literature, from nineteenth-century essays by European travellers to contemporary novels by Persian writers. Examples are sprinkled through this paper in italicized quotations.

It is a charming trait in Persia that anyone you meet understands the pleasures of a picnic and will make the best of all the trees and brooks and grassy places that they have.[2]

To Iranians, picnicking is both an art and a craft. The Iranian picnic has its own culture that has developed over the centuries – its own rituals, its own paraphernalia, its own choreography. The study of Iranian picnics requires one to consider many dimensions and viewpoints. Given the strong relationships and overlaps among the relevant dimensions, it is impossible to arrive at a unique grouping of such facets. The discussion in this paper is based on one such categorization as depicted in Figure 1 and outlined below:

- Social, Cultural and Anthropological Dimensions
    - **History** of Iranian picnics dating back to the ancient Persian empire
    - **Etymology** of words in the Persian language used to refer to picnics
    - Important **occasions** in Persianate societies where picnics are key elements
    - **Politics, regimes, wars, and religion** that have impacted Iranian picnicking rituals
    - References to Iranian picnics in Persian- and English-language **literature**
    - Picnics as a subject of classical Persian and contemporary Iranian **visual arts**

- Culinary and Logistical Dimensions
  - Typical **participants** in Iranian picnics
  - Familiar and unusual **sites** where Iranians hold their picnics
  - Key **paraphernalia**, equipment, and supplies taken to Iranians picnics, and how they are choreographed
  - **Food**, beverages, and culinary practices associated with Iranian picnics
  - **Activities** that take place during a typical Iranian picnic
- Philosophical Dimensions
  - The relationship between Iranian picnics and the **classical elements** of earth, water, air, and fire
  - Importance of all five **human sensory** organs in Iranian picnics
- The Unforeseen Dimension
  - A unique **unforeseen** peril for today's Iranian picnics

I will touch lightly upon some of these dimensions and go into more detail for others. The focus, however, will be on physical locations where the picnics are held, the paraphernalia, equipment, and supplies typically used to choreograph the picnic – and of course, the food which is the core of Iranian picnics.

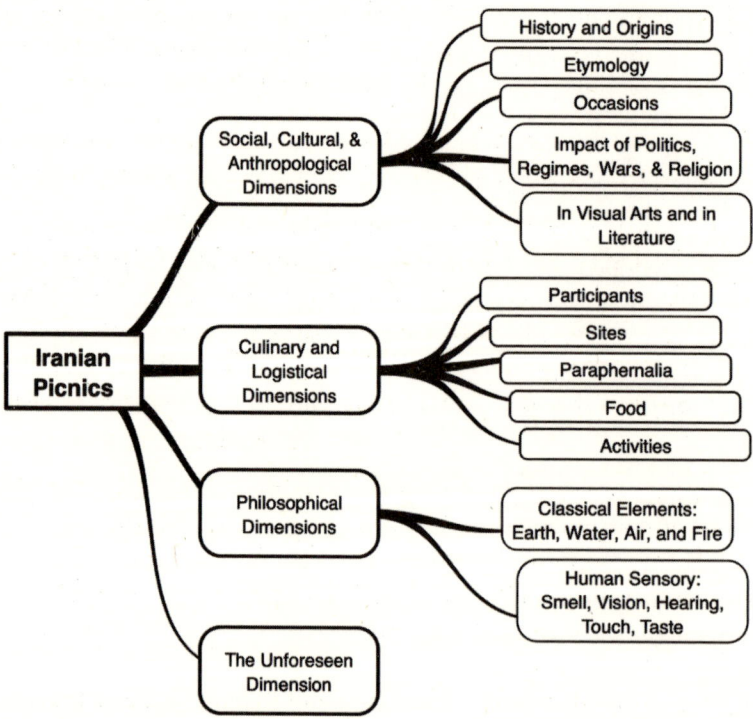

Figure 1. *Dimensions of Iranian Picnics.*

Picnicking in Iran

History and Origins

There are several secular and religious narratives about the origins of Iranian picnic rituals, all dating back to the first millennia BCE. Iranians' fascination with picnicking has roots in Zoroastrian beliefs as well as from the legendary Pishdadian King Jamshid of prehistoric Iran who celebrated outdoor festivals together with his people.[3]

One of the most well documented of these early picnics is grounded in the annual practice of *Sizdah-bedar* (Persian: سیزده بدر) which literally translates to Hi on the spring equinox. The literary masterpiece *Shahnameh*, the Persian Book of Kings, is the national epic of Greater Iran.[4] Written by the tenth-century Persian poet Abolghāsem Ferdowsi, it is of central importance to Persian culture. It narrates the mythology, legend, and history of ancient Persia from the creation of the world to the Muslim Arab conquest of Persia in the seventh century. It is stated in *Shahnameh* that the legendary King Jamshid, after having saved mankind from a severe winter, on the thirteenth day of spring advised his people that the weather was suitable and to spend a whole day happily in nature.[5] According to another legend, since thirteen is considered an unlucky number, people should leave their homes and spend that day outside for fear that something bad might happen to them inside.

There is a considerable amount of historical information – both textual and pictorial depictions – about the royal picnics of ancient Persian kings. This is because details of their activities – even the most minor ones – were well recorded. Examples of such records include the ancient expedition of Kāmbeez II (King of Persia 530-522 BCE – son of Cyrus the Great) to Ethiopia in search of the so-called Table-in-the-Sun; nineteenth-century picnics of Nasser al-Din Shah (King of Iran, Qajar dynasty, 1848-96 CE) which were more like outdoor banquets, often accompanied by hunting activities, where servants and cooks would go in advance to select and prepare a site and prepare the food; and modern era photographs of Mohammad Reza Pahlavi (Shah of Iran, 1941-1979 CE) and his second wife, princess Soraya, picnicking (Figure 2).[6]

Figure 2. Mohammad Reza Pahlavi (Shah of Iran, 1941-1979 CE) and his second wife, princess Soraya (1st from left), picnicking (photograph courtesy of Faryar Javaherian).

Etymology of Persian Words for Picnic

There are two different words for picnic in the Persian language – a classical (historic) word and a contemporary (popular, modern) word. The classical Persian word for picnic is 'تشگلگ' (Romanized: *golgasht*), which literally means recreating/spending time/enjoying/roaming in a flower garden.[7] This classical Persian word for picnic is rarely used in today's conversational language. It is this word, however, that points to the importance of gardens in Persian culture and to their relationship with picnicking in Iran.[8] Historically, Persian gardens are a representation of paradise which become a backdrop for picnics and associated feasts and activities.[9] The contemporary word for picnic in the Persian language is 'کین کیپ', which is simply pronounced *picnic*. It is a word that infiltrated into Persian from the French '*pique-nique*'.

Occasions

Although Iranians will use any excuse for a picnic, there are some key occasions which are the most popular. Without a doubt, the most important, the most historic, and the most popular of Iranian picnic occasions take place nationwide on the thirteenth day of the Persian New Year, as mentioned earlier. This day is referred to as the *Sizdah-bedar* which marks the end of the annual *Norooz* celebrations. *Norooz* (Persian: زورون), which literally means

Figure 3. Iranian picnicking on Sizdah-Bedar in Netherlands (photograph courtesy of Persian Dutch Network, CC BY-SA 3.0).

'new day', is the annual observance of the Persian New Year which begins on the spring equinox. Iranians begin preparing for *Norooz* several weeks before, and associated rituals and celebrations continue for two weeks afterwards, culminating with an epic picnic on *Sizdah-bedar*. It is such an important picnicking occasion that Iranians living in other parts of the world also picnic on this day (Figure 3). *Sizdah-bedar* picnics are all-day events preceded by days of preparation and cooking.

> *Everyone goes on the obligatory picnic ... everyone except the mullahs, that is, who'd just as soon do away with this pagan festival, which dates back over 3,000 years – well before the advent of Islam and its variants.*[10]

Aside from the *Sizdah-bedar*, there are other picnicking traditions driven by nature. In spring, when fruit trees begin to bloom (starting with almonds followed by citrus, cherry, pomegranates, stone fruit, and apples), Iranians like to picnic in the shade of these blooming trees and breathe in the aromas of the fragrant blossoms. In summer, when fruit trees are bearing fruit, Iranians have another seasonal excuse to linger in the shade of these trees and possibly participate in communal fruit picking activities. Most beautiful – and most delicious – of these is mulberry picking, when younger members of the group climb the huge mulberry trees, shake the branches, and encourage the super sweet mulberries to fall on sheets that have been laid on the ground underneath the tree.

In Iran, the workweek is from Saturday to Thursday in almost all public and governmental offices, and the one-day weekend is Friday – making Fridays, throughout the year, another very popular and desirable day for picnicking.

References to Picnics in Classical Persian and Contemporary Iranian Arts and Literature

Picnics have been a popular inspiration for the creation of exceptional works of art by Iranian visual artists – painters, photographers, mosaic artists, ceramic artists, and Persian miniaturists. There are well preserved artifacts, dating as far back as the fourteenth century, depicting picnics of royalty as well as common people. There are historic Persian visual art artifacts depicting Iranian picnics in permanent collections of museums around the world (Figures 4-8).

Figure 4 (top left). A depiction of friends picnicking by an Iranian contemporary painter (photograph courtesy of F. Javaherian). Figure 5 (top right). A depiction of a family picnic by an Iranian contemporary painter (photograph courtesy of F. Javaherian). Figure 6 (bottom). Seventeenth-century ceramic tile panel depicting picnic scene in Iran (photograph courtesy of Victoria & Albert Museum).

Figure 7 (left). Sixteenth-century Persian miniature painting depicting a royal picnic (photograph courtesy of Harvard University, Fine Arts Library). Figure 8 (right). black and white photograph depicting early nineteenth-century family picnic in Iran (photograph courtesy of Women's Worlds in Qajar Iran Collection, Harvard University).

In 2017, an extensive exhibition of such artifacts titled 'Iranian Picnic' (Persian: گلگشت ایرانی, Romanized: *Golgasht-é-Eerani*) was held in the Golestan Palace Museum in Tehran. The exhibition contained historic and contemporary visual arts showing Iranian picnics spanning a period of 3000 years. The site of the exhibition, Golestan Palace, is one of the oldest historic monuments in the city of Tehran and a UNESCO world heritage site. Two picnic-centric facts had contributed to the selection of the exhibition site. The name of the exhibition site, *golestan* (Persian: گلستان) which literally translates as 'flower garden', is directly related to the classic word for picnic in Persian language as was discussed earlier. Moreover, the museum already had a permanent collection of Persian miniatures depicting picnic scenes.

There are many occurrences of the classical word for picnic in Persian language 'گلگشت' (Romanized: *golgasht*) in Persian poetry dating as far back as 1100s.[11] These poets primarily used the word in the sense of its literal meaning for a most pleasant outdoor site for recreating, spending time, and roaming in a garden and often next to a stream:

'بده ساقی می باقی که در جنت نخواهی یافت / کنار آب رکن آباد و گلگشت مصلا'

O butler, bring what is left. Even in Paradise, you will not find the purity and freshness of this promenade.[12]

Familiar and Unusual Sites Where Iranians Hold Their Picnics

Selecting the site for picnics is not taken lightly by Iranians. Much thought and consideration go into the selection process – who is coming, who is not coming, the occasion, the season, the length of the outing, the predicted weather, the budget, the

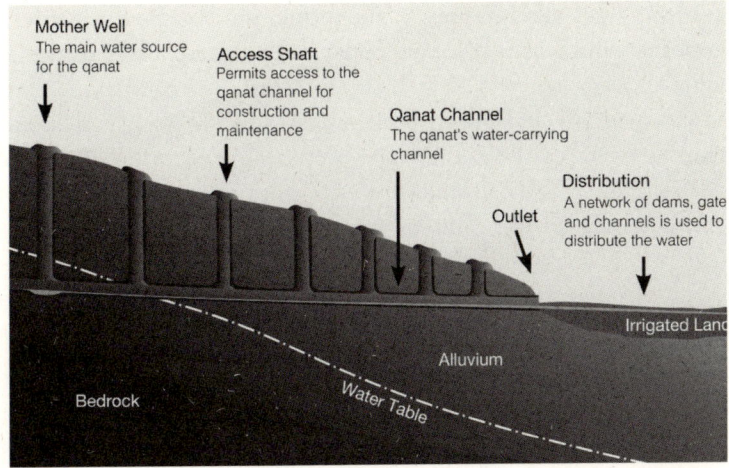

Figure 9. Cross-section of a typical Iranian qanat system (courtesy of Samuel Baily, CC BY 3.0).

distance to travel, the desired type of food, how much equipment to take, available transportation, etc. In fact, the selection process itself is a ritual of its own.

Iran is full of wonderful urban picnic areas, city parks, and roadside green spaces with their lawns, trees, and flowers faithfully taken care of by municipal gardeners and landscaping personnel. In the countryside, there are magnificent orchards, many open to the public for picnicking. There are also luscious oases in dry regions of the country. The existence of these green spaces is not only because of Iranians' love of outdoors and their obsession with picnicking but also due to ancient agricultural necessities.

On the Iranian Plateau, water is scarce and considered sacred. Sometime in the early first millennium BCE, the people of what would become Iran developed an ingenious method for transporting water over long distances in hot and dry climates without loss due to evaporation called *qhanāt* (Persian: قانات).[13] The underground structure of *qhanāts* is made up of a collection of vertical shafts, connected by a very low sloping underground tunnel which carries the water to the surface at a distance (Figure 9). Traditionally formal gardens and orchards were placed over *qhanāts* close to where the water would emerge from the underground system (Figure 10). The water would flow though the garden on its way to irrigate farms. Although the primary role

Figure 10. Shazdeh Garden outside the city of Kerman in Iran fed by water that comes from a qhanāt *originated 70 km away (photograph courtesy of S. H. Rashedi).*

of *qhanāts* is to supply drinking water for the people and irrigation water for agriculture, the resulting gardens and orchards became lovely oases for gatherings, recreating, roaming, and picnicking.

In addition to urban and rural green spaces and oases created around *qhanāts*, shores of natural rivers in the countryside are considered prime picnicking sites for Iranians – particularly in the summertime where the cool flowing water serves as a natural refrigerant. In fact, it is quite common for the picnickers to immerse beverage containers and whole watermelons in the flowing water to cool. The same is true not only for sites around natural rivers but also for areas around man-made water canals.

One of the most intriguing traditions of Iranians is to spread their picnic in cemeteries and the grounds of religious sites. One nineteenth-century visitor noted this long-standing practice: 'The shrine and tower are said to be 782 years old, and are kept in wonderful repair. And the place is often visited by Persians as an agreeable rendezvous

Figure 11 (top left). Picnicking at a cemetery (photograph courtesy of Zohreh Bayatrizi). Figure 12 (top right). Picnicking on the frozen grounds on the edge of ski slopes (photograph courtesy of Nader Mehravari). Figure 13 (bottom left). Picnicking in a deep river ravine where simple flat platforms are built into the side of the mountain (photograph courtesy Diego Delso, delso.photo, License CC-BY-SA). Figure 14 (bottom right). A family picnicking at a roadside truck stop (photograph courtesy of Lois Pryce).

for a picnic.'¹⁴ In such settings, picnics begin by with a prayer in remembrance for family members and friends who have passed away. It is then that eating and relaxing begins (Figure 11). Symbolically, eating nourishes not only the body but also the soul, and hence diminishes sorrow. Finally, food is offered to others who might be in the cemetery.

Frozen or muddy ground, lack of greenery, and other obstacles do not stop Iranians from having their picnics. Picnicking on frozen ground on the edges of ski slopes is quite popular in Iran where picnickers can rent a simple flat wooden platform to throw their picnic rug on, protecting them from the frozen ground (Figure 12). Wherever there are bodies of water, Iranians find a way to picnic. For years, entrepreneurs have constructed very simple flat platforms on the muddy grounds on the edges of rivers or in the deep rocky ravines to facilitate picnicking close to the water (Figure 13). When desperate, any out of the way piece of flat ground will become a picnicking site – traffic circles, caves, and even truck stops (Figure 14).

> *Or a picnic may be arranged for a moonlit night in the summer, when a cold supper or even a hot dinner may be served in a cave, of which there are many in the hills a few miles from the city.*¹⁵

Paraphernalia, Equipment, and Supplies Taken to Iranian Picnics

When it comes to paraphernalia taken to picnics, Iranians are not stingy. They often bring along a full set of equipment and supplies to provide day-long comfort, nourishment, and entertainment. Iranians are famous for not only fully loading their car trunks but also cleverly tying a slew of picnic necessaries to the roofs of their cars. Some of the key items are discussed below.

Carpet

The choreography of Iranian picnics begins with spreading a 'carpet' (rug, mat, blanket, tarp, a thick piece of cloth) on the ground (or on a simple wooden platform) – several of them if necessary. It could be as simple as a blanket or as precious as a handwoven Persian carpet. Iranian picnics don't exist without first laying out a carpet. The carpet represents the foundation on which the picnic is staged and where food and comfort are experienced. It softens the hardness of the ground. It is a representation of openness, hospitality, sharing, and reception. Since it is spread on the ground, it is also a sign of closeness of Iranians to nature.

> *It is considered improvident to travel without a carpet, which is an essential feature of a picnic, to lay the dust or cushion the roughness of the ground.*¹⁶

Sofreh

Before modern times, Persian society ate all their meals on the ground. *Sofreh* (Persian: سفره) is the term used to refer to a cloth on which the food is served on, and that people sit around to eat. It is equivalent to a tablecloth but can be spread on any surface

– not necessarily on a table. The word *sofreh* is both a noun and a verb. When used as a verb, it refers to the concept of setting-the-table, that is, getting ready for the food to be served. Typically, it is a thin (simple or luxurious hand-embroidered) square or rectangular piece of cloth dedicated for meals. Once the picnic carpet has been put on the ground, the *sofreh* is spread on the middle of the carpet. People can walk on the carpet but not on the *sofreh* – the only things that go on the *sofreh* are the food and the associated plates, bowls, glasses, and utensils. Normally, in one's home, a *sofreh* is spread right before eating and folded and taken away right after the meal is finished; *sofreh* is spread several times a day. For picnics, however, most families spread the *sofreh* once at the beginning of the picnic and let it stay out for the duration of the picnic. This is another indication of the centrality of food in Iranian picnics.

Cooking Gear

Although a good portion of food served at a typical Iranian picnic is cooked and prepared ahead of time at home, certain cooking gear is almost always brought along. Traditionally, the most popular of cooking gear is the Persian-style grill called *manghal* (Persian: منقل). *Manghals* are typically used to grill meat on skewers, and therefore, unlike western-style grills, do not have grates. In addition, there is also some type of heating gear to warm up prepared food or, more importantly, to boil water for making tea. These days, small portable propane tanks with a single burner on top serve this purpose. Such portable heating tools are now so popular now that a new word for it has infiltrated the contemporary Persian language: *gāz-é-picnicki* (Persian: گاز پیکنیکی) which literally translates into 'picnic gas'. Unlike western parks with picnic tables and elevated barbecue grills, regardless of whether it is a *manghal* or a modern-day portable propane-based burner, they are put on the ground where people squat down to use them – another sign that everything happens on the ground.

> 'Let's go for a walk in the woods', I suggested. 'What, a picnic?' she asked. 'No', I said, 'people go for walks here without taking half the kitchen with them'.[17]

Tea Making Provisions

Tea is the most popular and most consumed beverage among Iranians, so it is not surprising to see complete sets of tea making and tea drinking gear carried to picnics to make the gallons of tea that will be consumed. These could be as simple as a metal kettle to boil water and a ceramic teapot to brew the tea, all the way to a magnificent kerosene-based samovar that has been lugged into the picnic site even on the hottest days of the year.

> *At one time we lived in a house outside the city of Shiraz which had a lovely old garden [...]. It was a common thing for people to come and ask if they might spend the day in the garden, and we were only too glad to allow them to do so. They would come carrying a samovar and tea pot, a kalyan, a large bundle of lettuces and a bottle*

of sekunjabene, a syrup made of sugar, vinegar and mint; this they sprinkle over the lettuces, which they consider a necessary adjunct to a picnic. There was often a good deal of simple merriment among the people, but the great joy was to be sitting in a garden, and compared to the hot dusty, noisy road outside it was a veritable paradise.[18]

Comfort and Relaxation Necessities

Picnicking in Iran does not mean a lack of comfort. Although the carpet provides some level of softness, more comfort is often desired for the after-meal naps. It is not uncommon for picnickers to bring along a complete set of bedding including futon-type mattresses, large comfortable pillows, sheets, and blankets. For some – men and woman alike – full relaxation is not achieved without a few puffs from a hookah water pipe (Persian: قلیان, Romanized: *ghalyān*). Typically, hot coals from cooking *kabābs* over the *manghal* are used to put on the head of the hookah.

Other Equipment for Entertainment

Playing games is a common element of Iranian picnics – both those that can be played relatively quietly while relaxing on the carpet and those that require running around a field. Backgammon sets, playing cards, kites, a soccer ball or two, and small musical instruments are among the most common gaming paraphernalia that one finds used during Iranian picnics.

Food, Beverages, and Culinary Practices Associated with Iranian Picnics

The central focus of Iranian picnics is food that ranges from simple to elaborate. Picnic meals include a full range of food items: cold and hot; savoury and sweet; many prepared ahead of time at home, some bought on the way, and some cooked on-site, including snacks, starters, mains, desserts, and fruit. To illustrate the breadth of the food items that one might experience in an Iranian picnic, I have listed the most popular of such food items by categories:

- Snacks and starters:
  - *Ajeel* (mixtures of salted and unsalted nuts, seeds, and dried fruit)
  - *Kahoo-va-Sekanjebeen* (hearts of romaine lettuce dipped in a thick vinegar and mint-flavoured sweet syrup)
- Dishes prepared at home and served at room temperature:
  - *Māst-ó-Khiār* (yogurt, chopped cucumber, and mint)
  - *Kotlet* (pan-fried patties of ground lamb, mashed potatoes, and eggs)
  - *Kookoo-é-Sabzi* (chopped fresh herbs and egg frittata-like crustless-quiche-like dish)
  - Traditional Persian sandwiches using the Persian bâtard-like bread called *bolki*, stuffed with a variety of cold ingredients – the most famous being mortadella (*kālbās*) sandwich

- o *Salād-é-Olivieh* (one of the most popular salads with Iranian people comprised of a medley of chopped chicken, potatoes, eggs, and sour cucumber pickle mixed with a bit of mustard, some mayonnaise, and lots of olive oil)
- Dishes prepared at home and served warm (either kept warm during transport or warmed up at the picnic site):
  - o *Āsh* (an extensive class of thick porridge-like soup of chopped fresh herbs and a medley of legumes; in particular, *Āsh-é-Reshteh* which is a variety that has Persian wheat-based noodles)
  - o *Polow* or *Tahchin* (an extensive class of Persian steamed rice dishes where the rice is combined with other ingredients during the steaming phase.)
- Cooked onsite:
  - o Variety of traditional Persian grilled *kabābs*
- Accompaniments:
  - o Varieties of Persian flat bread such as *sangak, barbari, lavash*, or *taftoon* (in Iran, often purchased fresh from neighbourhood bakeries on the way to the picnic)
  - o Persian feta-like brined white cheese
  - o Plate of fresh herbs (mint, tarragon, watercress), scallions, and radishes
  - o Homemade pickles

Impact of Politics, Regimes, Wars, Religion, and the Risk of Unforeseen

Picnicking in Iran has successfully endured centuries of national upheavals and triumphs such as geographical expansions and shrinkages, winning and losing wars, economic expansions and stagnation, and personal freedoms and restrictions. In fact, picnics have always been a means to provide comfort during times of societal stress.

Figure 15. Contrasting images of Iranian picnics: 1976 on the left and 2019 on the right (left photograph © Bruno Barbey/Magnum Photos. Right photograph courtesy of Jan Cornall).

> *It [war] evokes an ominous sense of imminent danger, which makes the picnic all the more thrilling and exciting – just as during the war we would party every night because we did not know where or when the next bomb would fall in Tehran.*[19]

At the same time, however, there have been visible impacts on some aspects of Iranian picnics due to religious interpretations of the day. The most visible of such changes are due to the strict government-imposed social behaviours and codes of conduct since 1979, as contrasted in the two parts of Figure 15.

Picnicking for people of Iran has and continues to be more than just relaxing in nature, eating delicious food, and enjoying the company of family and friends. For some, the picnic is an analogy for returning to tribal and nomadic life. It is also a pursuit of a degree of freedom – regardless of how short-lived or risky that freedom might be. It is a freedom from the constraints of daily life, social and class barriers, religiously interpreted restrictions, guidelines of societal courtesies, and strict government-imposed social behaviours and codes of conduct – unless disrupted by an uncertain emergence of the moral police.

> *Picnics have an unforeseeable dimension in all cultures – rain might fall, wind might gather – but in today's Iran this is even more so, because the Basij, the militia that controls social behavior, might, like the lion in the savannah, spring up from any corner and disrupt the feast.*[20]

Closing

The act of picnicking is ingrained in the souls of Iranian people. It is one of the key characteristics of the Iranian people, one which has historically been, and which continues to be, independent of age, sex, income, position, religion, and ethnicity. It epitomizes the love Iranians have for nature, and some even believe it has been ingrained in Iranians from nomadic times. Iranian picnics and the accompanying feasts and activities are unique to the culture of Persianate societies. Iranian picnics celebrate long-held values ingrained in Persianate societies – friendship, family, nature, hospitality, and food. Eating in the open air is one of the Iranian customs, a tradition established thousands of years ago. In picnics, food becomes much more than the daily necessity of feeding the body. It becomes the central vehicle for enjoyment, relaxing, contemplating, and feeding one's soul.

Notes

1. Ella C. Sykes, *Persia and Its People* (London: Methuen & Co, 1910), p. 221.
2. Freya Stark, *The Valleys of the Assassins and Other Persian Travels* (London: J. Murray, 1934), p. 192.
3. Mahmoud Omidsalar, 'JAMŠID ii. In Persian Literature', in *Encyclopædia Iranica* (London: Routledge & Kegan Paul, 2008), vol. XIV, fasc. 5, pp. 522-28; an updated version is available online at <https://iranicaonline.org/articles/jamsid-ii> [accessed 3 May 2022].
4. Abolqasem Ferdowsi, *Shahnameh: The Persian Book of Kings*, trans. by Dick Davis (New York: Penguin Group, 2006).

5   Faryar Javaherian and Āzādah Shāhchirāghī, *Gulgasht-i Īrānī* (Tehran: Pizhūhishgāh-i Mīrās̱-i Farhangī va Gardishgarī, [2017 or 2018]).
6   Claudia Roden, *Everything Tastes Better Outdoors* (United States: Knopf, 1984); Faryar Javaherian, 'Picnic: Iranian Style', *Harvard Design Magazine*, 41 (2015), 118–19.
7   Alī Akbar Dihkhudā, Muḥammad Muʿīn, and Jaʿfar Shahīdī, *Lughat'nāmah* (Tehran: Dānishgāh-i Tihrān, 2000).
8   Faryar Javaherian, 'Iranian Picnic Culture and its Influence on Public Space Design and Social Culture', *Landscape Architecture Frontiers*, 4.6 (2016), 10–19.
9   Donald N. Wilber, *Persian Gardens and Garden Pavilions*, 2nd edn (Washington: Dumbarton Oaks, 1979); Faryar Javaherian, 'The Garden in the Carpet, the Carpet in the Garden', *2A Architecture and Art Magazine*, 47 (2022), 104–12.
10 Bernard Ollivier, *Walking to Samarkand: The Great Silk Road from Persia to Central Asia*, trans. by Dan Golembeski (United States: Skyhorse, 2020), p. 39.
11 *Ganjoor – Dar Dāneh-haye Adab-é Parsi* <https://ganjoor.net> [accessed 3 May 2022].
12 Shamsuddin Mohammad Hafiz Shirazi, *The Divan of Hafiz: Edition of Complete Poetry*, trans. by Henry Wilberforce Clarke (Dallas: Persian Learning Center, 2022), p. 8.
13 Edward Goldsmith, 'The Qanats of Iran', *Scientific American*, 218.4 (1968) 94–105.
14 Oliver B. St John and others, *Eastern Persia: An Account of the Journeys of the Persian Boundary Commission 1870-71-72*, Vol. 1 (London: MacMillan and Co, 1876), p. 161.
15 C. Colliver Rice, *Persian Women and Their Ways: The Experiences and Impressions of a Long Sojourn Amongst the Women of the Land of the Shah* (Philadelphia: J. B. Lippincott Company, 1923), p. 290.
16 Roger Stevens, *The Land of the Great Sophy* (Loindon: Methuen, 1962), p. 67.
17 Tara Kai, 'Mother Visit', in *A World Between: Poems, Short Stories, and Essays by Iranian Americans*, ed. by Persis M. Karim and Mohammad Mehdi Khorrami (New York: George Braziller, 1999), p. 210.
18 Rice, pp. 206-07.
19 Javaherian, 'Picnic: Iranian Style', p. 118.
20 Javaherian, 'Picnic: Iranian Style', p. 119.

Food on the Move: Commensality in an Indian Train Compartment

Shirin Mehrotra

Introduction

In one of the iconic scenes in English literature, Agatha Christie wrote:

> All around us are people, of all classes, of all nationalities, of all ages. For three days these people, these strangers to one another, are brought together. They sleep and eat under one roof, they cannot get away from each other. At the end of three days they part, they go their several ways, never, perhaps, to see each other again.[1]

The above quotation from 1934, to a certain extent, stands true for the Indian railways. In the complex rail network of the country where the longest train journey lasts close to eighty hours (a little over three days), the train compartment becomes a house on the move for many of its passengers. Being in such close proximity for a long duration makes way for the most intimate kind of social exchange, that of food. While anthropologist R.S. Khare believes that transit railway passengers eating together does not come under the purview of commensality, I continue to use the term in this paper to define the specific kind of dining inside the railway compartments since it sets the boundaries of inclusion and exclusion defined by social norms.[2]

Gender Roles

Growing up in the 1980s–1990s in the northern part of India, I vividly remember my train journeys that were defined by food. It was the time when not all trains had a pantry car, which meant carrying enough to last for dinner and breakfast next morning for the overnight journeys we took every summer to visit my grandparents. My mother would pack an extra bag dedicated to food for our family of five – *roti, pooris,* and *paneer-aloo ki subzi* (cottage cheese and potato curry) with pickle, neatly packed in multi-layer stainless steel tiffin boxes for dinner; bread, butter, and boiled eggs for breakfast the next morning along with biscuits and *dalmoth* (a spicy dry snack made with a mixture of fried lentils and tiny strings of gram flour) for evening tea.[3] There would also be whole tomatoes, cucumbers, boiled potatoes, and eggs in case someone fancied a sandwich; salt and pepper shakers, a knife, small stainless-steel plates, and spoons; and a cool keg to carry water. However, despite this perfect portable pantry that my mother would carry, there were regular snack breaks at various stations – samosas, *pakode, chana jor garam* (snack made with roasted and flattened Bengal gram), peanuts, and *ghughnee* (boiled chickpea tossed with onion, chillies and lime juice). Vendors would hop-on and hop-off at each station, waking up the sleepy travellers with their shrill cries.

A coupé in a sleeper (not air-conditioned) or three tier air-conditioned coach in an Indian train has eight sleeper berths, which means there are usually two families and a couple of solo travellers sharing this space. During meal times on such long journeys, the train compartment would turn into a shared dining area. In such scenarios the woman of the family would usually take care of unpacking, serving the food to everyone, collecting the dirty dishes, and washing them at basin located outside the compartment near the exit door of the coach. I have vivid memories of my mother pulling boxes of *paratha, subzi*, and pickles out of the bag she neatly arranged before leaving the house. She would spread out an old newspaper on the berth and lay out the food, serving each of us on little plates. She would then re-pack everything in the bag once we were done eating.

In sharing her experiences of travelling in a train – often from Chennai to Hyderabad, sometimes from Chennai to Delhi or Ahmedabad – freelance journalist Charukesi Ramadurai told me about the planning that would go behind the food that was carried along:

> During our annual trips to my grandparents' home in Hyderabad – it was usually my mother and I – we would always carry food from home like a meal for an overnight journey. Occasionally we would go on longer holidays with my parents or extended family. For instance if we would go to Delhi, which was more than a 30 hour trip, there were multiple meals packed carefully and labelled in large carriers. There was no disposable stuff back then. I am talking about 30–35 years ago, in the 1980s. So we would actually carry these stainless steel dabbas [boxes]. It was always my mother who would make the food, pack it and she was the one who knew which meal came first. It would start from planning; which food would spoil early, which would stay for a longer time. And it was up to the women, in this case my mother or aunt, to serve the food and wash the plates because as I said these were not disposable containers. And I remember seeing other women co-passengers doing the same. I was once travelling to Ahmedabad from Chennai [it is a 36 hour journey] and I saw a Gujarati family where women spread out containers, took a knife and started chopping onions etc. to make fresh *bhel puri* [a popular street-side snack in the western part of India]. They would literally cook, as much as one can cook on a moving train, and they would clean up and wash and put things away.

In an upper caste Hindu society, gender roles are defined. Historically men have been distant from household chores, especially those involving cooking and child care. As R.S. Khare writes, 'the central axis must always develop between the women of the

Food on the Move: Commensality in an Indian Train Compartment

house and the domestic hearth.'4 The rules of domesticity are replicated in an Indian train compartment. It is up to the women to plan, arrange, cook, and organize the meals for the commute – which means hours spent in a kitchen before embarking on the journey, while also taking care of serving and feeding the family and cleaning up after everyone has eaten. Men's responsibilities, on the other hand, lay outside of the hearth or kitchen duties, which translate in the railway compartment too. During our train rides it was always my father who would step out on the platform to buy food or fill water in the keg. As Pierre Bourdieu writes, 'man is the lamp of the outside, woman the lamp of the inside'.5 Like home, the railway compartment too gets divided into spaces – outside versus inside, men's space versus women's spaces.6

Food Politics in Indian Railways: A Short History

During the British rule Indian train compartments and their corresponding dining cars were segregated for first class, second class, and Indian passengers. While the first and second class were all about luxury, with silverware and multi-course dining menus, the one meant for the Indian passengers was a dilapidated room. In *The Great Railway Bazaar*, Paul Theroux describes it in much detail: 'the dining car, at the bottom rung of this Indian social ladder, was a narrow room of broken chairs and slopped-over tables'.7 And despite the intermingling of religions in the train compartments, Hindus and Muslims drank water from separate containers and bought tea from separate windows of the tea van at the railway platform. While the British officials portrayed this segregation as a respect to both the faiths, it was largely to appease the high caste Hindus who would not eat food touched by the Muslims and lower castes. This act of 'cultural accommodation' also ensured that the communal differences between the Hindus and the Muslims stayed intact to suit the divisive strategy of the British colonizers.8

However, with the railways opening up a new and expansive world for rural and small-town Indians, the Hindu code of consuming food outside of one's home began to dilute. As Lizzie Collingham writes, railways also became 'clandestine spaces for experimentation' especially for the upper caste Hindus.9 In his *The Great Indian Railways* Arup K. Chatterjee further elaborates that the railway platforms and compartments became spaces for forbidden dining, of forbidden hands and food items.10 Brahmins, who were 'pure vegetarians' at home made a beeline for the omelette and mutton chops served at the railway restaurant. The concept of caste was willingly shed at railway refreshment rooms while still devoutly maintained back home. For the diners there was no way to find out the caste or religion of the cook and the server and so it became convenient to overlook. The Hindu commensal rules changed based on place and context and the distance from home. With more and more Hindu men travelling outside of their village in search of work, the caste rules of commensality started to break down.

Caste in Contemporary India

In post-Partition India, the segregation between Hindu and Muslim water containers and tea stalls stopped. It was still the time when not all trains served food, and so people either relied on the food available at the platform or carried home-cooked meals along. Long journeys facilitated camaraderie between the co-passengers that often led to sharing of food. 'People were cooped up in that little box and for those few hours they were all together in that journey,' Ramadurai told me, talking about how barriers broke down between people of different communities, religions, and caste. But, the anxieties of caste and religion still simmered below the surface. Caste was inquired in indirect ways. In the northern part of India it is a common practice to ask a stranger their full name – first name and surname, with the latter identifying one's caste. Then, there are the direct questions.

I asked people on Twitter to share their experience of caste and religion while travelling on trains and a respondent who goes by @SereneinBoston replied to my enquiry saying:

> An A/C (air conditioned) coach attendant asked me 'aap bycaste kya hain. What's your caste?' [while I was] travelling from Patna, before offering to share his chai with me in an almost empty compartment once. The question has stayed with me for over 30 years.

Caste and religion are also assumed through various physical markers that separate high caste Hindus from the rest.

I interviewed Punita, a start-up founder living in Mumbai who shared about her travels as a kid:

> I used to travel every year with my mom from Mysore to Bombay in the 1980–1990s. We used to go from Mysore to Bangalore and then Bangalore to Bombay. There was once a Gujarati couple. They had two plates and they were eating in those plates. Back then there were no paper plates or disposable plates so they offered us food in the paper that it was wrapped in. My mom would immediately offer the food to the copassengers but they always refused saying 'nahi nahi hum chawal nahi khate. No no we don't eat rice' or some excuse like that. We would be travelling towards North India and usually there were very few South Indians in the train. Back then I did not know anything. My parents told me that we were tribal (adivasi) and that's all. And we were told that if someone asks your caste just say we don't have a caste. People would immediately assume that we were Christians because we would only wear dresses or frocks stitched by my grandmother. We would always wear shoes with socks. My mom was very particular about those things. South Indians also used to notice that we did not wear anklets which were common among the South Indian Hindus or that we did not wear a bindi, bangles or ash on our foreheads. So there were no markers. People would assume and sometimes ask if we were Christians. I would get very confused. And they would say 'ki nai aap Christian lagte ho. You look like Christian'.

Food on the Move: Commensality in an Indian Train Compartment

Cooked food in India creates social and cultural hierarchies, which then makes way for humiliation and discrimination.[11] In the Hindu caste practice, a person from an upper caste can offer food to the lower castes but is forbidden to receive food from the lower castes. In *Beef, Brahmins and Broken Men*, Dr B.R. Ambedkar writes about two kinds of food taboos of the Hindus, 'there is one taboo against meateating. It divides Hindus into vegetarians and flesh eaters. There is another taboo which is against beefeating'.[12] Vegetarian Hindus regard meat as rotting matter and equate this physical impurity to the moral impurity of the person consuming meat, and beef eaters are regarded as the most socially and morally corrupt among all meat-eaters.[13] This dividing line is what marks Touchables from Untouchables. And since a high population of the Untouchables (Dalits) has converted to Christianity, receiving food from them falls under the purview of taboo for the Hindus.

While caste and religion are easy to identify among fellow passengers – through inquiring surname, family history, or physical markers – what happens when food is being purchased from the platform or being served in the train? There are two theories. In the upper caste Hindu homes in North India (mainly Uttar Pradesh and nearby regions) there exists the concept of *kachcha* (uncooked or raw) and *pucca* (cooked) *khana* (food). *Kachcha khana* is food which is cooked in water, like *dal* (lentils), *khichdi* (one pot meal of lentils and rice), rice, and vegetables that are boiled in water. *Pucca khana*, by contrast, is where the food is cooked, fried, or deep-fried in ghee or oil. During festivals and rituals where brahmins are fed, the food that is served is what qualifies as *pucca khana* – *pooris*, fried or stir-fried vegetables, *halwa* (a sweet dish made of semolina), all cooked in ghee and thus purified – the only foods the brahmins deemed fit to eat at a lower caste home. The food served on the railway platforms is often of the latter kind – *pooris* with potato *subzi*, samosa, *kachauri* (deep-fried bread), *pakoda* (fritters) – making it pure enough for the upper caste Hindus to eat without the fear of pollution.

The second theory pertains to the caste of those cooking and serving food inside the train. A few years ago one of the vendors contracted by IRCTC – the railways ticketing and catering subsidiary – posted an advertisement in a newspaper calling applicants from the Agrawal Vaish community to apply for the posts of railway food plaza manager, train catering manager, base kitchen manager and store manager (all posts related to cooking and handling of food).[14] Agrawals are Baniyas – an upper caste who are traditionally employed as cooks in households as well as cooks across industries. While the ad was taken down after a public uproar, it is telling of the caste preferences in hiring behind closed doors.

Food and Identity Politics

In post-Partition India the hierarchies of food do not exist in the Indian railways the way they did during Colonial rule, but only on a surface level. The rules of dining inside a train compartment continue to be dictated by the upper caste Hindus. The minorities and the lower castes still cannot exercise their right of eating the food of their choice without the fear of being questioned, frowned upon, or – far less often but far worse – lynched.

Ibrar, a journalist from Kolkata, told me over a Zoom interview:

> My train journeys have been about food. And since I am a Muslim it's also about how to negotiate the space when there's a vegetarian sitting next to you. So I would always ask '*main non-veg kha raha hoon aapko problem toh nahi hai?*' (I am eating non-veg hope you don't have a problem). I would do that when I was travelling alone. I'd ask because I know my place as a Muslim…I don't know if it's the right thing to say. And I also know that I should be accommodative and respectful of someone else's dietary habits. This is the reason my family would not carry a lot of cooked food from Kolkata thinking what if the smell comes out and somebody has a problem. Kolkata has a lot of Marwaris (upper-caste Hindus who are traditionally vegetarians) so we'd always think that it might be an inconvenience if there are a few Marwaris in the compartment. Not because of the fear of fight, but more because why should we inconvenience someone. The brain works in a way that we think, what if someone is eating pork and we have a problem? Even though Muslims don't know what pork looks like [*toh koi kha bhi raha hai toh pata nahi chalega*] we won't know if someone's eating it. But non-veg is non-veg.

The majoritarian control over what one eats in public places upholds the myth about India that it is largely a vegetarian country. However, according to the data from National Family Health Survey (NFHS), close to 80% of Indian men and 70% of Indian women consume eggs, meat, fish – either one or all of these. According to a demographic analysis done in 2018 which looked at the decadal shift in food preferences in India, there hasn't been a significant change in vegetarianism; only 23–37% Indians are vegetarians.[15]

However, there has definitely been an increase in the assertiveness of vegetarianism among the upper caste Hindus. In November 2021 IRCTC announced that it will be serving only vegetarian meals on certain routes – those connecting to the Hindu religious sites – to promote vegetarian-friendly travel; a *sattvik* certification by an NGO called Sattvik Council of India has been proposed for the trains connecting Hindu religious destinations.[16] In Hindu scriptures a *sattvik* diet refers to meal that is devoid of any meat, onion, or garlic and follows the rules of the brahmanical ritual purity. Consumption of non-vegetarian foods has negative caste-linked associations (low intellectual ability, sexual desire, and tendency for violent behaviour) among upper caste Hindus, which is perpetuated in the Ayurveda texts too. In 2020, IRCTC tried to replace the dishes (egg curry, *appam*, etc.) on the trains in Kerala with vegetarian north Indian dishes like *chhole bhature* (deep-fried bread with chickpea curry) and *kachauri*.[17] A public uproar led to a change in this decision, but it goes to show the creation of a homogenous idea of a vegetarian India, which is not limited to just the railways. Currently twenty-three states of India have banned beef, fifteen states do not allow eggs in their mid-day meal programmes, and Air India – a government-run airline till recently – has largely turned vegetarian.[18] Being a government-owned operation that comes under the constitutional mandate, not allowing non-vegetarian food in trains, airlines, or mid-day meals is an act of discrimination.

This state-led discrimination, however, is not a recent development. The seed of creating vegetarian-only trains (with the larger agenda of creating a vegetarian Hindu nation) was sowed way back in 2001 in Gujarat when Narendra Modi – the current prime minister of India – became the chief minister of the state.

In *Pogrom in Gujarat: Hindu Nationalism and Anti-Muslim Violence in India*, Parvis Ghassem-Fachandi writes:

> From the moment he assumed political office in 2000, [...] Modi engaged in a micropolitics of ahimsa, which he deployed whenever it seemed convenient. One of his first promises as acting chief minister was to clamp down on illegal slaughterhouses, which in Ahmedabad were incidentally all owned by Muslims.[19]

Right after the Gujarat pogrom, the Modi government proposed to develop ahimsa tourism and to open a modern Ahimsa University. Ironically, the practice of vegetarianism in India is not connected with the idea of ahimsa or non-violence. In fact, it's rooted in religion and the ideas of purity and impurity.

Nascimento Pinto, a Mumbai-based journalist, told me over a chat:

> I was recently travelling with a Jain family whose daughter is a baker. Since her daughter is a baker, they were also talking about how their food is very 'pure' and 'clean' because they're from the community. I remember this other time, I was travelling, we got to know that the co-travellers were vegetarian and so that they don't feel uncomfortable (can't remember if they told us or we opted), we went to other seats where our family was and ate it there. We usually eat meat. Not beef. It was pork I think. Like *sorpotel* (a Goan Portuguese dish made with pork) or something.

These ideas are perpetuated by the Bhartiya Janta Party, the current ruling party, that made cow protection a part of their election manifesto making way for militant right-wing groups that attack Muslims and Dalits on the suspicion of selling, carrying or consuming beef. The cases of violence against those consuming or trading in beef have increased since 2014, the year Narendra Modi came to power. According to Human Rights Watch, a New York-based NGO, the cow vigilante groups have gained more power since the beginning of Modi's reign. At least 44 people were killed in the name of *gau raksha* (protection of cows) between 2015 and 2018, out of whom 36 were Muslim.[20]

In the past decade, the food landscape of the Indian Railways has changed vastly. Most long-distance trains now have a pantry car attached; trains like Rajdhani and Shatabdi Express include food – a snack and a meal – in the cost of ticket.[21] And those who neither carry home-cooked food nor want to eat the unappetizing meal catered by IRCTC can order food from a restaurant and get it delivered right to their seat at the station of their choice. The options have certainly made the train journey convenient for women; the availability of food means fewer hours spent labouring in the kitchen before the trip and time off from managing food and feeding everyone during the trip.

However, the identity politics and discrimination associated with food has only increased. In the past few years there has been a crackdown on meat shops and road-side stalls selling non-vegetarian food across the country, especially in Gujarat and Gurugram, stating that they hurt religious sentiments of vegetarian Hindus.[22] Incidentally, most of these shops are run by Muslims. In the Indian Railways, apart from demands for vegetarian only trains, there have been other instances too which showcase the intolerance of upper caste Hindus towards meat-eating communities. In 2018 a PIL (public interest litigation) was filed in the Gujarat High Court demanding a segregation of seats based on people's food preference.[23] The political climate inside the train compartment has started mirroring that of the country.

Notes

1. Agatha Christie, *Murder on the Orient Express* (New York: William Morrow, 2011 [1934]), p. 24.
2. R.S. Khare, *The Hindu Hearth and Home* (Durham, NC: Carolina Academic Press, 1976).
3. Tiffins are three or four round stainless-steel compartments stacked and sealed with a tight-fitting lid to avoid spillage.
4. Khare, p. 184.
5. Pierre Bourdieu, 'The Kabyle House or the World Reversed', in *Algeria 1960: The Disenchantment of the World, The Sense of Honour, The Kabyle House or the World Reversed*, trans. by Richard Nice (Cambridge: Cambridge University Press, 1979), pp. 133-153 (p. 133).
6. Sarovar Zaidi, 'Homing and Unhoming: Taxonomies of Living', *Chiragh Dilli*, 7 August 2020 <https://chiraghdilli.com/2020/08/07/homing-and-unhoming-taxonomies-of-living/#_edn1> [accessed 20 January 2022].
7. Paul Theroux, *The Great Railway Bazaar: By Train through Asia* (Boston: Houghton Miffin, 1975), p. 114.
8. Lizzie Collingham, *Curry: A Tale of Cooks and Conquerors* (Oxford: Oxford University Press, 2006), p. 201.
9. Collingham, p. 204.
10. Arup K. Chatterjee, *The Great Indian Railways: A Cultural Biography* (New Delhi: Bloomsbury, 2017).
11. Gopal Guru, 'Food as a Metaphor for Cultural Hierarchies', in *Knowledges Born in the Struggle*, ed. by Boaventura de Sousa Santos, Maria Paula Meneses (New York: Routledge, 2019), pp. 146-61.
12. B.R. Ambedkar, *Beef, Brahmins & Broken Men: An Annotated Critical Selection from The Untouchables*, ed. by Alex George and S. Anand (New York: Columbia University Press, 2020), p. 147.
13. Julia Twigg, 'Food for thought: Purity and Vegetarianism', *Religion*, 9.1 (1979), 13-35.
14. Anisha Dutta, 'Railways Caterer Sought Upper-Caste Candidates, Pulled Up by Government', *Hindustan Times*, 7 November 2019 <https://www.hindustantimes.com/india-news/railways-caterer-pulled-up-for-upper-caste-bias-in-recruitment-fires-hr-manager/story-iWduPYmgMLOXp5kBzRryIL.html> [accessed 26 May 2022].
15. Prannv Dhawan and Sabah Gurmat, 'As Higher-Caste Influential Hindus Turn Vegetarian, States, Govts Try To Enforce Vegetarianism', *Article 14*, 19 July 2021 <https://article-14.com/post/as-higher-caste-influential-hindus-turn-vegetarian-states-govts-try-to-enforce-vegetarianism--60f6447a8b2e7> [accessed 26 May 2022].
16. Devanjana Nag, 'Good News for Vegetarian passengers! Now Enjoy Fresh & Pure Veg Food in Indian Railways', *Financial Express*, 2021 <https://www.financialexpress.com/infrastructure/railways/good-news-

for-vegetarian-passengers-now-enjoy-fresh-pure-veg-food-in-indian-railways/2370939/> [accessed 27 May 2022]; 'Indian Railways Will Soon be Serving Only Vegetarian Food in Some Trains', *Mint* (14 November 2021 <https://www.livemint.com/news/india/indian-railways-will-soon-be-serving-only-vegetarian-food-in-some-trains-irctc-aims-for-sattvik-certificate-11636848545617.html> [accessed 26 May 2022].

17 Navya Singh, '"We Associate Our Identity with Food," Railway Restores Kerala Delicacies On IRCTC Menu After Massive Outrage', *The Logical Indian,* 23 January 2020 <https://thelogicalindian.com/exclusive/railways-food-menu-outrage-19412> [accessed 26 May 2022].

18 The Midday Meal Scheme is a school meal programme in India designed to better the nutritional standing of school-age children nationwide.

19 Parvis Ghassem-Fachandi, *Pogrom in Gujarat: Hindu Nationalism and Anti-Muslim Violence in India* (Princeton: Princeton University Press, 2012), p. 156.

20 'Violent Cow Protection in India', *Human Rights Watch,* 18 February 2019 <https://www.hrw.org/report/2019/02/19/violent-cow-protection-india/vigilante-groups-attack-minorities> [accessed 26 May 2023].

21 The Rajdhani Express is a series of passenger train service in India operated by Indian Railways connecting the national capital New Delhi with the capitals or largest city of various states. Shatabdi Express trains are a series of fast (called Superfast in India) passenger trains operated by Indian Railways to connect Metro cities with other Major cities. Shatabdi Express are day-trains and they return to the station of origin the same day.

22 Shuma Raha, 'The Culture War Against Meat-Eaters', *Deccan Herald,* 16 November 2021 <https://www.deccanherald.com/opinion/the-culture-war-against-meat-eaters-1051289.html> [accessed 27 May 2022]

23 Anwesha Madhukalya, 'Separate Seats for Vegetarians and Non-Vegetarians on Trains? Someone's Filed a PIL in Gujarat High Court', *Business Today,* 1 October 2018 <https://www.businesstoday.in/latest/economy-politics/story/separate-seats-vegetarians-non-vegetarians-on-trains-someone-filed-pil-gujarat-high-court-153507-2018-10-01> [accessed 26 May 2022].

An Army Marches on Its Stomach: Portable Foods in War and Peace

Johanna Mendelson Forman

The early twenty-first century has not been peaceful. From the wars in Afghanistan and Iraq, to the ongoing conflicts in Syria and Yemen, to the frozen conflicts in Africa, the common challenge in all of these crises has been access to food. For militaries engaged in fighting and for civilians who are in harm's way, how to get people fed is both a logistical and a humanitarian problem.

As I write this paper, we are witnessing the brutal fighting in Ukraine after the unprovoked invasion by Russia in February 2022. This conflict involves thousands of foreign troops and local forces that need adequate calories to fight. Civilians fleeing this conflict must have enough food and water to survive.

One thing that has not changed over time is the concept of portable food. Since organized warfare began, in addition to weapons, food has played a leading role in whether a military force will be able to prevail. Without the requisite calories, success in battle is jeopardized. Napoleon Bonaparte's oft repeated refrain that 'an army marches on its stomach' reflected this challenge. He demanded new technologies for preserving foods to make them portable.

The story of food portability in modern times is really a history of technologies that ensured that armies were well-fed when on the battlefield rather than left to forage for their dinner. Starting in the early nineteenth century, the continued efforts to transform our daily bread into a feast with a three-year shelf life has driven research and enhanced the notion of a portable meal.

Ironically, the invention of portable meals for the military has also become a part of more popular culture related to war – the predilection for portable foods by survivalists. Starting with the Cold War, when citizens filled their nuclear fall-out shelters with military meals, to more recent times, when people are preparing for increased natural disasters due to climate change, a growing industry of portable meals has become a feature of online advertising and YouTube videos.

This paper will explore the origins of portable meals for the military, examine the technological advances that have made it possible for longer field engagement in conflicts, and review the expansion of portable meals to the civilian population. Meals Ready-to-Eat (MREs) have become the gold standard of portable meals used by armies around the world, in part because of the investment that the US government made in the twentieth century to create foods that were both nutritious and easy to carry.

Today, food portability has expanded to the needs of civilians in conflict regions, with the creation of humanitarian rations to help offset starvation and malnutrition. It is an important component of the portable meal industry, but also a reflection of the ongoing global impact that migration from war and climate has on daily access to food. The war in Ukraine has brought renewed attention to the challenges that modern armies, even a great superpower like Russia, have in provisioning their troops.

I will touch upon this subject, albeit briefly, because of the anecdotal evidence that the failure to adequately supply the Russian army has contributed to its inability to achieve military objectives, despite its numbers and weaponry. Reverting to plunder and foraging, the Russian troops demonstrate how difficult logistics can stymie an army.

Finally, I will examine another dimension of portable meals – their attraction in popular food culture, where consuming meals over social media feeds has become a way to demonstrate what I would call 'battlefield terroir', with different nations' portable meals gaining a following in yet another dimension of our contemporary food-crazed culture.

The Origins of Portable Military Meals

Chinese warrior and philosopher Sun Tzu (500 BCE) advocated foraging for food or stealing it from an enemy as the way to feed his armies. Ancient Egyptians also encouraged foraging, even as they introduced a strategy of food storage depots along the campaign route. But foraging and plunder alone was not a sustainable means of providing warriors with adequate calories, even if they had enough weaponry.[1] It took the Romans to create a better organized supply and logistical system for their army. The Romans' success in imperial conquests is a result of a much more sophisticated way to provision their soldiers.

The Romans also discovered the importance of portability: 'The load carried by Roman legionaries when marching to battle has been estimated at about 20 kg. with soldiers expected to maintain a marching pace of about 4.5 kph for distances of about 30 km.' In addition to their military equipment, 'each Roman legionary would typically carry several days of food, as well as cooking equipment and a bladder of water', and also some salt, dried bread, wine, and salted meats. What is clear is that Roman armies ate a varied diet of proteins, legumes, and cheese. Grain was believed to make up the bulk of a Roman soldier's diet.[2] The typical Roman ration contained about 3400 kcl and weighed about 1.3 kg.[3]

Rome's technological contribution to portable meals was the addition of salting to preserve foods. This made it possible to have more variety. Garum, a fish paste that was used as a salting ingredient, was one source of preservation, but it is not clear whether this substance was carried with legionaries into battle.[4]

The Technologies of Portable Food

From the end of the Roman empire through the Middle Ages, 1500 years, there was no major advance in food technology that would allow for greater food portability.[5] It was not until the nineteenth century and the ascendance of Napoleon as emperor that the quest for greater technology to enhance portability emerged as an essential part of military strategy.

During the nineteenth century, militaries usually travelled with their supplies and cattle. In the American Civil War, for example, herds of beef cattle followed the army. Milk also made it to the battlefield, thanks to the invention of the condensed variety by Gail Borden in 1856. And desiccated potatoes and vegetables, a precursor to dehydration, made these items portable.[6] Coffee, an indispensable drink and stimulant, was given to Union soldiers just before a battle, as a way to get them ready to fight. The Confederacy was deprived of this drink because the Union blockade of the South made it difficult to procure supplies for soldiers.[7]

The twentieth century saw vast advances in food technology that allowed for different forms of portable foods. But it was not until World War II that the rations we know today were developed in laboratories dedicated to military use. The goal was to have foods that would meet emergency needs in the trenches (Trench Rations), leading eventually to the MREs that are now the gold standard for modern militaries – a product of large investments in the search for foods with a long shelf life.

The invention of retort pouches for packaging food was the most significant technological advance after canning. It allowed for a lightweight, easily transportable means of carrying food to the battlefield. The Natick food labs developed this invention in the 1970s. These pouches allowed for the mass production of MREs as they could keep food preserved for long periods of time, were easy to sterilize once filled with food, and, most important, took up little space in a knapsack.[8]

The MRE programme was a transitional moment in the history of military feeding. It eliminated the need for field kitchens in remote areas while also allowing a soldier to carry enough nutritious food to sustain fighting. MREs were highly portable. According to Cara Eisenpress, who researched this shift in feeding, 'its success rested on two facts: MREs weighed less and contained more variety and nutritional benefit than World War I's Trench Rations (bread and sardines), World War II's K-rations (canned meat, chocolate bars, chewing gum) and C-rations (bread and coffee), and Vietnam's Long-Range Patrol (freeze-dried chicken stew and escalloped potatoes)'.[9]

The road to MREs was prepared by a variety of portable rations developed for combat during World War II:

- C rations. These were intended for 3-21 days in the field, each containing 3700 calories. Each one had six 12 oz cans: three meat and three bread. Sugar, instant coffee, and candy were in each pack.
- D rations. These were for survival only. They contained 'unpalatable chocolate bars' in gas-proof wrappers that could withstand up to 120 degrees Fahrenheit.
- K rations. These were a 'lightweight' nutritional blitz ration for paratroopers, tank soldiers, and others who needed something more substantial than D rations or less bulky than C rations.[10]

The lessons learned from World War II, where the United States found itself feeding 11.4 million people in 1944 when the allies invaded Germany, provided expansion of the Combat Feeding Directorate, and ultimately resulted in a dramatic shift in the way militaries were fed in the twentieth and twenty-first centuries.

From the Korean War to Vietnam, military food continued to benefit from technological advances and improved logistical capacity. The old role of the Quartermaster Corps, the logistical and supply centre for any modern military, has expanded to a more efficient and interservice effort to deliver food to the field. While this paper focuses on portability, I would be remiss if I did not mention the importance of the quartermaster in history. In the United States the role of quartermaster started with the American Revolution when Congress authorized a quartermaster for the Continental Army. These soldiers become the chief procurement officers, responsible for everything from clothing to food to weapons supplies. Today, however, so much of what is done by the Quartermaster Corps revolves around contracts with private sector contractors that support the military food programme in war and peace.

During the Vietnam War, portable combat rations – Meal Combat, Individual, or MCI – were used as a balanced nutritional descendent of the C rations of World War II and Korea. This was an improvement that took advantage of new dehydration techniques. But the real revolution in portability was the Long Range Patrol ration, or LRP. Lightweight and compact, it had a flexible package because no cans were used. There were eight freeze-dried, precooked entrees that could be eaten by adding water or dry, like popcorn. A LRP also contained instant coffee and creamer, sugar, toilet paper, matches, and a plastic spoon. It was enough food for ten days.

During the Gulf War, Meals Ready-to-Eat were the food that transformed portability, not because of the variety of meals offered, but because of a new technology that changed the safety of food packaging – retort technology that used polymers filled with nanoscale particles to replace laminated retort pouches.[11] This was a major advance in the safety of packages as it avoided the dangers of consuming plastics when a meal was heated, or when food was stored in all-plastic containers. Thus, the dangers of contamination of foods by plastics was resolved through another technological advancement.

There will continue to be challenges to getting food into the field. But each new conflict will also benefit from the most advanced technologies in food preservation that continue to make meals portable, nutritious, and palatable.

Beans and Bullets: The Natick Food Labs and Combat Ready Foods

From World War II until today, Natick, Massachusetts, remains the place where the US government has invested the most in food research and development programmes. It has also been the source of contracts for feeding millions of people serving in the military. The largest food producers along with the government have one of the most lucrative long-term private-public partnerships in recent history.[12]

Academia also has played a supporting role. The learning arising from partnerships with universities has allowed the application of scientific techniques like freeze-drying and irradiation to support greater portability and food safety. The result has been thousands of inventions and patents that solidified the tie between the military, the Natick food centre, and private sector producers.

'When you go to war, you've got to bring your beans and your bullets. And those beans and bullets have to be shelf stable, high quality, and ready to go': here Anastacia Marx de Salcedo encapsulates the work of this operation-central of portable eating for the US military.[13] Marx de Salcedo's excellent volume about this subject is among the most comprehensive. I do not plan to repeat what her research says, but rather to address the object of her study: how the Natick Labs became the hub for military food portability.

The labs continue to research and implement a multifaceted approach to modern military food operations that provides nutritious, palatable, durable, and portable meals. Natick is also a technology hub for food preservation and innovation and has developed processes to allow food to go from US farms to the mountains of Afghanistan and to the space station.

Marx de Salcedo's research concludes that the connection between the quest for portable military food and our current love affair with processed junk food is an important by-product of the work done at the Natick food labs. The science of food preservation crossed civil-military boundaries.[14] The granola bars we buy for a quick energy fix, or artificially flavoured meat, or chips are really only one step removed from the MREs that sustain soldiers fighting in the field.

Portability Marches Forward

We know that modern military forces train to be lean, mean fighting machines. From the introduction of the MRE as a combat ration in 1975 to the present versions, the contents of these rations have evolved to reflect changing tastes, new food preservation technologies, and a greater awareness of sustainability and food waste. Biodegradable spoons were added in 1994. Starting with three menus for these meals, today there are twenty-four different choices, including the addition of a vegetarian meal. There are also kosher and halal meals that reflect the ethnic and religious composition of the armed forces and demonstrate respect for cultural food preferences.

For the wars of the twenty-first century, coupled with evolving strategies to use smaller groups of highly trained soldiers in combat, the need for portable, highly nutritious food continues to be of paramount concern. In an article about a visit to the Natick Labs' Combat Feeding Center, one food technologist described the mission of food optimization technology this way: 'There's a big emphasis on load. The less a soldier carries, the fewer opportunities his muscles have to collapse beneath the weight of his sack.' Thus, even today the goal of making highly nutritious foods as small and light as possible recalls centuries of war-fighting experience 'in which an army least burdened by either starvation or food could win the fight'. The development of the

MRE and ongoing refinements to it 'enable a soldier to survive with the lowest possible logistic burden. It is lightweight, heavy-duty, and requires no cooking, though soldiers can warm their entrée pouches with a water-activated magnesium heating packet'.[15]

First Strike Rations (FSRs) are a light version of the MRE that reflected a need to have food in an intense battle situation in the first 24 hours. They were introduced in 2007 as a practical response to what soldiers were doing with their MREs – stripping them down to the bare minimum contents to lighten the load in the field. Eventually, the FSR were created to address this experience. Once again, portability took precedence.[16]

Nutrition is another aspect of military food that remains a central part of food research. How do you get a fighting soldier enough calories to be sustained in the field? This has been an ongoing challenge, but the 1200-calorie MRE, and a larger packet First Strike Ration, which is 3700 calories, have been tested to allow a soldier the energy needed while in the field. The continuing refinements to the MRE enable a soldier to survive with the lowest possible logistic burden.

Finally, MREs are a form of soft power. It is often the case that meals are given away to civilians during military exercises abroad. A US military MRE is highly prized, not so much for its culinary value, but because it can be an additional source of food in many regions where militaries undertake training exercises abroad.[17]

Portable Meals of Other Militaries

These portable meals are a reflection of the broader foodways of the respective countries that produce them. They support the caloric needs of armies but also provide familiar tastes of home to soldiers serving in the field.[18]

We often speak about the importance of food as a means of transporting someone back to their home when they are travelling, or tasting the food of another cuisine as a means of introducing a person to different cultures. But what can be said of food that is produced in laboratories and so highly processed as to have no terroir, let alone anything other than chemicals that simulate flavours of popular foods? This is the challenge for MREs when efforts are made to have them represent present-day tastes.

Other nations that prepare MREs for their militaries are more sensitive to some of their cultural needs, if only because the contents of these packages are really used to recreate national taste preferences. Significant are the continued use of tinned meats in the French MREs, or the inclusion of cigarettes in some of the Asian packages.

While lots of thinking goes into the contents of international MREs, it is unclear whether those who consume their field rations are actually thinking about their family meals, or more concerned with merely their survival and quelling their hunger pangs.

Survivalist Eating

The demand for civilian portable meals began after the Vietnam War. Returning veterans who had eaten Long Range Patrol rations created a market for these products for use in recreational camping. Backpackers also learned about these easy-to-carry foods and spawned an industry for these products in the 1970s.[19]

The COVID-19 pandemic spawned a new demand for military-style portable meals. But even before this latest global plague, survivalists had purchased MREs to stock their bomb shelters. In early 2020, the market for these shelf-stable aluminium-pouched foods expanded when basics like flour and sugar were in short supply at supermarkets and people grew fearful of diminishing food supplies. A growing business in private vendors selling military meals made it profitable to search for unused MREs. It is technically illegal to sell MREs produced under a government contract, but internet auction sites are replete with boxes of these foods that are no longer being used.[20]

During the pandemic, many who bought MREs considered them a way to avoid possible exposure to COVID when shopping at grocery stores. These new consumers joined the ranks of individuals eating MREs as part of a culinary survivalist subculture that thrives on the exotic.

One thing that has become clear is that with natural disasters increasing due to climate change, stockpiling MREs will grow more common. In the United States, the Federal Emergency Management Agency (FEMA) is already purchasing meals from the same vendors developing food for combat. With natural disasters increasing, this trend will likely grow.

Humanitarian Crises

Portable foods have become a standard part of the humanitarian assistance packages that militaries use to support people in conflict zones. They are a spinoff of MREs and manufactured by the same companies that have contracts with the Department of Defense.

Since the beginning of the twenty-first century, humanitarian daily rations, or HDRs, have been used to support refugees and the internally displaced. They were designed to provide a full day's calories to moderately malnourished people who are often on the run or living in encampments. Like MREs, they are contained in aluminium pouches, and the nutritional contents of the foods are intended to address different needs.

The meal is designed as a complete day's supply of food; a minimum of two entrees is provided in each meal bag. Complementary components are included to provide the balance of the daily nutritional requirements that call for not less than 2200 calories. HDRs are always vegetarian to accommodate different cultural food preferences. Delivery of these emergency packets is usually done by helicopter drops, with foods attached to large palettes.[21]

Portable Food in Popular Culture and Public Art

For the modern age, portable military meals have become a form of cult popular culture. The proliferation of YouTube stories about eating military meals suggests that there is more to these aluminium packages than just food for survival. What explains this fascination with meals that are meant for survival in combat among the civilian population? Stories vary, but one common thread in the narrative of those who consume these meals is related to portability – ease of preparation and having them nearby.

Among the unusual participants in the MRE food subculture are those who make it their regular business to taste test these military meals from both the United States and other nations. Steven Thomas, a well-known taster with a regular YouTube channel, likes to think of opening MREs as a form of 'time-travel'. When you open a meal that is quite old – 50-100 years old – it provides a window into tastes of other times.[22] For those who eat older MREs, these packages contain a form of culinary archaeology.[23]

MRE packaging has also gained a place as public art.[24] During the Milan Food Expo in 2015, an exhibit of K-ration packaging was featured 'to demonstrate the value of essentiality as the key to dealing with precarious situations, using design to meet genuine and primary human needs in admirable fashion', according to the curator Giulio Iacchetti.[25] The display highlighted how the containers, from food that was consumed, were transformed into other objects that had aesthetic value.

MRE pouches also have found their way into the fashion world, with the food pouches converted into pocketbooks. The aluminium pouches have been used for practical applications as well, such as rain booties that Haitian children wore after the US military entered the country in 1993 to restore the legal government of President Jean Bertrand Aristide and again in 2010 after an earthquake. The aluminium pouches were also used in Haiti as wallets. The applications cited represent yet another twist on the notion of portability – shoes and handbags.[26]

The War in Ukraine

In ways unlike other military experiences, the Russian invasion of Ukraine has created a new focus on food portability and shelf life. Social media has exposed this challenge, with soldiers posting on Twitter and on videos how unprepared Russia was to provide for its fighters. Journalists report a massive failure to provision for thousands of soldiers in cold weather. Morale is always affected by the lack of food. Ukraine has presented the perfect storm for this type of neglect.

During the War in Afghanistan, the Soviet Union provided a very basic, awful-tasting combat ration called the 'Sukhoi Pak' (Dry Ration). Consisting entirely of preserved bread and canned foods, the three choices were: (1) a can of preserved meat, several hard crackers, a small can of cheese, a tea bag, and a large cube of sugar; (2) two cans of meat mixed with buckwheat groats and hard preserved bread; and (3) a can of meat stew and a second can of vegetables or fruit. This did not work out well. After that war, the Soviets developed an 'improved' ration called the Mountain Ration.[27]

Today, the Russian Federation uses a ration called the 24-Hour Individual Food Ration, or *Individualnovo Ratsiona Pitanee – Povsedyen* (IRP-P). It contains a day's worth of food (most of it canned) and a hexamine-based folding stove. Complaints from soldiers document Russian MREs that have expired years ago. And social media shows commissary trucks that accompanied the convoys lacked any real provisions except onions, potatoes, and porridge. Furthermore, Russian troops have had to contend with low supplies of their MREs. At the core of these problems is 'the military's lack of concern for the lives and well-being of its personnel'.[28]

Ukraine's security service claimed that Russian soldiers were eating dogs instead of the ration packs they were given. Soldiers are 'sick' of the ready-to-eat meals they've been provided, according to a 45-second conversation between a serviceman and his family. An audio recording of the call was uploaded on Twitter after it was captured by Ukraine's security services. 'Are you eating okay at least?' the soldier is asked. 'It's not that horrible', he says, adding, 'We had Alabay [a kind of sheepdog found in Central Asia] yesterday. We were looking for some meat'.[29]

Foraging to find food, and meat in particular, has underscored how the practices used by Russian soldiers in Ukraine today are more reminiscent of the methods of ancient armies than the practices of a modern military quartermaster system.

Some Final Thoughts and Conclusions

What does portability mean today? As we see fighting in Ukraine and fleeing refugees, we are confronted by the urgency of portable food for war-fighting and for refugee survival. Some things over the course of history never change. Armies need to eat. Civilians who are war's victims also require food to survive. The Romans were the first to recognize that an army did travel on its stomach, even if that famous quote was attributed to Napoleon centuries later. Portability made it possible for an army to march and to conquer efficiently.

Technology goes to food preservation but also to right-sizing meals so that they are light and easy to carry. What was created at the Natick Labs to provision soldiers has become some of the most popular civilian portable foods, from granola bars to protein drinks.

We are the beneficiaries of the portability science. But we are also witnesses to the challenges that humanity faces in light of the geopolitical crises of the world in which we live. Humanitarian rations reflect the terrible state of refugees and the displaced, not only from war but also from climate change. We also understand the importance of portability in emergencies arising from natural disasters. There still remains a need to have the basics of survival – food and water – to provide relief in these crises.

What does this all mean for the future? In the United States, Natick Labs and our civil-military complex will continue to improve the foods available for our armed forces. Similarly, the contracts that advance research and development of techniques that allow for greater portability, improved nutrition, and better taste will continue. We civilians will also benefit as the world continues to demand safe and portable snacks whose origins were not in the field, but in the laboratory.

Notes

1 Chris Forbes-Ewan, Terry Moon, and Roger Stanley, 'Past, Present and Future of Military Food Technology', *Journal of Food Science and Engineering*, 6 (2016), 308-15 (p. 309).
2 B.J. Whipp, S.A. Ward, and M.W.C. Hassall, 'Paleo-Bioenergetics: The Metabolic Rate of Marching

Roman Legionaries', *British Journal of Sports Medicine*, 32 (1998), 261-64, as cited in Forbes-Ewan, Moon, and Stanley, pp. 309, 310. See also R.W. Davis, 'The Roman Military Diet', *Britannia*, 2 (1971), 122-142.

3   J.P. Roth, *The Logistics of the Roman Army at War (264 BC-AD 235)* (Boston: Brill, 1999), as cited in Forbes-Ewan, Moon, and Stanley, p. 310.
4   See entries on Garum in *Wikipedia* and *Encyclopedia Britannica*.
5   Forbes-Ewan, Moon, and Stanley, p. 310.
6   Paul Dickson, 'Combat Food', *Oxford Encyclopedia of Food and Drink in America*, ed. by Andrew F. Smith, 2 vols. (Oxford: Oxford University Press, 2004), vol. 1, pp. 275-279 (p. 276).
7   Jon Grinspan, 'How Coffee Fueled the Civil War', *The New York Times*, 9 July 2014 <https://opinionator.blogs.nytimes.com/2014/07/09/how-coffee-fueled-the-civil-war> [accessed 15 May 2022].
8   AskUSDA, 'What Is a Retort Pouch?', *US Department of Agriculture*, 10 November 2022 <https://ask.usda.gov/s/article/What-is-a-retort-pouch> [accessed 23 May 2022]; skUSDA, 'What Is the History of Retort Pouches?', *US Department of Agriculture*, 10 November 2022 <https://ask.usda.gov/s/article/What-is-the-history-of-retort-pouches> [accessed 23 May 2022]. See also the overview of retort packaging in *Combat Index* <http://www.combatindex.com/> [accessed 23 May 2022].
9   Cara Eisenpress, 'Eat to Kill', *Gastronomica*, 14:2 (Summer 2014), 68-77 (p. 70) <https://www.jstor.org/stable/10.1525/gfc.2014.14.2> [accessed 19 May 2022].
10 Dickson, p. 278.
11 Anastacia Marx de Salcedo, *Combat-Ready Kitchen: How the U.S. Military Shapes the Way You Eat* (New York: Current, 2015), p. 166.
12 The Combat Capabilities Development Command (CCDC) Soldier Center was formerly the US Army Natick Soldier Research, Development and Engineering Center. CCDC SC is a military research complex and installation in Natick, Massachusetts, charged by the US Department of Defense with the research and development of food, clothing, shelters, airdrop systems, and other servicemember support items for the US military. The installation includes facilities from all the military services, not just the Army, and is so configured to allow cross-service cooperation with the many academic, industrial, and governmental institutions in the Greater Boston Area.
13 Marx de Salcedo, p. 12.
14 Marx de Salcedo, pp.172-73.
15 Eisenpress, pp. 76, 71.
16 'First Strike Ration', *MRE Info* <https://www.mreinfo.com/other-us-rations/current-us-rations/first-strike-ration/> [accessed 30 May 2022].
17 Marx de Salcedo describes the way US military troops on training exercises use their MREs to provide additional food to young kids in the field (p. 234).
18 Jeanne Kim, 'Soldiers Rations Kits Tell Us a Lot About Their Nations', *Quartz*, 22 January 2015 <https://qz.com/330590/a-look-inside-the-worlds-military-ration-kits/> [accessed 30 May 2022].
19 Dickson, p. 278.
20 Priya Krishna, 'Unloved by Generations of Soldiers, the M.R.E. Finds a Fan Base', *The New York Times*, 6 June 2021 <https://www.nytimes.com/2021/06/08/dining/mre-meals-read-to-eat-html> [accessed 23 May 2022].
21 'Humanitarian Daily Rations', *MRE Info* <https://www.mreinfo.com/other-us-rations/current-us-rations/humanitarian-daily-ration/> [accessed 30 May 2022].
22 Krishna.
23 Steve Thomas, *Steve1989MREInfo* <https://www.youtube.com/channel/UC2I6Et1JkidnnbWgJFiMeHA/about> [accessed 26 May 2022].
24 Andrea Borghini and Andrea Baldini, 'Cooking and Dining as Forms of Public Art', *Food, Culture and Society*, 25.2 (2022), 310-27 <https://doi.org/10.1080/15528014.2021.1890891>.
25 Domitilla Dardi 'Giulio Iacchetti, K-Ration Triennale, Milan', *Klat*, 5 February 2015 <https://www.klatmagazine.com/en/design-en/giulio-iacchetti-razione-k/36351> [accessed 31 May 2022].

26 The Smithsonian Museum has a collection of repurposed MRE containers found after Hurricane Maria: see search results for 'MRE', *Smithsonian Institution* <https://collections.si.edu/search/results.htm?q=MRE&fq=online_visual_material%3Atrue&gfq=CSILP_1> [accessed 2 May 2022].
27 'Russian IRP', *MRE Info* <https://www.mreinfo.com/international-rations/russian-irp/> [accessed 28 May 2022].
28 Dara Massicot, 'The Russian Military's People Problem', *Foreign Affairs*, 18 May 2022 <www.foreignaffairs.com/printnode/1228915> [accessed 21 May 2022].
29 Aparna Shandilya, 'Russian Troops Eating Dogs Instead of Rations Provided, Ukrainian Security Service Claims', *Republic World,* 31 March 2022 <https://www.republicworld.com/russia-ukrain-crisis> [accessed 23 May 2022].

Going to the *Gahambar*: Tradition and Invention in Parsi Community Feasts

Meher Mirza

Imagine a thrum of diners sitting at rows of trestle tables, each one scarved in plates of palm leaves. A pageantry of wait staff clomp down the rows in orderly fashion, spooning a roll call of Parsi dishes onto the leaves. Fish dunked in a vinegary sauce made with egg. Chicken under a thatch of *sali*, crackle-crisp shards of fried, salted potato. Bronzed *masala dal* eaten with mutton *pulao* rice. Stacks of *rotlis* cooked on a *lohri* over firewood. Sometimes there is *boi* fish, smeared with turmeric, chilli, salt, then shallow fried to gold. Sometimes, there is gently simmered chicken and potato. A babble of Gujarati unspools around you. The afternoon sun slants its rays down from an azure Mumbai sky.

This is a typical *gahambar* feast, the holy congregational feast of the Parsis, a miniscule Indian community with a peripatetic past. But in order to fully understand the shifting narrative of the Parsi *gahambar*, it is crucial to place the community within its unique, millennium-long history in India.[1]

Parsis: A Primer

Zoroastrianism likely dates to pre-651 CE in Iran and Central Asia. Zoroastrianism became the bedrock of three mighty empires – the Parthian, the Sassanian, and the Achaemenian, perhaps the first superpower the world has ever seen; at its height, its realms stretched from the Indus River to the Balkans in Europe.

But in the seventh century, an epoch ended. Repeated Arab invasions splintered an already-weakened realm, leading to the collapse of the empire in 641 CE. Waves of Zoroastrians took the sea route to India – the proximity of India to southern Iranian ports and centuries-old trade ties between the two peoples made it an obvious choice.

At the Gujarati port they landed at – Sanjan – they were allowed to practise their religion but not to proselytize. (This holds true to this day. Only Zoroastrians may enter their fire temples in India, and weddings may only take place after sunset). In return for shelter, the refugees forsook their arms, yielded their native Persian for the local Gujarati language, and dressed according to the Gujarati custom. These became the Parsis, a reference to their place of origin in the Pars, or Fars, region of Iran.[2] They began to function as agriculturalists, artisans, merchants, revenue collectors, and, from the seventeenth century onwards, traders and shipbuilders, occupations which held them in good stead under colonialism.

It was during the British raj that Parsis truly flourished. Many sallied forth to Mumbai, blazing trails in the fields of cotton and opium mercantilism, banking, and trade. They helped to build the infrastructure of the educational system; inspired political and social reform movements; built up its banking sector, insurance services, and the newspaper industry (the city's oldest newspaper begun in 1822, *Bombay Samachar,* was run by Parsis); and assisted in the pioneering of the industrial revolution in western India. There followed a cultural collusion where elite Indians constructed themselves within the image of the British and adopted an Anglophile ethos, one of several minorities at the time to do so. Undeniably, a certain amount of collaboration would have ensured a presence that would otherwise have been precluded to such a minority community.[3] The city would become (demographically, at the least) the centre of the Parsis.

No wonder then that the culture is a puzzle-board of syncretism; it is heavily indented with the passage of Parsi peregrinations. Parsi cuisine, for instance, is chequered with influences from Iran, Gujarat, Goa, Maharashtra, England, and even the Netherlands. The Parsi *garbo* is drawn from Gujarati folk music and dance traditions. Under colonialism, a few wealthy Parsis hired English governesses, who taught the children of the household to play the piano and the violin.

In an independent India, the imperatives of development and progress have caused the attrition of conventional occupations, with the old systems of land ownership, mercantilism, and political mobilization altering unrecognizably. This, to an extent, has led to the trope of the dying Parsi, in numbers if not also in cultural capital.[4]

The Food of the *Gahambar*

The Zoroastrian *gahambar* festivals (spelt variously as *gahanbars, ghambars,* etc.) are a throwback to a time when Iran and Central Asia were still roved by pastoralists; their roots lie in the agricultural tradition. The Maidyozarem Gahambar, the Maidyoshahem Gahambar, the Paitishahem Gahambar, the Ayathrem Gahambar, the Maidyarem Gahambar, and the Hamaspathmaidyem Gahambar — these were the holy days of obligation enjoined on the Zoroastrians by Prophet Zarathustra, celebrating the agricultural cycle as well as the stages of evolution (depending on which religious or academic authority you ask).[5] A *gahambar* may also be held in honour of the dead.[6] Four days of liturgy and prayer were followed by a congregational meal, the feast enriched by the participation of rich and poor alike. *Gahambars* were always free, and traditionally open only to Zoroastrian men; they were meant as affirmations of community and commensality, a means of charity.[7]

Here is the late Parsi priest, Ervad Soli Dastur, describing the *gahambars* of his village:

Other chefs were busy cooking *vaghaarela chaawal* [brown rice] for the *dhanshak,*

Gahambar: Tradition and Invention in Parsi Community Feasts

dhanshak dar [dal] with the freshly cut goat meat, *papeta-ma-gosh* [meat and potatoes], *ravo, malido* [Parsi sweets], *kachubar* [salad], and other savoury meals.

A specialist team then used two *choolaas* [stoves] to prepare lots and lots of *rotlis*. Perhaps the term 'production line' started with this team! What a pleasure it was to see them work in unison kneading the dough, making *rotli* balls, and rolling the *rotlis* with 'singing metal rolling pins' [hollow metal rolling pins filled with small stones, which when rolled make a pleasing sound. I still have my mom's singing rolling pin!]. An expert chef baked the rolled *rotlis* on a huge, round *lohri* [a slightly concave, circular metal utensil] 6 to 8 at a time, deftly switching them around its periphery, while cooking one in the centre of the *lohri* until it puffed up [*foolkaa*]. When done, he would neatly arrange them on a *saadri* [large rectangular woven mat from dried coconut palm fronds].

Helpers collected the cooked *rotlis* and stacked them on utensils. What a sight to see!

[…] After all dishes were cooked, it was time for the noon *Satum* prayer, which we all silently attended.

Afterwards, my father led us in the *Jamwaani Baj* prayer [a prayer made before a meal]. We then partook in the solemn *gahambar* feast, where we were served in *panghats* [group of people served at a time depending upon the seats available] on *paatraa* [dried leaves woven together with wooden pins]. Depending on the number of people to be served and the seats available, two or more *panghats* might be needed. The last *panghat* was reserved for the chefs – they were served by the children.

After everyone was served, the big job of cleaning started and everyone pitched in. Leftovers were distributed among families; the utensils were cleaned, dried, and placed into the *balad gaallis* [a cart pulled by two oxen]; and hauled away to be stored in the library go-down [storage room].

Our family was the last to leave. By then, it was time to perform the *Uziran Geh Bui* ceremony by one of our family *Mobeds* [priests]. After that, everything was locked up and we headed home. One more *gahambar* became history!⁸

The doyenne of Parsi cookery, chef and recipe writer Bhicoo Manekshaw, writes that '[e]ach person may contribute to the function whatever little he has, and rich and poor sit down to eat together. If a person is too poor to contribute anything, he can help manually in the preparation of the food'.⁹ Participants pitched in with whatever they could afford to contribute – rice, vegetables, lentils, meat. Those who couldn't afford anything, offered their time and labour; preparations for a *gahambar* could begin the night before since they were usually held at lunchtime. Everyone pitched in with the cleaning, and leftovers would be distributed to families.

Traditionally, the food served was unfussy, and mutton was once the normative meat (scriptures suggest the sacrifice of a goat was required). Indian liquors such as *mhowra* (drawn from the *mhowra* flower) or toddy, sap slicked from the palmyra tree and left to ferment until its flavours deepened, was also meant to be an intrinsic part of the meal. Manekshaw writes, 'The meal is simple – *masala ni dar ne chawal* (spiced dal and rice) served with *kachumber*. Sometimes, a generous person may contribute an additional dish, which is usually *papeta ma gosht* (meat cooked with potatoes)' – thus gesturing towards the staunch Indian heart of the meal.[10] The *dar* was the beating heart of a *gahambar* meal, but there were plenty of outliers. In Bharuch, a *gahambar* served only *mithai*. In Gholvad, a *gahambar* served *bhaji ma gos* (greens cooked with mutton). And in the Gujarati hamlet of Variav, an annual *gahambar* celebrates brave Parsi women who stood up to the exorbitant taxation policies of their king; it's said the only dish served every time was the bitter *vaal* (a type of pulse), a metaphor that elevated plain food to parable.

Gahambars and Charity

> At their solemn Festivals [...] each Man [...] brings with him his Victuals, which is equally distributed, and eat in common by all that are present. For they shew a firm Affection to all of their own Sentiments in Religion, assist the Poor, and are very ready to provide for the Sustenance and Comfort of such as want it. Their universal Kindness, either in implying such as are Needy and able to work, or bestowing a seasonable bounteous Charity to such as are Infirm and Miserable; leave no Man destitute of Relief, nor suffer a Beggar in their Tribe.
> — John Ovington, *A Voyage to Surat in the Year 1689*[11]

Zoroastrianism is striking in that it is a convivial faith, one that doesn't espouse arduous peregrinations as pilgrims, nor fasting nor deprivation of any sort. It emphasizes the 'pleasant duty to be as merry as possible, since in the doctrine joyfulness is a positive virtue, a weapon to defeat sorrow and care'.[12] And so via the twining of food and religion, the *gahambars* came to be seen as one of the central sites of Parsi 'authenticity', espousing the Parsi virtues of charity and social cohesion. In this way, these congregational feasts were central to the way in which the Parsis defined their identity – as a charitable, community-forward, and socially-adaptive people who nonetheless safeguarded an orthodox faith.[13] It helped articulate their *Parsipanu* – the we-consciousness of the Parsis. Various interviews I conducted (with Mumbai and diaspora Parsis) for this paper reiterated exactly this, emphasizing that this was a 'typical Parsi thing' to have a 'big, open feast for the good of the community', especially since 'so many of our traditions are tied to food'.[14] In keeping with this sentiment, moneyed Parsis began to provide for *gahambars* in their wills, both within India and to their co-religionists as far afield as Iran and China.

For a long time in the eighteenth and nineteenth centuries, Mumbai got the benefit of hefty trust funds towards *gahambars* by Mumbai's first baronet, Jamsetjee Jeejeebhoy, the enormously wealthy opium merchant-benefactor, who hoped to popularize them as a means to construct a shared group identity. In 1838, he is said to have donated Rs 1,68,500 (approximately £1700) to turn them into annual features. His charity later extended to the towns and villages in Gujarat in which his co-religionists were found in fairly large numbers, i.e. Udwada, Valsad, Navsari, Bharuch, Billimora, Chikli, and Gundevi, amongst others.[15] There was purportedly a *gahambar* in which both Parsis and those of other communities received gifts of cash and clothes from Avabai, Jamsetjee Jeejeebhoy's wife. After the humans, even the stray dogs were fed *khichdi* and ghee!

'Decline' of the Urban *Gahambar*

Yet, towards the tail end of the nineteenth century and onwards, the community and, subsequently, the *gahambar*, began to change. Many Parsis were involved in the food and hospitality business at this time, and consequently the community began to experiment with a number of reforms in the fields of dress, dining, and domesticity. Women's education was lofted by reformers, and as a result, in the 1860s, ladies gradually began to attend family meals with the men and children, anathema up until then.[16] In 1859, the 'young class' began to accept the foreign ways of the chairs and tables, the glasses and the plates, the knife and the fork, and this was reflected in the *gahambars* as well. In 1916 and 1922, a fund was begun for a women's *gahambar* in Mumbai.

Parsi prosperity vaulted thanks to burgeoning business opportunities, and more Anglicized, wealthy Parsis began to thicken the exclusive parts of the city. Naturally, class striations began to appear. Consequently, *gahambars* gradually devolved to becoming mere social dinners with people of the same income level.[17] For instance, wealthy Parsi cloth merchants in Mumbai held their own *gahambar*. In 1892, some Parsis set up a *gahambar* fund for Zoroastrian traders in Hong Kong and Guangdong.[18] And ever the diplomats (as an economically prominent but numerically insignificant group), sometimes prayers were proffered for imperial-driven objectives, such as by the Parsis of Solapur who invoked Divine Aid on behalf of British arms in Afghanistan.[19]

These became 'food-forward' functions, a religiously-sanctioned means of showcasing the wealth of more privileged Parsis who had now begun to occupy important public positions in the city. The much-vaunted simplicity of the Mumbai *gahambar* had begun to dwindle.

Fractious letters flew in to newspapers, fanged with outrage at what the letter-writers believed to be wanton excesses in the *gahambars*, fearing that privilege, coming in the wake of immense successes under colonial rule, had unmoored Parsi merchants from the pastoral and religious origins of the *gahambars*. Certain influential Parsis believed that extravagance in such feasts enervated the community's morals, as well as draining resources that might be put to better use elsewhere: this to the extent that limits were put into place on how much food and how many guests could be served at functions. There was some fear that Parsis, especially those of considerable fortune, would begin to think themselves beyond the bounds of conventions since they were free from the tight social patterns of the village.[20]

One such letter was published in the Bombay *Times* in 1844, by the intriguingly named Q (later revealed to be Maneckjee Cursetjee, Parsi reformer and educationist):

> The Parsees have been accustomed to vie with each other without consulting their means, to keep up appearances among the caste – by giving feasts, and incurring other extravagant and unnecessary expenses, for the mere sake of propitiating that monster – Custom, and to be thought great; and many of the most wealthy and distinguished families among them, by such displays, or rather by committing such folly, have dissipated their fortunes, and the poor by adopting similar modes of proceeding have added debt to debt, and at last become so embarrassed as to end their days in absolute penury.[21]

John Hinnells, scholar of Zoroastrianism, wrote of this divide too, explaining that by the early twentieth century, the urbane, urban Parsi had lost most of his connections with the agricultural roots of their religion. At least in Mumbai, the *gahambar* had devolved into a largely social, secular custom, up until the 1939-45 war when all public banquets were entirely proscribed by the Government as a measure of economy. The custom was revived in the 1970s, with full participation of women.[22]

Contemporary *Gahambars*

Lavish Mumbai *gahambars* continue to the present day. In some cases, guests are even required to purchase tickets (a limited number of free passes may be kept aside for the economically underprivileged). In 2014, the writer Bachi Karkaria described a Kolkata *gahambar* in which 'curried mutton and *masala dal-chawal* [was served by Parsi volunteers] to the seated rows of partakers – including the Hindu and Muslim workers'.[23] Professional caterers now provide viands that rival wedding meals. At a Navroz *gahambar* I attended a few years ago, tables were groaning with the weight of catering doyenne Tanaz Godiwalla's spread – *dabba gosht*, great scoops of mutton *pulao dar, sali marghi,* foil chicken, *akuri,* custard, even ice cream and chocolate lava cake.

In the wake of such opulent meals follow the jeremiads bridling at the superficiality of a contemporary generation. 'Today, many people have the misguided thought that we gathered in these bygone *Gahambars* just to eat and drink,' wrote Ervad Dastur. '[...] Sure we had big feasts, but there were also prayers [...].'[24]

Elsewhere, the priest of a prominent fire temple observed, 'they go to eat and make merry, but they have forgotten all rituals and customs.'[25] My interviews reflected this too, with several members of an older demographic (those over 55) lamenting the loss of piety, religion, and a 'community feeling'.

Some of these points of view remain anchored in an assumed prelapsarian Zoroastrian epoch, one that was premised on the construct of a utopic past; they reflect certain Parsi preoccupations with the decline of the community both demographic and economic.

However, rural *gahambars* and those celebrated in Parsi colonies (enclaves) refute this charge of Parsi ossification into elitist enterprises; the performance of community and commensality, prescribed as crucial elements of a traditional *gahambar*, were still privileged by the residents.

A Navsari-dwelling respondent who I interviewed for this paper described regular afternoon *gahambars* in Navsari where someone would personally go round to the *baugs*, i.e. Parsi enclaves, inviting the men to attend (a *sheri* or women's *gahambar* was held separately). The meal was thinned to simplicity itself – rice, dal, and *kachumber*, and leftovers were packed away for the infirm and elderly who were unable to attend.²⁶ At least this resident had never contributed monetarily.

Another village resident, Mrs O, offered reminiscences that reiterated the pastoral and religious bedrock of the meals: 'The Borkharivalas from Malesar used to hold a *gahambar* where *vaal* would be served, since it was the time of the new crop. A *jashan* was done in the evening, and people were given *vaal ne rotli* [*vaal* and roti] [...]. As funds increased, a side-dish was sometimes served, or *vaal ne khariya* [*vaal* and trotters].'²⁷

Diasporic Repositionings

Some of the largest changes have come through Parsis outside India, many of whom grapple with which community boundaries to preserve, both in principle and practicality. Most diasporic Parsi-Zoroastrians live in cities. This makes the celebration of a community festivity an even greater expression of collective religious practice and repossession of the past. The *gahambars* therefore become the most prevalent means of bringing adherents – who may be scattered across a wide area – together.

For a tiny minority that has been doubly displaced, T.M. Luhrmann contends that, '[t]here seems to be an acute awareness that if Zoroastrians do not forge some new sense of commonality, they will literally disappear'.²⁸ Through the *gahambars*, therefore, there is the sculpting of a new community consciousness and a new community conscience.

Like in all communities, there are orthodox and progressive members, but there appears to be an elasticity to the mingling of cultures. For instance, the old forms of the *gahambar* are sometimes leashed to that of North American Thanksgivings. Zoroastrian children may carve pumpkins in autumn. The Zoroastrians of Montgomery County in Maryland, USA, held a *gahambar* in solidarity with Christchurch (New Zealand's deadly shooting).²⁹ There are even *gahambars* purporting to be environmentally-friendly. At one such, the meal included *lagan nu achar* (dried fruit pickle), sago crisps, Parsi stew with roti, egg curry rice and *kachumber* – entirely vegetarian (if you eat egg) and with the addition of *gulab jamun* (a most un-Parsi sweet).

The festivals have shifted shape in other ways too. Both in Mumbai and in the diaspora, some programmes are fringed by song and dance recitals, comedy skits, Parsi *naataks* (plays in Gujarati), and even Bingo games. Religion is sometimes mingled with business; in 2019 in Dubai, a *gahambar* showcased the work of local Zoroastrian entrepreneurs, selling jams, biscuits, chocolates, and Parsi *vasanu*, *malido* and *badam pak* (sweet-savoury fudges).

And in the East Asian communities of Hong Kong, Macao, Guangdong, and elsewhere, the Parsi-only rule of attendance is waived by re-naming memorial *gahambars* as 'Dinners in Memory Of' so that non-Parsi spouses may gather as part of the congregation (the liturgies are performed earlier to a Zoroastrian-only audience).

Elsewhere, no re-christening is deemed necessary. Feroza Mistry Nusbaum, Secretary, ZAPANJ Board of Trustees, wrote the following in an email interview with me:

> *Gahambars*, like all of our gatherings and functions, are open to all individuals who would like to participate. We have mostly Zoroastrians who attend as well as many non-Parsi and non-Zoroastrian spouses who attend with their children as well. My spouse, for example, is not Zoroastrian, and he has been welcomed by the community and is well loved, as are our children. We also occasionally have children or families who bring their neighbours or close friends who are not Zoroastrian to an event from time to time and I think this is very beautiful to include others and help them learn about Zoroastrians, who we are, and what is important to us. I think there have been times in my childhood when a particular *dasturji* (priest) may not feel comfortable doing a *jashan* when there was a non-Zoroastrian person there, but I do not see this occurring much in our community any longer. My feeling is that to do so would be to move backwards, to exclude and would encourage many people away from Zoroastrianism and the community – which is the opposite of what we are trying to do.[30]

Even amidst this pluriformity, diaspora Parsis adhere to the ontological barebones of the *gahambars*. Fresh generations are introduced to the religion through short talks and lectures. Smaller Zoroastrian communities (such as in Ontario) still band together to host the feasts with food and labour offered by individual members.[31]

In these ways, the *gahambar* is slowly amplifying its secular gustatory identity. An older Mumbai generation might view these changes through an elegiac lens – an aunt dismissed it as 'tomfoolery' – but in fact these fresh narratives mirror the dynamic nature of Parsi culture throughout its history. As a diasporic people, they have perfected the art of existing in a state of liminality, partaking of different cultures yet ultimately retaining for themselves the refuge of their formative ethno-religious identity. There is now the sense of forging a brand new Parsi identity, seesawing with the sway of several cultures while staying within the bounds of respect for their scriptures.

Notes

1. Mary Boyce, scholar of Iranian languages and religion, believes Zoroastrianism to be the oldest of the revealed credal religions and 'some of its leading doctrines were adopted by Judaism, Christianity and Islam… so named because its prophet, Zarathustra, was known to the ancient Greeks as Zoroaster' (*Zoroastrians: Their Religious Beliefs and Practices* (London: Routledge, 1979), p. 1).
2. The apocryphal legend handed down to Parsi children today is one of assimilation and accommodation. The tale is told that when the Zoroastrian refugees first landed in Gujarat, the local ruler, Jadhav Rana, sent a bowl of milk filled to the brim to connote that there was no space for more people in his kingdom. The visitors supposedly then added sugar to the milk and returned the bowl. The unspoken

3. message was that the Parsis, like the sugar, would ensure the sweetening of the milk (and thus symbolically of their adoptive land) without displacing anything.
3. The population of the community in India has dipped by 22% to 57,264 in 2011 from 69,601 in 2001; according to the most recent census data, the total Parsi/Zoroastrian population in 2011 stood at 57,264 (Press trust of India, 'Parsi population dips by 22 per cent between 2001-2011: study', *The Hindu*, July 26, 2016 <https://www.thehindu.com/news/national/other-states/Parsi-population-dips-by-22-per-cent-between-2001-2011-study/article14508859.ece> [accessed 27 May 2022]).
4. Tanya M. Luhrmann, 'We Are Not What We Were', in *The Good Parsi: The Fate of a Colonial Elite in a Postcolonial Society* (Cambridge, MA: Harvard University Press, 1996), pp. 126-57.
5. Mary Boyce, 'Gahanbar', *Encyclopaedia Iranica*, 2012 <https://www.iranicaonline.org/articles/gahanbar> [accessed 25 May 2022].
6. They are also held for multiple other reasons today: to celebrate a birthday, even that of a fire temple, to commemorate long-ago events, even just to offer thanks.
7. In the early years of colonial India, the *gahambars* were rife with rule and ritual; for instance, only Parsis could enter orthodox Parsi households during the holy days; Parsis would not pare their nails on holy days; dogs were (and are) meant to be fed after the people's stomachs have been glutted.
8. Ervad Soli Dastur, 'A Gahambar Retrospective – at the Saghdi', *FEZANA Journal*, 28 (Fall/September 2014), 61. Here and elsewhere, bracketed clarifications are by the author.
9. Bhicoo Manekshaw, *Parsi Food and Drinks and Customs* (New Delhi: Penguin, 2000), p. 85.
10. Manekshaw, p. 85.
11. Nora Kathleen Firby, *European Travellers and Their Perceptions of Zoroastrians in the 17th and 18th Centuries* (Berlin: D. Reimer, 1988), p. 325.
12. Mary Boyce, 'Festivals', *Encyclopaedia Iranica*, 2012 <https://iranicaonline.org/articles/festivals-i> [accessed 25 May 2022].
13. Jesse S. Palsetia, 'Partner in Empire: Jamsetjee Jejeebhoy and the Public Culture of Nineteenth-Century Bombay', in *Parsis in India and the Diaspora*, ed. by John Hinnells and Alan Williams (London: Routledge, 2008), pp. 81-98.
14. The trope of the Parsi *bon vivant* is inscribed in Mumbai's popular culture.
15. Jehangir Mody, *Jamsetjee Jeejeebhoy – The First Indian Knight and Baronet – 1783–1859* (Bombay: RMDC Press, 1959), pp. 81-82.
16. Delphine Menant writes, 'two years earlier [1859], when an attempt had been made to admit them to the drawing room and the dining room, the innovation had been described as "dangerous". The chief argument was drawn from the (alleged) disastrous effect of this custom in England, such as the numerous cases of divorce and domestic troubles, which were attributed to the mutual intercourse between the sexes' (*The Parsis: Volume 2*, trans. by M.M. Murzban (Bombay: Danai, 1994), p. 321).
17. Mary Boyce, *Zoroastrians*, p. 208.
18. John Hinnells, *The Zoroastrian Diaspora: Religion and Migration* (Oxford: Oxford University Press, 2005), p. 181.
19. 'Commercial Notes', *The Times of India*, 27 September 1879, p. 2.
20. John R. Hinnells, 'Changing Perceptions of Authority Among Parsis in British India', in *Parsis in India and the Diaspora*, ed. by John Hinnells and Alan Williams (London: Routledge, 2008), pp. 100-18.
21. Cursetjee Manockji, *The Parsee Panchayet: Its Rise, its Fall and the Causes that led to the same; being a Series of Letters in the Bombay Times of 1844-45, under the signature of Q in the Corner* (Bombay: L. M. Souaz Press, 1860). Together with anti-colonial activist and pioneer of Indian nationalism, Dadabhai Naoroji, Cursetjee helped introduce social reforms into the community such as inter-dining between men and women; an unheard-of concept previously.
22. Hinnells, *The Zoroastrian Diaspora*.
23. Bachi Karkaria, 'Hakka Doodles', *The Times of India*, 5 February 2014 <https://timesofindia.indiatimes.com/blogs/erratica/hakka-doodles> [accessed 3 October 2022].
24. Ervad Soli Dastur, 'A Gahambar Retrospective – at the Saghdi', *FEZANA Journal*, 28 (Fall/September

2014), 62.

25 Anirudh Raghavan and others, 'Circuits of Authenticity: Parsi Food, Identity, and Globalisation in 21st Century Mumbai', *Economic and Political Weekly*, 50.31 (2015) 69–74 <http://www.jstor.org/stable/24482166> [accessed 30 May 2022].

26 My father remembers being taken for village *gahambars* as a child – residents living in Parsi enclaves would rush back to their homes after the prayers, only to change into loose flowy garments such as night gowns and pajamas, in order to enjoy the meal more fully!

27 Philip G. Kreyenbroek, 'Mrs O', in *Living Zoroastrianism: Urban Parsis Speak about Their Religion* (London: Taylor & Francis, 2013), pp 67–68.

28 Tanya M. Luhrmann, *The Good Parsi: The Fate of a Colonial Elite in a Postcolonial Society* (Cambridge, MA: Harvard University Press, 1996), p. 213.

29 Tejinder Singh, 'Zoroastrians Express Solidarity with Gun Violence Victims, Observe Gahanbar', *The America Times*, 16 March 2019 <https://www.america-times.com/zoroastrians-express-solidarity-gun-violence-victims-observe-gahanbar/> [accessed 30 May 2022].

30 Personal email to the author. ZAPANJ is the Zoroastrian Association of Pennsylvania and New Jersey.

31 'Religious Events September December', *Vision OZCF Newsletter*, 2019, p. 20 <https://ozcf.com/resources/Documents/Newsletters/2019/OZCF%20September-2019%20NL.pdf> [accessed 30 May 2022].

The Perils and Promises of Portability: Sweets and Snacks in Modernizing Japan

Tatsuya Mitsuda

Sweets and snacks, including confectionery, invariably transcend temporal and spatial constraints.[1] Candies, chocolate, or cookies are almost always eaten in between meals – sometimes in lieu of them – and are neither fixed for consumption at home such as at the family dinner table nor anchored to commercial venues such as restaurants, cafés, or eateries. Chewing gum, jelly sweets, or potato snacks can be taken on journeys, munched on at work or at school, devoured during leisure activities, or simply eaten at home to fill time, stave off boredom, or satisfy an immediate craving – often while playing games or watching television. Because sweets and snacks can move relatively freely across time and space, societies, especially those that have become saturated with confectionery, have grown suspicious of their movement. Societies have, as a result, sought to intervene in who makes them, how they are made and sold, when and how they are consumed, and who should eat them and at what age.[2]

Social commentators and academics alike have over the years lamented the rise of sweets and snacking. Shortly before the publication of his ground-breaking work on sugar and empire, Sidney Mintz lambasted the practice of snacking. He pinned the blame for the ubiquity of sweets in the post-war American diet on the 'food technologist' who was 'interested' merely in making profits, thus obliterating 'the social significance of eating with other people'.[3] More recent investigations have continued in the same vein, seeing the rise of eating in between meals as a symptom of a society 'in which the collective rules organizing temporal, social and spatial dimensions of eating are disappearing'.[4] As Mary Douglas pointed out many years ago, precisely because of their shape-shifting characteristics, the practices inherent in the eating of sweets and snacks have led to attempts to restrict their consumption based on arguments that variously combine health, order, or manners.[5] Sweets and snacks pose threats because they contain health-harming ingredients; placed in the wrong hands they could quickly turn into germ-ridden vectors of diseases; devoured without discipline they could destabilize meal routines; and the ways in which they are consumed – on the go, slouched on the sofa, or away from the table – could offend adult, class, or intercultural sensibilities, especially when eaten at inappropriate occasions and venues.

At least part of the problem with sweets and snacks, it seems, lies in their portability, which this paper shows is a recently acquired characteristic. For much of human history, carrying food over long distances risked spoilage. In the absence of adequate preservation techniques, perishable foods were limited to the places in which they were produced. In her book *Fresh: A Perishable History*, Suzanne Freidberg demonstrated that it was from the late nineteenth-century that technological innovations such as refrigerators made it possible for foods such as meat, eggs, fruits, milk, and vegetables to be transported across the world.[6] Reducing the 'perils of portability' was thus a major ambition of mankind, the attainment of which permitted the infiltration of foods into times and places that had previously been off limits. Even though Freidberg was mainly concerned with producers and suppliers operating on national and international scales, consumers acting on smaller scales were also wary of how long they could carry food with them before external influences, both human and non-human, threatened edibility. Focusing on sweets and snacks as they made their way into Japanese foodways, this paper investigates how and why confectionery in the late nineteenth and early twentieth centuries became portable. By shedding light on the perils as well as the 'promises of portability' that certain confections brought, this contribution illuminates the various factors that not only restricted 'improper' sweets and snacks to certain spaces and places but also helped launch 'proper' sweets and snacks into new ones.

Snacks, Travel, and Tourism

During the Edo period (1603-1867), the political stability brought about by the Tokugawa Shogunate allowed for the kind of economic prosperity that underwrote the explosion of travel, including for recreation and tourist purposes. Much of the

Figure 1. *The sign on the left is for* abekawamochi, *a local confectionery delicacy associated with the Abe River.* Fifty-three Stations on the Tōkaidō: The Abe River near Fuchū, *c. 1840 (reproduced with permission from Hiroshige Museum of Art, Ena).*

infrastructure that was laid out – such as the five highways constructed to link the capital, Edo (Tokyo), with the provinces – made it possible for regional feudal lords, *daimyō*, to fulfil the requirement to reside in the capital for extended periods. Facilities offering lodgings, rest, and refreshments sprang up in post stations to service such needs, where the travelling entourage could look forward to treating themselves, among other things, to sweets and snacks, perhaps while sipping on tea (Figure 1).

By the beginning of the eighteenth-century, commoners were also using these facilities. They travelled long distances to make pilgrimages to shrines and temples, including to the Grand Shrine in Ise, located in central Japan, which represented a once-in-a-lifetime experience for most of the population. For these people, a common practice was to choose a representative who would be entrusted with money not just for worshipping at shrines and temples on their behalf but also to bring back mementos of their trip. Oftentimes, merely coming back with gifts such as trinkets or amulets was deemed insufficient: souvenirs from post stations and products from local regions were also carried back and distributed in the community.

Confectionery formed an important part of this burgeoning travel culture. Shops sprang up around shrines and temples keen to cash in on the commercial opportunities, and it thus comes as no surprise that sweets and snacks became heavily associated with specific religious venues. In his book on the history of gifts and souvenirs, Kanzaki Noritake, the foremost authority on early modern travel, showed that Dazaifu Tenmangu in Fukuoka was known for its plum flower-looking rice cakes (*umegamochi*), Kinkakuji in Kyoto for its charbroiled rice cakes (*aburimochi*), or Fushimi Inari Shrine in Kyoto for its gelled confection (*yōkan*).[7] At Ise Grand Shrine, rice cakes with red bean paste (*akafuku*) became the signature confection. It is important to point out that buying mochi, dumplings (*dango*), or *yōkan*, especially the sweet variety, was not cheap. They tended to be luxuries for most of the population – it was a treat that travel made possible and for which pilgrims saved up and looked forward to. Equally important is that the confections, most of which were perishable, were not – unlike today – carried back as souvenirs because of the risks of spoilage. Travellers preferred instead to bring back durables such as handicrafts that could withstand the rigours of time and distance.

Even though a burgeoning confectionery culture had already developed in early modern Japan, which reflected a maturing consumer society on a par with Europe, sweets and snacks continued to be produced and consumed locally.[8] Such restrictions on portability did not necessarily mean that the value of buying and eating confectionery diminished for travellers. In fact, consuming the local sweets and snacks *in situ* became part of the appeal: it attained value as an act that could only be experienced there and then. Near to temples and shrines, travellers would sit down to enjoy their sweets and snacks while sipping on tea. In the travelogue *Ise Sangū* (1848), the writer described, following his visit to Osaka Tenmangū Shine, that he went in search of the famous flour-based confection of *manjū* made by the confectioner, Toraya.[9] With a splendid view of Osaka castle, he took out his *manjū*, sat down and drank his tea, and relaxed his feet before setting off into the sunset. Not being able to bring back the famous local sweets and snacks – referred to as *meika* – was thus not necessarily a hindrance to their fame. Memories of them were likely brought back and transmitted by word-of-mouth, while published guides spread the word nationwide, whetting the appetites of future pilgrims before it was their turn to go and experience themselves what they had heard and read about.

Not a great deal changed after the Meiji Restoration (1868), when Japan, faced with threats to open the country, decided to modernize: sweets and snacks continued to be produced and consumed locally. Despite the development of railroads, which in theory made it possible to carry goods and people farther and more swiftly, this innovation was slow to bring *meika* benefits. At Ise Grand Shrine, the representative *meika*, *akafuku*, because it was made with fresh mochi, continued to be bought and eaten near the shrine. In 1897 the Sangū Railway Company and the Kansei Railway Company, both of which operated lines into Ise, published tourist guides, but little mention was made of *akafuku* as a potential souvenir.[10] They continued instead to recycle early modern recommendations about durables. In fact, only cans make an appearance as food-related gifts. And it was not for a lack of commercial effort that *meika* failed to reach a broader geography. Following the opening of Nihon railway line, which opened the shrines and temples of Nikkō for mass tourism, *yōkan* makers were quick to exploit the commercial opportunities to sell their products to visitors. However, complaints and concerns about the gelled confection's perishability were never far away: an investigation conducted by the Tochigi Prefectural Hygiene Institute revealed that Nikkō *yōkan* became unfit for consumption ten days after it was produced.[11]

Eventually, *meika* did become relatively mobile in the first few decades of the twentieth century. At Okayama, famous for its *kibidango*, confectioners saw an opportunity to market the sweet dumplings to soldiers returning from conflict. They exploited the association the confection had with the folklore peach boy (Momotaro) – a story about a young boy who successfully defeated ogres – to conduct a brisk trade selling the product to soldiers disembarking from the port of Ujina in Hiroshima. As part of its sales pitch, *kibidango* vendors attempted to lure customers by appealing to how *kibidango*, because it was reconstituted using *gyūhi*, a softer variety of mochi that preserved better, could be taken back home as a souvenir.[12] By the beginning of the twentieth century, *kibidango* confectioners who set up shop in Okayama station were selling an average of 600 a day. A similar story developed in Ise, where *akafuku* confectioners, experimenting with new ingredients and packaging material, opened shops in railway concourses. They promised travellers that it could be carried by 'passengers making trips hundreds of kilometres away'.[13] In the case of *akafuku*, children on school outings appear to have helped confectioners shift their wares by the 1910s. However, attempts to sell either *kibidango* or *akafuku* beyond the regions for which they were famous faltered, suggesting that the value of these confections was firmly rooted in the places with which they had become indelibly associated. Purchasing *meika* away from these places was not the same as buying it there.

The Perils and Promises of Portability

Military Conflict and the Importance of Portability

Reflecting on the rise and struggles of *meika* is important: it helps show the limits to the portability of indigenous confectionery and sets the context in which progress towards confectionery portability took place in the second half of the nineteenth century. Most of the confectionery encountered so far – mochi, *dango*, or *yōkan* – are Chinese-inspired indigenous sweets and snacks. Referred to increasingly by the turn of the century as Japanese-style confectionery, or *wagashi*, these confections shared similar ingredients of rice, flour, beans, sugar, and agar, rendering them mainly plant-based products. Yet the impetus towards the 'portabilization' of sweets and snacks came from Western-style confectionery, or *yōgashi*. Although the Portuguese had introduced Iberian confectionery back in the sixteenth century, the introduction of modern Western confectionery such as biscuits, caramel, and chocolate, all of which invariably contained animal-derived ingredients such as milk and butter, proved to be politically and economically more significant.

In relation to the political context, news that the British had forced China to open its market to free trade, for which the Opium War (1839-1842) was fought, sent shockwaves through Japan. Sparking moves to modernize and westernize the country's defences, military reforms included overhauling the rations warriors carried, resulting in the introduction of ship's biscuits.[14] Unlike rice balls, the staple for combatants in the past, ship's biscuits were more durable: they could be kept and stored for long periods and carried across long distances without fear of spoilage. Rice balls were, by contrast, a liability: their preparation involving the steaming of rice could give away locations to the enemy. Nor were they particularly portable because they could not be carried over long distances.

One of the first to successfully make ship's biscuits was the local governor of Nirayama Izu, near Tokyo, Egawa Tarozaemon (1801-1855). As a pioneering student of Western military science and technology, Egawa experimented with Dutch, American, and Russian methods of ship's biscuit making, eventually adopting the Dutch method and setting up a production facility in his home province in 1843. Through his military academy, Egawa is also thought to have handed down his knowledge of ship's biscuit making to the 300 or so domains that made up Japan.[15] Even though the ship's biscuits he produced did not see action in his time, prominent domains such as Chōshu and Satsuma deployed them in the various civil wars that erupted following Japan's decision to open up to the West: they eventually became an integral part of rations in both the navy and the army.

Military imperatives, sparked through a change in the international political context, had wide-ranging implications.[16] First, it introduced and popularized, for the first time since the Portuguese, Western-style confectionery: biscuits became the gateway confection that readied native palates for other types of *yōgashi*. Second, and connected to the first, military deployment created commercial incentives for confectioners to invest in the kinds of machinery – typically imported from British manufacturers such as Huntley & Palmers – that allowed for mass production, thus making it possible to churn out sweets and snacks on an industrial scale and for national, even international, markets. Third, the association with military conflict elevated the apparent utility of *yōgashi*. They were seen to have a practical use and, as butter and milk were later accepted as ingredients, *yōgashi* were seen to be more healthy and helpful in strengthening the nation than *wagashi*.

From the 1880s, when Japan began to extend itself militarily in East Asia, biscuits marketed to the civilian population were advertised as vital food in maintaining state security, empowering soldiers, and extending the empire. Keen to exploit the new medium of newspapers, Western-style confectioners showed off the military orders they had received and emphasized how well their products travelled and kept, despite the distances and the duration involved. During the midst of the Sino-Japanese War (1894-1895), the Asahi newspaper carried an advert for salmon biscuits. Boasting that as suppliers of biscuits to the army they were contributing to nation-building, the confectioner also emphasized their portability.[17] Their biscuits were not only suited for soldiers, but they were also touted as beneficial to hunters and travellers because they could be carried on ship and rail and survive exposure to extremes of heat and cold. Fūgetsudō and Kimuraya, both pioneering Western-style confectioners, were also quick, following victory in the War, to associate themselves with the outcome, creating biscuits in the shape of military caps, medals, flags, and guns to commemorate the triumph of Imperial Japan.[18] In such ways, by the end of the nineteenth century *yōgashi* began to find its way into places and spaces to which *wagashi* had hardly any access.

Sweets, Children, and the Perils of Portability

These moves towards confectionery portability took place at a time when sweets and snacks were becoming a ubiquitous feature of everyday urban life. In the fin-de-siècle, sweets and snacks became not just affordable to most adult inhabitants of towns and cities – they also began attracting custom from children who became increasingly conspicuous consumers of confectionery. Penny sweet shops (*dagashiya*) began to pop up in increasing numbers to cater to this juvenile demand, and the fact that they required little capital to operate made them particularly attractive to impoverished households wanting to generate a side income (Figure 2).[19] For this reason, the *dagashiya* were typically located in the entrance of tenement houses in working class districts and attracted

Figure 2. A reconstructed penny sweet shop (dagashiya) *that was typically set up in the entrance of small tenements (reproduced with permission from Shitamachi Museum, Ueno Park, Tokyo).*

opprobrium because children from respectable families appeared to be sucked into them after school. Much of adult society's concerns revolved around the cheap, unhygienic, and unhealthy street sweets (*dagashi*) that penny sweet shops notoriously sold. They also focused on the tricks and chicanery that shopkeepers supposedly sprang on innocent children to rob them of their allowances and which, it was feared, encouraged gambling. Even children from well-to-do households would be lured in to the *dagashiya*'s orbit. The fear among middle-class society was that its children would mix too freely with children from the lower classes.

Between the end of school and the beginning of supper, children would typically encamp at the penny sweet shop and go on collective crawls looking for the best deals and entertainments offered at various *dagashiya*. Sculptured candy (*amezaiku*) appears to have been a big draw: the product could be individualized, and the price included the entertaining spectacle of the candy being moulded into shapes resembling fruits and animals. Sucking slowly on the candy meant the enjoyment could be extended further. Invariably, the sweets and snacks that children bought at the *dagashiya* would be carried in unpredictable and risky ways. They could be taken from one penny sweet shop to the next, left around on the floor, shared among friends, exposed to the elements (a particular concern during the hot and humid summer months) and carried around for hours on end, especially among children whose parents worked in trades that meant children needed to feed themselves supper. Left in the hands of young children, sweets and snacks could quickly turn into vectors of disease, bad habits, and undesirable company. They embodied, in short, the perils of portability.

Restricting the movement of both children and 'bad' sweets was thus touted as the solution. Making them play with children from the same social milieu within the confines of the home was one way in which children could be prevented from becoming too mobile. Feeding them 'proper' confectionery – preferably Western-style confectionery – at fixed points in the afternoon could also minimize the risks of children carrying around improper and unvetted sweets and snacks. A further solution proposed was to take competition to the street sweets. Mothers would be instructed to move children back home and keep them there through concocting sweets and snacks that would maintain their children's interests. Magazine and newspaper articles not only showed that home-made sweets and snacks had the advantages of being cheap, healthy, and hygienic, but they also had the benefit of being original and tailor-made to each family's needs. Even though expressions were still embryonic, home-made sweets and snacks were promoted as something that commerce could not replicate, injected as they were with motherly affection.

Western-Style Sweets and the Promises of Portability

The increasing ubiquity of sweets and snacks among children brought into sharp relief the perils of portability and ignited efforts to develop confections that could be safely carried. Most of these attempts came from Western-style confectioners; but rendering sweets and snacks more portable proved difficult. In the trade journal

Kashi Shimpō, the correspondent lamented how the biscuits and drops exhibited at the domestic industrial exposition, held in 1907, came up short.[20] He charged that Kimuraya, already famous as a baker producing sweet red bean bread, lagged ten years behind those of foreign manufacturers and despaired that the drops presented by Matsuya and Matsukazedō were also inferior because they failed to withstand the summer heat. There were distinct advantages to taking drops outside and eating them in between work, he pointed out, but that Japanese confectioners lacked the technological know-how to turn this promise into reality.

Such obstacles were largely overcome by the Western-style confectioner Morinaga. Unlike the likes of Fūgetsudō and Kimuraya, Morinaga Taichirō (1865-1937), the founder, received his training as a confectioner in the United States. Following his return from California in 1899, he experimented, albeit unsuccessfully, with selling milk caramel to the Japanese public. Reticence about the smell of dairy, it appears, constituted a major obstacle.[21] After the Russo-Japanese War (1904-1905), Morinaga sensed that milk was becoming increasingly accepted and tried again, putting them in a tinned can and retailing 80 pieces for 40 *sen* – not a cheap price tag. Sales remained flat until Ōgushi Shōji, who was responsible for the product, detected a gap in the market when observing people in public places.[22] At the Imperial Theatre – a Western-style theatre that opened in 1911 – Ōgushi sensed the unease the audience felt because they could neither smoke nor eat their lunch boxes during performances. Bringing in sweets and snacks to this public venue was also problematic on account of the rustling noise the packaging would make. Adapting milk caramel to these sorts of public spaces, Ōgushi thought, would help liberate people from hunger in between meals and from thinking about the nuisances the smell or noise would cause fellow citizens.

To combat these problems, Ōgushi and his team came up with a cardboard box – deliberately designed to resemble a cigarette box – in which pieces of caramel were individually wrapped in paper. Unlike the tin cans that had previously served as the container, the whole package was lighter, and the caramel pieces packed closer together, thus helping to reduce noise. This also meant cheaper costs, and the number of caramel pieces was decreased to 20 and retailed for just 10 *sen*. One major concern was how well the caramel kept. Given that it contained milk – a highly perishable ingredient – fears were high that the product could struggle to attract custom in the rainy season and during the sweltering summer months. Fears that milk caramel would not last as the company had hoped for provided the fillip to Morinaga to strengthen its marketing offensives. They included sending out its employees, equipped with milk caramel, on train journeys to show off the product and enact how it could be eaten on the go – an early example of stealth marketing.[23]

During the 1910s, when milk caramel took off, adverts followed the well-trodden path of Western-style confectioners in the past to show customers how it could become indispensable in activities that involved some travel, recommending taking them to the office, to the theatre, or to sports days (Figure 3). To address scepticism about the suitability of milk caramel in the summer, Morinaga reassured readers that its products were also appropriate for making the sweltering summer months more bearable and persuaded them to take milk caramel to their summer retreats.[24] Milk caramel, which was also being referred to as pocket caramel, achieved its breakthrough at the 1914 Tokyo Exposition: visitors lapped them up, bringing them back home as souvenirs of their visits. Paradoxically, the habit of bringing back souvenirs, a practice that pilgrimages had given birth to, helped spread the fame of the product.

By the late 1920s and early 1930s, the pleasures of portability became central to marketing campaigns – a message that chocolate took over from milk caramel. Like milk caramel, where the paper wrapping became the iconic symbol of the product's portability, the silver wrapping in which chocolate was packaged became another marker of chocolate's portability. Juxtaposed with images of women and men engaged in hiking, tennis, golf or dancing, the adverts produced by Morinaga as well as by arch rival Meiji depicted spaces and places that extended beyond the type of places and spaces that had been imagined for caramel in the 1910s. Chocolate could be taken not just on public transport, in theatres, and the office – it could be carried much farther afield and turn weekday accompaniments elevating labour productivity into weekend companions of pleasure. Now accepted as a staple of school outings, adverts by Morinaga showed children as well as women moving freely and amusing themselves in modern

Figure 3. *Morinaga's milk caramel advert recommending the product as indispensable in leisure activities such as sport.* Tokyo Nichi Nichi Shimbun, *13 May 1915 (courtesy of Morinaga & Company).*

Figure 4. *Morinaga's milk chocolate adverts showing children engaged in sports and picnics. There were also free tickets to amusement parks.* Asahi Shimbun, *31 March 1933 (courtesy of Morinaga & Company).*

recreational public spaces such as gardens, amusement parks, the seaside, or zoos (Figure 4). Depicting chocolate protruding from inside lunch boxes, the adverts impressed upon the onlooker how chocolate enabled such outdoor activities. Even though Morinaga and Meiji's marketing ambitions of transcending seasonality by selling chocolate over the hot summer months did not necessarily succeed, carrying sweets and snacks around became a common practice. No longer did consumers need to concern themselves about issues of perishability that might hamper the movement of the sweets and snacks. As long as consumers were buying the right type of confectionery, it was implied, consumers did not need to worry.

Conclusion

In common with other foodstuffs, carrying sweets and snacks presented problems of perishability. Before Western-style confectionery entered Japan in a major way, sweets and snacks were largely produced and consumed locally. Exposed to the vagaries of the elements and the environment, it simply did not make sense to carry them long distances. Despite tourism and technological developments, such as railways, Japanese confectionery remained largely confined to the areas in which they were produced. Within an everyday context, indigenous sweets and snacks, while becoming mass consumer products, also tended to be confined because of issues of perishability. In the case of street sweets favoured by children, the perils of portability manifested themselves most clearly. Starting with biscuits and then moving to caramel and chocolate, however, *yōgashi* manufacturers were able to increasingly present their products as promising a better portable experience. Gaining from an association with the military and the apparent health benefits of milk and butter, Western-style confectioners sunk substantial capital into marketing campaigns, convincing consumers to take them on transport, to work, and then on to recreational venues. Eventually targeting women and children, the likes of Morinaga and Meiji succeeded in portraying their products as solutions to the perils of portability. As a result, the spaces and places in which modern snacks and sweets could be consumed were largely extended. No doubt this was successful in allaying the fears of certain households who otherwise had to devise strategies to limit the movements of problematic sweets and snacks as potential vectors of disease and bad morals. However, physical restrictions to portability did not necessarily pose limits to the human imagination. Local confections that became famous nationwide because of tourism may not have been carried back, but their fame nonetheless spread across the country. People who heard about them wanted to experience eating them *in situ*, commemorating their visits to shrines and temples and bringing back the memories packaged with confectionery consumption. It is a practice that still lives on.

Notes

1 There has been little scholarly attempt to provide a workable definition of sweets and snacks. In the recent *Oxford Companion of Sugar and Sweets* (2019), the 600 or so contributions suggest that sugar

would loom large in any definition. Here the terms sweets and snacks will be used interchangeably to refer to the act of eating in between meals. While these tended to be sweet confections, this investigation incorporates unsweetened snacks to reflect more accurately the historical diversity that characterizes confectionery culture.

2 For examples of this in western countries, see Samira Kawash, *Candy: A Century of Panic and Pleasure* (New York: Faber and Faber, 2013); Bee Wilson, *Swindled: The Dark History of Food Fraud, from Poisoned Candy to Counterfeit Coffee* (Princeton: Princeton University Press, 2008); Wendy A. Woloson, *Refined Tastes: Sugar, Confectionary, and Consumers in Nineteenth Century America* (Baltimore: The John Hopkins University Press, 2002).

3 Sidney Mintz, 'Meals Without Grace: Sugar & Fat in American Eating', *Boston Review*, 9.5–6 (1984), pp. 6–7.

4 Jukka Gronow and Lotte Holm, *Everyday Eating in Denmark, Finland, Norway and Sweden: A Comparative Study of Meal Patterns 1997-2012* (London: Bloomsbury, 2019), p. 4.

5 Mary Douglas, 'Food as a System of Communication', in *In the Active Voice* (London: Routledge 2011), pp. 82–124; Mary Douglas and Michael Nicod, 'Taking the Biscuit: the Structure of British meals', *New Society*, 19 December 1974, pp. 744–47.

6 Susanne Freidberg, *Fresh: A Perishable History* (Cambridge: Belknap Press, 2010).

7 Kanzaki Noritake, *Omiyage: Zōtō to Tabi no Nihon Bunka* (Tokyo: Seikyūsha, 1997), p. 148. In keeping with East Asian convention, names will be rendered with surnames first, and given names second, unless they appear in English-language publications.

8 Penelope Francks, *The Japanese Consumer: An Alternative Economic History of Modern Japan* (Cambridge: Cambridge University Press, 2009), p. 30.

9 Kanzaki, *Omiyage*, p. 158.

10 Suzuki Yūichirō, *Omiyage to Tetsudō. Meibutsu de kataru Nihon Kindaishi* (Tokyo: Kōdansha, 2013), pp. 70–2.

11 Suzuki, p. 108.

12 Suzuki, p. 52.

13 Suzuki, p. 75.

14 Shibukawa Michio, '*Bakumatsu no hyōryō pan ni tsuite*', *Gunji Shigaku*, 51 (2015), 4–21.

15 Adachi Iwao, *Pan to Nihon Shi. Shokubunka no Seiyōka to Nihonjin no Chie* (Tokyo: Japan Times, 1989), p. 35.

16 The following arguments are based on Tatsuya Mitsuda, '"Sweets Reimagined": The Construction of Confectionery Identities, 1890–1930', in *Feeding Japan: The Cultural and Political Issues of Dependency and Risk*, ed. by Andreas Niehaus and Tine Walravens (Cham, Switzerland: Palgrave Macmillan, 2017), pp. 53–82.

17 *Asahi Shimbun*, morning edition, 6 January 1895, p. 8.

18 *Asahi Shimbun*, morning edition, 6 April 1895, p. 5.

19 The points in this section are mainly based on Tatsuya Mitsuda, 'Consumed by Sweets: Children, Snacking, and Parents in Modern Japan', *Journal of the History of Childhood and Youth*, 14 (2021), pp. 63–84.

20 '*Kashi daikan*', *Kashi Shimpo*, 1 May 1907, p. 5.

21 Morinaga Taichirō, *Kachiku Shōrei Ron* (Tokyo: Takahama Jirō, 1924), p. 8.

22 Kobayashi Gizaburō, *Kashi 30 nen Shi* (Tokyo: Kashi Shimpō Sha, 1936), p. 455.

23 Morinaga Seika, *Morinaga 55 nen Shi* (Tokyo: Morinaga Seika, 1954), p. 101.

24 *Yomiuri Shimbun*, morning edition, 5 August 1915, p. 1.

A Rational Approach: Portable and Practical Eating 'Beyond the Seas'

Jacqui Newling

The 'hungry' Years

On 26 January 1788, eleven ships carrying over 1000 British colonists – 700-odd convicts, 300 marines and civil servants and their families – weighed anchor in Sydney Cove, Port Jackson (Sydney Harbour). Now known as the 'First Fleet', the ships had departed England in May 1787, stopping *en route* at Tenerife, Rio de Janeiro, and the Cape of Good Hope. The voyagers ate fresh produce in port, successfully alleviating scurvy, and acquired additional supplies for the intended colony 'beyond the seas'.

A reliable food supply system was integral to the colonizing mission. The Administration opted for a longstanding victualling model designed to feed government workers and dependents in distant and off-shore locations, distributing a weekly allocation of salt-provisions from a commissary store. The ration scale was based on that issued to troops then stationed in the West Indies; seven pounds of flour, four pounds of salt pork or seven of salt beef, three pints of dried peas (pease), six ounces of butter, and a half-pound of rice or an additional pound of flour per man per week. The women's scale was two-thirds and children's one-third of the men's. The fleet carried what was calculated as two years' allowance, with replenishment supplies to be delivered within that time. The ships also carried breeding stock (sheep, cattle, pigs, goats, and poultry), and plants and seeds to cultivate in government farms and colonists' kitchen gardens. Native vegetation would suffice until cultivated gardens became productive, and fish, shellfish, and small game would offer fresh alternatives to salted meat while imported livestock could multiply.

Such was the plan, but little of this eventuated. Grain crops struggled in the sandy harbourside soils, and most of the sheep and cattle that survived the voyage met with accident. Native game was elusive and fish catches were inconsistent. Contrary to popular mantra, native vegetation was readily consumed, but the types that suited British tastes and cooking techniques were quickly exhausted by voracious foraging in the vicinity of the settlement. Garden vegetables augmented the diet, but the ration remained the principal source of food.

By late 1789 Governor Arthur Phillip was concerned at the reduced state of the stores, and that replenishment provisions may not arrive. (A storeship, HMS *Guardian*, had been sent from England in 1789, but it struck disaster in the Southern Ocean, and could not complete the voyage). Phillip reduced the ration by one-third in November 1789 as a precautionary measure. With no ships appearing by March 1790, he sent a third of the people to the small 'satellite' settlement on Norfolk Island, some 1400 km north-west of Sydney. HMS *Sirius* – one of two ships left to service the colony – was wrecked on the island's shore in the process. The colony's remaining ship, HMV *Supply*, was quickly dispatched to the nearest trading port, Dutch Batavia (now Java, Indonesia) seeking provisions. In April Phillip reduced the ration to less-than-half allowance. Hungry and fearful, the colonists eked out their remaining supplies, supplementing the ration with whatever they could glean from their gardens and local land and waterways. Relief finally arrived with the Second Fleet of transport ships, in late June. Although carrying another thousand convicts, many of them chronically ill, their provisions were enough to restore the ration to full allowance until additional supplies arrived from Batavia in December.

Food security remained tenuous for several years, with ration stores running critically short on various occasions. The continuing arrival of convicts, many unable to work effectively due to poor health from mismanaged transportation conditions, further drained the colony's limited resources. References to short rations and general inadequacy abound in colonists' commentaries and the first five years of settlement have become known as the 'hungry', 'starving', and 'starvation' years. Careful analysis of the primary record provides a much more nuanced understanding of the colonists' foodscape, however. Certainly there were periods of shortage, but the ration was not the only source of food. Nonetheless, tropes of paucity and deprivation have dominated the colonial narrative from 1788 to the present. The colonists are also criticized for their inability to feed themselves in a land that sustained the First Nations for millennia. I have detailed elsewhere the First Fleet colonists' responses to the foods available to them, but this paper discusses the ways that colonists across the different tiers of society cooked and dined with limited resources and rudimentary facilities, so far from the 'mother table' in England.[1]

The Colonial Kitchen

The British were experienced in mobilizing people en masse to distant locations, and were adept at installing necessary infrastructure for temporary use. Along with key structures such as storehouses, a hospital, and a prefabricated house for the governor, a sketch map of the colony dated April 1788 (Figure 1) shows cooking places (Figure 1.1) and a bakehouse and oven (Figure 1.2). These were common features in military encampments and, while seemingly makeshift, provided catering facilities for large numbers.

Drawn as bullseye-like markings – one near the marines' and convicts' camps and another near the hospital (Figures 1.1 and 1.2) – the 'cooking places' were field or trench kitchens, designed for a number of cooks to prepare meals simultaneously (Figure 2). Described in military treatises, they were relatively quick to build, convenient to use, and fuel efficient.[2] Similarly, domed 'beehive' baking ovens could be assembled from bricks or clay (Figure 3). Cooking was also performed on open campfires, and in time more commodious kitchens were installed at the troops' barracks and officials' residences, while the governor's house had its own detached kitchen.

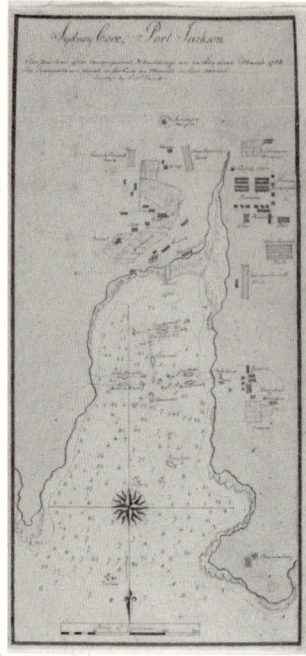
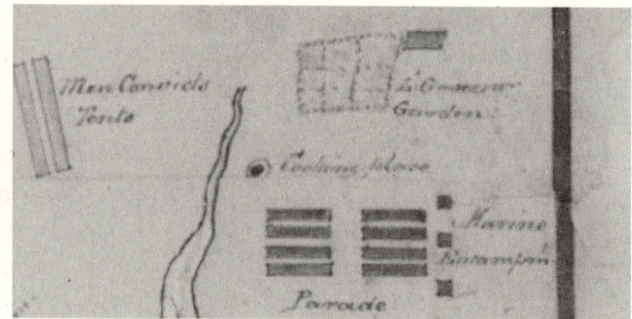
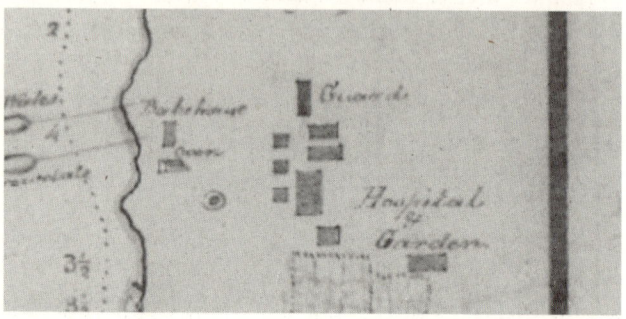

Figure 1 (left). Sydney Cove, Port Jackson [map], 'The position of the encampment & buildings are as they stood at March 1788', by William Bradley, chart from his journal A Voyage to New South Wales, 1802. Mitchell Library, State Library New South Wales, Safe 1/14. Figure 1.1 (top right). Sydney Cove, Port Jackson [map], William Bradley March 1788 (detail showing cooking place). Figure 1.2 (bottom right). Sydney Cove, Port Jackson [map], William Bradley March 1788 (detail showing bakehouse and oven near the shoreline, and additional cooking place symbol).

Marines cooked and ate in mess groups, with members taking turns to cater for their fellows. Convicts may have banded together in similar fashion, or in family groups, and officers had convict servants to tend their meals. Marines and convicts were provided with wooden bowls, platters, and spoons, and each was issued a knife, a hatchet, and small felling axe for firewood. Cookware included billy-style tin kettles, iron pots, and frying pans, distributed among mess groups. The governor complained that the bowls and platters were 'not larger than pint basons [sic]' and made regular appeals for more cooking pots and frying-pans, noting receipt of the latter would be 'a saving of spades'.[3] Other adaptations were made in the absence of conventional utensils, such as 'a couple of ramrods' from the marines' muskets being used to broil over a campfire.[4]

Officers' tables were dressed with cloths adorned with pewter and ceramic dinnerware of English and Chinese origin, some of it quite fine and decorative. At Government House, wine and punch were imbibed from delicate, footed drinking glasses, and tea was poured from porcelain teapots. The governor and his guests enjoyed the services of a French cook,

Portable and Practical Eating 'Beyond the Seas'

while convict servants attended the dining room. Polite table etiquette was observed, and on some occasions guests were entertained with music and song. On New Year's Day in 1789, 'during dinner-time a band of music played in an adjoining apartment; and after the cloth was removed, one of the company sang in a very soft and superior style'.[5] These social niceties seem incongruous with the rustic nature of the settlement (Figure 3).

The colony's first celebratory dinner was held on 7 February 1788 following an official ceremony by the governor to confer the public orders and take formal possession of the colony. After the declarations, senior officials were invited to dine on a 'cold collation' laid out in a marquee erected for the purpose. To the diners' dismay and disappointment 'the Mutten which had been kild yesterday morning was full of maggots [...] nothing will keep 24 hours in this country'.[6] The colonists had much to learn in their new locale. The king's birthday on 4 June 1788 provided the next opportunity for celebration. Again, the officers convened with the governor:

> About 2 oClock we sat down to a very good Entertainment, considering how far we are from Leaden-Hall Market, it [consisted] of Mutton, Pork, Ducks, Fowls, Fish, Kanguroo, Sallads, Pies & preserved Fruits, The Potables consisted of Port, Lisbon, Madeira, Teneriffe, and good old English Porter, [which] went merrily round in Bumpers.[7]

While this meal was for a privileged few, it illustrates the range of imported and local foods that were available just five months after the colonists' arrival. The mutton, pork (fresh, rather than salted, for this illustrious occasion), and likely the fowls were brought with the fleet, while the fish, ducks, and kangaroo were caught locally. The 'sallads' were potentially fresh, young cultivated greens such as mustard and cress, mâché, and endive, alongside a variety of wild species that colonists likened to celery, sorrel, and parsley.

Figure 2 (left). Recreated trench kitchen, Yorktown Discovery Museum, Virginia, USA (Jacqueline Newling, 2016). Figure 3 (right). 'Sydney Cove, Port Jackson. 1788', William Bradley, drawings from his journal 'A Voyage to New South Wales', 1802+ (Opp. p. 84), State Library New South Wales Safe 1/14. The baking oven is depicted as a domed structure next to the bakehouse on the western foreshore (lower right).

Wood cabbages, the fleshy pod which grows in the apex of cabbage palms (*Livistona Australis*), made 'very good eating either as a Sallad or just as it comes out of the Tree'.[8] Pies could have been savoury or sweet, the latter perhaps with fillings of lily pilly, native currants, or native cherries which were readily used in tarts and jellies.

More generally, the colonists were free to manage, cook, and consume their food as they saw fit (though selling or trading their ration was strictly forbidden). The colonists brought with them in their 'invisible luggage' their British food culture: 'an ensemble of shared knowledge, attitudes and practices that people bring to selecting, preparing and eating food'.[9] Entrenched in what sociologist Pierre Bourdieu termed *habitus*, these culinary sensibilities informed the colonists' understanding and expectations of 'good' food, which included the form it should take and how it should be prepared and consumed. These informed the colonists' views of the ration, native produce, and the food habits of local people.

The Ration

The ration and the ways it was cooked demonstrate the portability of Georgian food culture. The Navy Board set the standard scale. The full ration provided an estimated 13,500 kilojoules per day, and the women's allowance, about 9,000. These are consistent with today's recommended intakes for very active or labouring men and women. No distinction was made between the classes, but for a daily half-pint per day of 'grog' – a privilege denied to convicts. The ration could be altered at the governor's discretion, when components were in short supply, or withheld as punishment for poor conduct. Phillip understood that food was fuel and important for morale, and little could be gained from a hungry workforce. Much to the marines' chagrin he adhered to the Navy Board's parity policy, but he did make some concessions. Recognizing that flour was preferred to rice, Phillip allowed the option of the pound of flour to be issued in its place. He believed four pounds of pork per week was too little, and issued five pounds instead. He increased the children's allowance to half rather than one-third of the men's allowance, and when he reduced the ration by one-third to conserve supplies in November 1789, he did not reduce the women's: all adults received the same issue. Subsequent reductions and substitutions were implemented on equal bases.[10] While these decisions benefited the colonists' energy levels and overall health, they more quickly depleted the two-year supply of rations.

The ration is often disparaged by commentators today, and certainly to modern tastes it seems crude and monotonous. But in context of the times, it was quite acceptable, very practical, and, all going well, relatively reliable. The ration components brought with the First Fleet were good quality; however later shipments, particularly those from Asian ports, were often substandard – meat was often bony, 'dholl' from India 'boiled hard' even after lengthy cooking, and bread made with 'soujee' (semolina-like flour) quickly soured.

Meat rations had to be soaked overnight to leech out some of the salt. Pork was often cooked with the pease (pea and ham soup-style), and beef boiled (corned beef style), with any vegetables at hand flavouring the broth which could be drunk as soup for supper. Extended periods in brine broke down the meat's structure, and by 1790 the meat 'shrunk away to nothing' in the cooking.[11] Instead of boiling, colonists 'toast[ed] it on a fork before the fire, catching the drops which fell on a slice of bread, or in a saucer of rice' taking full advantage of the flavour and caloric value in the drippings.[12]

The flour (probably wholemeal rather than a refined white grade) was generally used to make bread. People could exchange an agreed quantity of flour at the public bakery for a ready-made loaf, or make their own, cooked over coals on a shovel or in an iron pot, or buried in hot ashes in the damper tradition. Flour was mixed with butter or fat to make dumplings and cloth-boiled 'hasty' puddings, or boiled with greens for a porridge-like gruel dish not unlike colcannon.

Pease were boiled in cloth to make pease pudding – a highly nutritious and filling dish, often enriched with butter or fat. Forming a dense mass once cold, it could be sliced and fried in a pan for supper.

Rice, which replaced the more traditional oatmeal, could also be made into puddings, sweetened with currants or sugar when available, or made into a congee-style gruel with a nub of salt-pork for flavour.[13] Like pease and flour, or biscuit aboard ships, rice was used to add bulk to soups and stews.

Cultivated Produce

The prevailing view of cultivation in the early colony depicts barren, impoverished, and unyielding soils in a hostile land. Certainly, European grain crops such as wheat, barley, and oats failed to thrive due to sandy soils, low rainfall, a lack of manure, and heat-affected or otherwise travel-damaged seedstock. Colonists were encouraged to grow their own garden produce, and convicts were given time off on Saturday afternoons to tend their kitchen gardens. Early cultivation efforts proved frustrating, but by springtime gardens were flourishing.[14] Those who had made the effort were rewarded with turnips, potatoes, corn, cabbages, and other greens 'in abundance', and pumpkins, melons, and cucumbers grew with 'unbounded luxuriancy' in their respective seasons. Unfortunately for many industrious gardeners, thefts were common – hardly surprising in a convict colony.

Richer agricultural ground was found a few miles west of Sydney at Parramatta where by the end of 1789 vegetables at the government farm were 'plentiful' and 'luxuriant'. The grain harvest yielded two hundred bushels of wheat and thirty-five of barley which were kept for seed for the following year.[15]

Few histories acknowledge the many successes colonists made of their gardens and the contribution of cultivated produce to the diet, choosing to focus instead on the initial struggles experienced in the first few months of settlement and the ongoing difficulties growing European grain crops. Maize thrived, and although regarded as inferior to wheat and producing coarsely textured bread it was the staple for convicts and poorer settlers for decades to come.

Chickens and other domestic fowl provided eggs and perhaps the occasional cockerel for the pot. Fresh meat remained a luxury, however, as 'prudence forbade us' to kill the imported breeding stock.[16] In June 1788 the few head of cattle escaped into the bush, so ewes and she-goats were the only source of milk.

Native Produce

Despite claims to the contrary in many histories, colonists displayed adventurous tastes and experimented widely – indeed, 'The pot received everything we could catch or kill'.[17]

Native birds and game animals were greatly relished. On an inland excursion a surveying party enjoyed a 'kettle of excellent soup out of a white cockatoo and two crows'.[18] They later enjoyed 'ducks picked, stuffed with some slices of salt beef, and roasted, and never did a repast seem more delicious'.[19] Greyhounds were trained to catch emus and kangaroos which proved difficult to hunt with guns. Emu was likened to mutton and beef. Kangaroo, when young, was comparable to veal and mutton though much leaner. Older ones were 'more tough and stringy than bull-beef'; although their flesh was eaten 'with avidity', 'the tail is accounted the most delicious part, when stewed'.[20] Possums also 'eat very well'.[21]

Colonists waxed lyrical about the range of fine fish in the harbour, 'which are extremely delicious'.[22] Vast quantities of shellfish were obtainable from the harbour foreshores, where 'Oysters, Cockles & Muscles are to be got for a little Trouble'.[23] These were, of course, staple foods for Aboriginal people. Although fish catches were inconsistent and rarely made any saving to the meat-ration, there were occasions when seine hauls produced hundreds and occasionally thousands of fish. The local people, who maintained sustainable fishing practices to ensure a constant supply, were astonished, distressed, and at times hostile about the volumes colonists' caught. To avoid conflict the governor ordered they be given a portion of each catch.

Native greens augmented the supply of cultivated vegetables, bringing necessary freshness and nutrients to the salt-heavy ration. Foraging expeditions revealed native alternatives to spinach, sorrel, parsley, sage, sea celery, purslane, samphire, cabbages, and 'sev'ral wholesome unknown vegitables'.[24] A range of small fruit and berries were also welcomed into the colonists' diet.

The colonists adopted a native alternative to tea, which, although enjoyed by people of all classes in England by the mid-1780s, was not included in the ration. When boiled and left to infuse, the leaves of a wild sarsaparilla (*smilax glyciphylla*) made a 'pleasant' and 'wholesome' drink. It had 'much the taste of Liquorish & serves both for Tea & Sugar'.[25] Aboriginal people chewed these leaves for medicinal purposes, and it is possible that colonists were aware of this practice. The colonial surgeons believed the so-called 'sweet tea' to be a 'powerful tonic' and, erroneously, a scurvy preventative. Whether for perceived health benefits or the enjoyment of its taste, it was 'drank universally'.[26] Demonstrating a level of discernment, some colonists blended the sarsaparilla with leaves from a melaleuca 'Tea Tree, so called because a little of the leaves being put into the Native Tea gives it a pleasant spicey taste'.[27] Like other useful resources, when

Figure 4. 'Three natives attacking a sailor' artist unknown (Trustees of the Natural History Museum, London: reproduced under licence).

first discovered the creeping plants were found close to the settlement, but within a few months 'the great consumption had [...] rendered it scarce'.[28]

Many native plant foods require careful detoxifying processes before consumption, and without this knowledge several colonists suffered 'violent wretchings' from eating ill-prepared specimens. Whenever possible, colonists paid attention to the ways local people procured, prepared, and cooked their food, but many of these processes were time-consuming and labour-intensive. The resources that colonists found appealing were quickly depleted, and foragers and hunters had to venture further afield, putting themselves at risk of attack from Aboriginal people, who objected to their lands and waterways being invaded, and their food and medicines raided (Figure 4).

The colonists are often accused of being neophobic, refusing or fearing the unfamiliar, and accepting native produce only 'under stress of great necessity'.[29] Certainly, colonists were selective in their choices, but often for practical reasons: many foods were time-consuming to source and process, supply was inconsistent or insufficient, and they were unsuitable for broadscale collection and distribution.[30] There is ample evidence, however, that First Fleet colonists actively sought and enjoyed native foods.

A Shared Table

The British had little appreciation of Aboriginal people's careful resource management and regarded their food habits as coarse and primitive; however they were keen to know 'whether or not the country possessed any resources, by which life might be prolonged'.[31] In the absence of formal interaction with local leaders (who purposefully chose not to engage with the newcomers), the governor held captive at his house two Aboriginal men who were kidnapped by force. The first man taken was Arabanoo, a Cammeraygal man, held from December 1788 until his death from smallpox in May 1789. Six months later a Wangal man, Bennelong, was captured to fill his place, residing at Government House until he escaped in May 1790. It appears these men learnt more of the colonists' practices than their captors learnt of theirs. They were taught to dine and take tea in the company of the officers, and were commended for mastering social etiquettes. Arabanoo was 'as much at his ease at the tea-table' as the Englishmen, managing his cup and saucer 'as though he had been long accustomed to such entertainment'.[32] Bennelong regularly joined the governor for meals and coffee, and learnt to toast the King's health with a glass of wine.

By March 1790 it was evident that the portable food system that underpinned the colonizing mission was failing. With the loss of the *Sirius* and the fate of the *Supply* unknown, the weekly ration was reduced in April 1790 to less than half its original standard: two-and-a-half pounds of flour, two pounds of pork, and two pounds of rice. Per day, this equated to about 160 g (1 cup) of flour, 130 g of pork, and 130 g (2/3 cup) of rice, delivering perhaps 6,000 kJ (see Figure 5). Public works were stopped and all hands turned to food production and procurement – fishing, hunting, foraging, and gardening. To assist with responsible husbandry the ration was issued daily, one week's allowance served to groups of seven people. Even at the governor's table, guests were asked to bring their own bread: 'Every man when he sat down pulled his bread out of his pocket, and laid it by his plate.'[33] The colonists made efforts to conceal their food insecurity from Bennelong and his people, fearing they would take advantage of the colony's weakened state. Their senses of Imperial certainty and cultural superiority were severely challenged; they were hungry, isolated, and vulnerable.

Myth-Making

Reliance on the ration proved precarious, but as we have seen it was not, and was never expected to be, the colonists' only source of food. The food insecurity crisis in 1790 was something of a perfect storm, caused partly by colonists' initial hubris and confidence in their portable food system and a readily-available local food supply, but also by incidents beyond their control. Medical historians established in the 1970s that at no time were the colonists actually 'starving', and that death rates were minimal during episodes of food shortage.[34] Rather, substandard conditions aboard convict transport

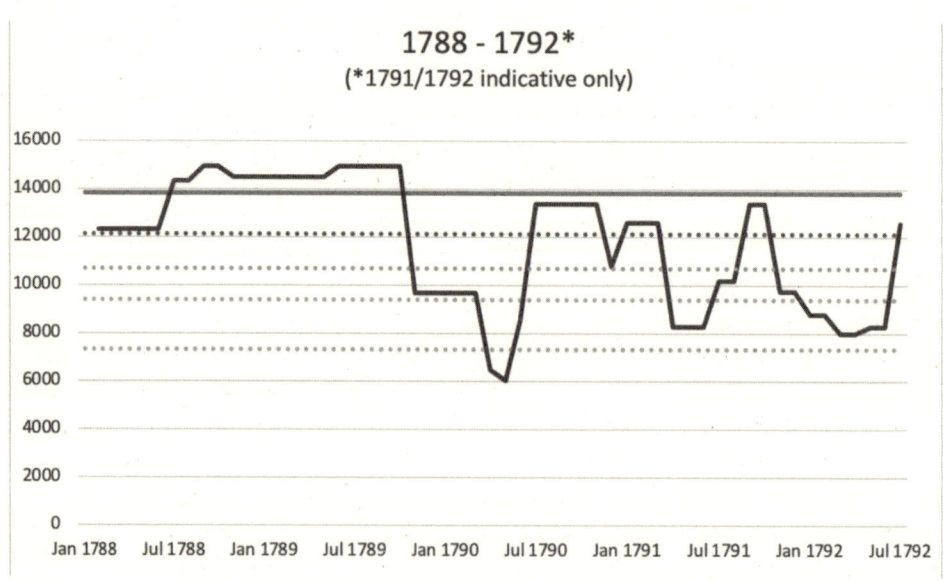

Figure 5. Ration variation 1788-1792 (estimated kilojoules) compared with the original Navy Board standard and recommended benchmark levels (Jacqueline Newling).

ships caused mortality spikes in the months after arrival. And yet, mantras of starvation persist. Temporary ration reductions were imposed when the stores ran low, but these episodes of dearth are often conflated or misinterpreted as permanent restrictions and presented as the norm. Such interpretations perpetuate notions of hapless and ignorant British colonists subsisting on 'starvation-sized rations' in a colony where 'famine conditions prevailed'. Preferring 'starvation to experimentation', they doggedly insisted on 'familiar English foods'.[35] Careful, considered analysis of the primary record reveals many examples that challenge or contradict these negative stances. Claims of First Fleet colonists' abject ignorance and neophobic rejection of native foods, barren soils and unproductive gardens, perpetually short rations and privation, all resulting in hunger are commonly repeated 'myth-information'.

This paper has offered a more reasoned understanding of the First Fleet colony's foodscape. It invites readers to consider the qualitative as well as quantitative value of the colony's various food sources, including the social and emotional aspects of the colonists' culinary encounters. The foods and cooking facilities in the First Fleet colony may today seem basic and limited, but they offered enough scope for colonists to make choices and exercise agency in the ways they cooked and ate. The colonists drew from past experiences to make choices according to what was available to them, improvising and making concessions, adapting their transported tastes and skills to the peculiarity of their circumstances. They found inventive ways to satisfy personal preferences and maintain established tastes, cultural norms, and social sensibilities in their cooking, consumption, and eating practices. They maintained beloved rituals and cultural constructs such as social dining and taking tea, despite but perhaps because of their situation, perched in a makeshift settlement on the edge of a vast and strange land owned by others. Rather than nostalgic yearnings of 'home', these traits and customs were deeply embedded in the First Fleet colonists' identities. Prior experience, preconceived notions, and cultural norms are integral aspects of people's abilities to adapt to new environments or changed circumstances.[36] Readily judged as signs of ignorance, inflexibility, and narrowmindedness, they should be recognized as forms of resilience, demonstrating the strength – and portability – of food and food culture in shaping the colonists' adaptive abilities in a foreign environment, providing a sense of order, comfort and civility against strangeness and uncertainty.

Acknowledgements

This paper was prepared on Wangal country – unceded lands of Australia's First Nations. I pay respects to Elders and knowledge holders, past and present, and acknowledge their people's deep and continuing connection to the country they have nurtured for 60,000 years.

Notes

1   Jacqueline Newling, 'First Fleet Fare: Food and Food Security in the Founding of Colonial New

South Wales, 1788-1790' (unpublished doctoral thesis, University of Sydney, 2021) <https://hdl.handle.net/2123/24785>.

2. See Humphrey Bland, *A Treatise of Military Discipline, In which is Laid Down and Explained the Duty of the Officer and Soldier, Thro' the Several Branches of the Service* (London: Daniel Midwinter, 1743), pp. 244-45 <https://books.google.com.au/books/about/A_Treatise_of_Military_Discipline.html?id=xHtUAAAAYAAJ&redir_esc=y> [accessed 22 May 2022].

3. Phillip to Nepean, 28 September 28 1788, *Historical Records of Australia*, Series 1, Volume 1, Edited by Frederick Watson (Sydney: Library Committee of the Commonwealth Parliament, 1914), p. 86; Phillip to Nepean, 18 November 1791, *Historical Records of Australia*, p. 308.

4. Watkin Tench, *A Complete Account of the Settlement at Port Jackson* (London: G. Nicol and J. Sewell, 1793), p. 27. <http://adc.library.usyd.edu.au/view?docId=ozlit/xml-main-texts/p00044.xml> [accessed 18 May 2022].

5. Tench, *Account*, 1 January 1789, p. 14. See also, Newling, 'Phillip's Table: Food in the Early Sydney Settlement', *Sydney Journal*, 5.1 (2017), 69-83 <https://doi.org/10.5130/sj.v5i1.5730>.

6. Sydney, Mitchell Library, State Library of New South Wales, Safe 1/27a, Ralph Clark, 7 February 1788, *Journal on the Friendship During a Voyage to Botany Bay and Norfolk Island* <http://archival-classic.sl.nsw.gov.au/_transcript/2017/D00018/a262.html> [accessed 10 May 2022].

7. George Worgan, 4 June 1788. *Journal of a First Fleet Surgeon* (Sydney: University of Sydney Library, 2003 [1788]), p. 5 <http://adc.library.usyd.edu.au/view?docId=ozlit/xml-main-texts/worjour.xml> [accessed 10 May 2022].

8. Fowell, Letter 10, p. 20.

9. Colin Bannerman, 'Making Food History in Australia', *Australian Humanities Review*, 51 (2011).

10. The parity policy was abandoned under later administrators: convicts often issued with inferior substitutes.

11. David Collins, *An Account of the English Colony in New South Wales* (Sydney: University of Sydney Library, 2003 [1798]), vol. I, p. 109 <http://adc.library.usyd.edu.au/view?docId=ozlit/xml-main-texts/colacc1.xml> [accessed 18 May 2022].

12. Tench, *Account*, footnote p. 40.

13. The butter was expended by September 1789, when sugar was issued instead, for an undisclosed period (Collins, p. 81).

14. Phillip to Sydney, 28 September 28, 1788. *Historical Records of Australia*, p. 75; Sydney, Mitchell Library, State Library of New South Wales, Newton Fowell Papers and Letters, Letter 11, 5 January 1789 <http://adc.library.usyd.edu.au/view?docId=ozlit/xml-main-texts/fowjour.xml> [accessed 18 May 2022].

15. Collins, p. 88.

16. Tench, *A Narrative of the Expedition to Botany Bay* (Sydney: University of Sydney Library, 1998 [1789]), p. 106 <http://adc.library.usyd.edu.au/view?docId=ozlit/xml-main-texts/p00039.xml> [accessed 22 May 2022].

17. Hunter, John. *An Historical Journal of the Transactions at Port Jackson and Norfolk Island* (Sydney: University of Sydney Library, 2003 [1793]), p. 70 <http://adc.library.usyd.edu.au/view?docId=ozlit/xml-main-texts/hunhist.xml> [accessed 22 May 2022]

18. John White, *Journal of a Voyage to New South Wales* (Sydney: University of Sydney Library, 2001 [1790]), p. 148 <http://adc.library.usyd.edu.au/view?docId=ozlit/xml-main-texts/p00092.xml> [accessed 18 May 2022].

19. White, p. 149.

20. Tench, *Account*, pp. 128, 171.

21. A letter from Sydney Cove, Port Jackson, 24 November 1791, *Dublin Chronicle*, 25 September 1792, p. 2.

22. Tench, *Account*, p. 176. Sharks and stingrays were also readily consumed.

23. Worgan, 12 June 1788, p. 14.

24. Daniel Southwell, May 27, 1788, *Historical Records of New South Wales, Volume 2: Grose and Paterson, 1793-1795*, ed. by Frank Murcott Bladen (Sydney: Charles Potter. 1893), p. 667.

25. Sydney, Mitchell Library, State Library New South Wales: Safe 1/14, William Bradley, *A Voyage to New*

South Wales, p. 136 (October 1788).
26 Tench, *Account*, footnote p. 17.
27 *White,* insert between pp. 232-33.
28 Collins, p. 57 (March 1789).
29 Lionel Stone, 'Australian Food Culture Originated in Britain', *Food in Motion: Proceedings of the Oxford Symposium on Food and Cookery 1983*, ed. by Alan Davidson (London: Prospect Books, 1983), pp. 99-121 (pp. 102, 106).
30 Jacqueline Newling, 'Dining with Strangeness: European Foodways on the Eora Frontier', *Journal of Australian Colonial History*, 13 (2011), 27-48.
31 Tench, *Account*, pp. 33-34.
32 Hunter, p. 132.
33 Tench, *Account*, p. 42.
34 Bryan Gandevia, 'Socio-Medical Factors in the Evolution of the First Settlement at Sydney Cove 1788-1803', *Royal Australian Historical Society Journal*, 61.1 (March 1975), 1 -25 (p. 7).
35 Stone, p. 103, 104, 106.
36 Knut Oyangen, 'The Gastrodynamics of Displacement: Place-Making and Gustatory Identity in the Immigrants' Midwest', *Journal of Interdisciplinary History*, 39.3 (2009): 323-48 (pp. 324-30); Amartya Sen, *Poverty and Famines: An Essay on Entitlement and Deprivation* (Oxford: Oxford University Press, 1981), pp. 16-17, 25.
37 American Food and Nutrition Board, *European Recovery Program: Hearings Before the United States Senate Committee on Foreign Relations, Eightieth Congress, Second Session* (Washington, DC: US Government Printing Office, 1948) pp. 905-09 <https://books.google.com.au/books?id=p-RJAQAAIAAJ> [accessed 22 May 2022].

Hoosh, Dogs, and Seal Meat: The Role of Food in the Race to the South Pole

Diana Noyce

'The fate of nations depends on the way they eat.'
— Jean Anthelme Brillat Savarin[1]

In 1911, Britain's Robert Falcon Scott (1868–1912) and Norwegian Roald Amundsen (1872–1928) set out on a quest to be first to reach the South Pole – the last frontier to be conquered. The British planted the first flag on the Antarctic continent (*Southern Cross* expedition 1898-1900), and it was expected Scott would win the race and raise the British flag at the South Pole.

Antarctica is a vast wilderness of ice and mountains. This southernmost continent is the coldest, driest, windiest, and most isolated place on Earth. For the privilege of being the first to tread this inhospitable yet so desirable spot, both Amundsen and Scott were prepared to drag themselves 1600 miles (2575 km) across a frozen wilderness and face extreme suffering and danger. Most people know that Scott reached the Pole but died heroically on the return journey. For the victorious Norwegian, who on 14 December 1911 planted the Norwegian flag at the geographic South Pole, many regarded his achievement as just plain good luck, as well as being a little devious. Amundsen wrote in his account of the expedition:

> I may say that this is the greatest factor – the way in which the expedition is equipped – the way in which every difficulty is foreseen, and precautions taken for meeting or avoiding it. Victory awaits him who has everything in order – luck, people call it. Defeat is certain for him who has neglected to take the necessary precautions in time; this is called bad luck.[2]

With vastly different approaches to conquering the Pole, the reasons for Amundsen's success and Scott's failure have always been controversial. Explanations for Scott's disaster have tended to dwell on his struggle against overwhelming odds, including terrible weather conditions, vital time and energy expended on scientific observation, and man-hauling 16 kg of geological specimens, as well as human and tactical errors.[3] This explanation overlooks the major difference between Scott and Amundsen in the matter of food – both quantity and quality – especially given Scott's reliance on the heart-breaking work of man-hauling sledges to and from the Pole, rather than using dogs like Amundsen. The contrasting fates of two teams seeking the same prize at the same time indeed invite comparison. Was it Norway's fate to be first at the Pole because of good luck, or was Amundsen's team better prepared – and more importantly better fed – than Scott's team? Good food and reliable transport are fundamental components to the success of any expedition whether a military campaign or polar exploration. Amundsen's expedition had both. Moreover, in a life-or-death journey the ability to eat your transportation is essentially efficient. It is nutrition on four legs.

The Heroic Age of Antarctic Exploration 1897–1922

Antarctica was like no other continent. Unlike previous continental explorations, once an exploring party left the coast it was completely reliant on portable provisions. As the South Pole is a great distance inland, food needed to be transported by the expedition members. It had to be not only portable but also durable.

A great deal is known today about food and clothing required for basic survival in extreme conditions. Much of this knowledge was discovered the hard way, by men suffering from cold, starvation, and nutritional deficiencies while exploring polar regions. Extreme cold and hard work whether travelling by dog sledge or especially by man-hauling a sledge uses a great deal of energy. As such, explorers require a sufficient, balanced diet including all seven food groups needed by humans: carbohydrates, fats, and proteins – to supply energy – and vitamins, minerals, fibre, and water – to ensure bodies run smoothly. Furthermore, a balanced diet depends on the individual – how big they are, how old they are, what biological sex they are, and the activity in which they are engaged.[4]

Early explorers underestimated the required amount of food which often left men fiercely hungry. Personal diaries written during the period reveal frequent and disproportionate attention to food and in particular sledging rations.

The key problem in sledging calculations was balancing food and weight. With nutritional science in its infancy, the impact of extremes of cold on the body imperfectly understood, the variable effects of altitude unknown, and even the causes of scurvy (the great horror of expeditions on both sea and ice) misunderstood – it is easy to see why the optimum balance was not often achieved.

Attempts to Reach the South Pole

It was on the sledging journeys to the South Pole where deficiencies in vitamins and calories or kilojoules were most apparent.

The *Southern Cross* expedition, a British expedition led by Norwegian Carsten Borchgrevink (1864–1934), was the first attempt to reach the Pole. Despite leaving late in the season, the party of three – the first to use dogs and sledges – managed to reach 78°50'S, a new Farthest South record. Next was the National Antarctic Expedition (1901–1904), known as the *Discovery* expedition, led by Scott. An unknown Royal Navy Lieutenant, Scott had no predilection for polar exploration but was to lead two attempts at the Pole. Scott, Dr Edward (Bill) Wilson, and Ernest Shackleton set out on skis with nineteen dogs, five sledges, and supporting parties on 2 November 1902 to try to reach the Pole.

However, Scott and his team were not proficient on skis and were inexperienced with dogs: progress was slow. Moreover, the dogs' food lacked calories and was tainted causing the dogs to sicken. Wilson and Shackleton were forced to kill the dogs. The men, too, were hungry and sickening, struggling with snow blindness, frostbite, and signs of scurvy. Nevertheless, Scott's team continued southward, man-hauling until the shortage of food forced them to turn back, but they set a new Furthest South record of 82°17'S. Troubles multiplied on the home journey, when Shackleton collapsed with scurvy. Wilson's diary entry for 14 January 1903 stated, 'we all have slight, though definite symptoms of scurvy'.[5] This was an understatement. Scott and Wilson struggled on, while Shackleton, short of breath and at times coughing up blood, needed to be carried. The party eventually reached the *Discovery* on 3 February 1903, and Shackleton was invalided home.

Expedition diaries yield clues but not hard evidence as to what occurred on polar expeditions. As authors Roland Huntford and Chris Turney reveal, many diaries of expedition members, particularly Scott's and Wilson's were censored before publication.[6] However, after the *Discovery* expedition Wilson published a detailed account of the medical aspects of the expedition in the *British Medical Journal*. He recounted they were so hungry that they were unable to sleep – and when asleep were disturbed by food dreams. Moreover, scurvy was evident not only in the polar team but in all expedition members. However, once fresh seal meat was eaten regularly scurvy symptoms dissipated.[7]

In 1907, the 33-year-old restless and ambitious Shackleton, determined to try for the Pole again, this time under his own command, set sail on the *Nimrod* for Antarctica. On 29 October 1908, Shackleton, Jameson Adams, Frank Wild, and Eric Marshall, with four sledges, four ponies, and provisions for ninety-one days, set out for the South Pole along what became known as the Great Beardmore Glacier. After the disastrous experience with dogs on the *Discovery* expedition, Shackleton chose ponies, but they proved unsuccessful, and when all four died the team resorted to man-hauling. Within 97 miles (180 km) of the Pole, Shackleton made the agonizing decision to turn back. Again, the amount of food required for such a journey was underestimated. Although Shackleton was aware of the 'need for fatty and farinaceous (carbohydrates) foods in fairly large quantities', Shackleton's narrative of the return journey reveals a different story.[8] The men were 'appallingly hungry', obsessed with food, and Marshall was very ill. Hunger propelled them: 'Our food lies ahead, and death stalks us from behind,' wrote Shackleton.[9] Shackleton set a new record, reaching 88°23'S.

Final Attempts at the South Pole

Fired by his failure to capture the Pole, Scott, now 42 years of age, launched the British Antarctic Expedition in 1910. Scott's expedition set out from Cardiff, Wales, on 15 June with a crew of 65 aboard the *Terra Nova*. On board the ship were 4 Wolseley motor sledges, 19 Manchurian ponies, 33 Siberian dogs, materials to build a hut, and provisions. A well-insulated icehouse on the upper deck held 162 carcasses of frozen

New Zealand mutton, 3 beef carcasses, and sweetbreads and kidneys.

The Norwegian Antarctic Expedition 1910–1912 was led by Roald Amundsen, a 38-year-old, powerfully built man over six feet in height. As a youth he insisted on sleeping with open windows even during Norwegian winters to help condition himself for a life of polar exploration. An experienced professional explorer, Amundsen had spent much time in both the Arctic and Antarctic. As a member of the *Belgica* expedition (1897–99), the first to overwinter in Antarctica, he witnessed the horrors of scurvy. Frederick Cook (1865–1940), the expedition doctor who would soon claim the North Pole, described scurvy as 'polar anaemia'.[10] Cook taught Amundsen the curative powers of fresh seal and penguin meat. Nutritional analysis confirms that seal meat is a reasonably good source of ascorbic acid (Vitamin C), providing two milligrams per one hundred grams of meat; seal liver provides nearly ten times that amount. Men over nineteen years of age need a minimum of 90 mg daily.[11] Moreover, when in 1903–06 Amundsen discovered the elusive North-West Passage from Atlantic to Pacific, he spent two winters on King William Island, where he learnt Arctic survival skills from the local Netsilik Inuit people. He learnt to drive a dog sled, eat raw fish and meat thus preserving Vitamin C, and swap heavy woollen clothing for furs.

Amundsen departed from Oslo 9 August 1910 with a crew of eighteen aboard the *Fram*. However, Amundsen was devious about his intentions. The crew set sail believing they were bound for the North Pole. Not until reaching Madeira did Amundsen inform his crew he planned to go south instead of north. He then sent a telegram informing Scott he was heading south. The race was on.

On board the *Fram* was a prefabricated wooden hut and provisions for two years, including live pigs, fowls, and sheep which were killed within the Antarctic Circle and frozen in the ice. Amundsen also took 97 Greenland dogs. He chose these dogs for their powerful bodies and heavy coats, but also because, unlike some other animals, they will eat one another. He intended to use the dogs as transport, then kill some along the way to provide fresh meat for the men and remaining dogs. Dogs synthesize their own Vitamin C, which would help to maintain Vitamin C levels in both the dogs and the men.

Base Camp

Expeditions needed bases on the continent to overwinter while preparing to attempt the South Pole in the summer. Scott established his base at Cape Evans. Amundsen landed at the Ross Sea's Bay of Whales about 400 miles (644 km) from Scott's base, and 70 miles (113 km) closer to the Pole.

After constructing huts, it was time to lay depots of food along the individual routes to be taken to the Pole. The aim of the first season's depot-laying for Scott's party was to place a series of depots on the Barrier from its edge (Safety Camp) to 80° S, for use on the polar journey beginning the following spring. The final depot would be the largest, One Ton Depot, so the expedition team would not have to carry all its supplies. However, the ponies were struggling, so Scott decided to lay One Ton Depot at 79°29'S,

35 miles (56 km) north of its intended location.¹² It was to be a fatal tactical error. Amundsen formed three depots and laid three tons (3048 kg) of provisions, including 22 hundredweight (1118 kg) of seal meat and 165 litres of paraffin oil to feed eight men. In Antarctica, winter lasts from around April to September, so after the depot laying, they had about five months to wait before they could start for the Pole.

During winter both Scott and Amundsen made further preparations, including sorting and packing sledging rations. As Scott intended his team to man-haul the sledges for the last 150 miles (242 km) as well as the return journey, he made careful calculations to minimize weight. Amundsen's men also prepared themselves physically by maintaining an exercise regime and building up body reserves of nutrients to sustain them. Scott's men did neither.¹³

The cook was the most important person to these would-be conquerors. Amundsen's expedition cook was an overweight but experienced 'polar' cook named Adolf Lindström. He rose each morning at six to prepare hot buckwheat cakes spread with whortleberry and cloudberry preserve, traditional Norwegian antiscorbutics. Amundsen claimed Lindström's cakes 'slipped down with fabulous rapidity'.¹⁴ Wholemeal bread enriched with wheat germ and leavened by fresh yeast which Lindström brewed was also served with butter and cheese. This provided Vitamin B-complex, the importance of which, according to Huntford, has been overshadowed by the focus on Vitamin C.¹⁵ Seal meat was served twice a day, lightly cooked, preserving most of the Vitamin C. A thick,

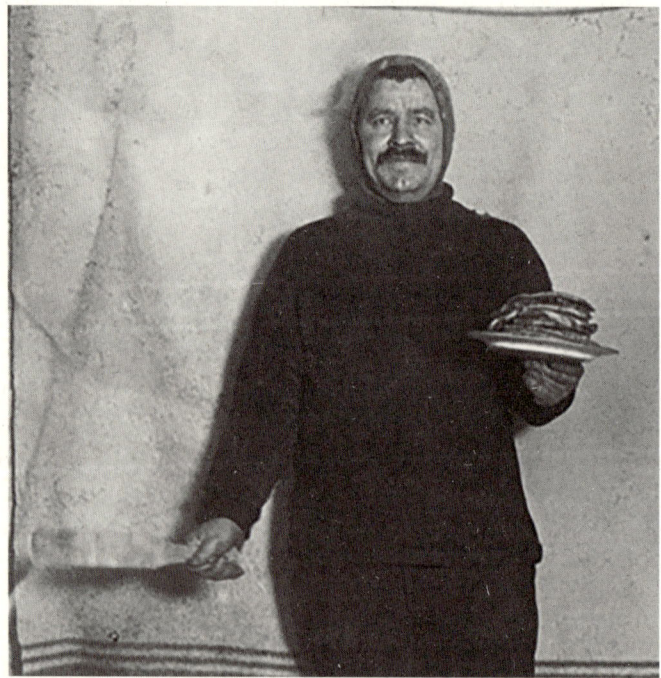

Figure 1. *Adolf Lindström with hotcakes (Wikimedia Commons).*

black seal soup made with potatoes, carrots, cabbage, turnips, peas, celery, prunes, and apples was one of Lindstrom's signatures dishes. Before the winter set in Amundsen had 60 tons (60,000 kg) of seal meat as well as penguin meat in their winter quarters, which he thought was enough for the men and dogs.[16]

For dessert, the Norwegians ate green plums, tinned California fruits, and cloudberries, all rich in Vitamin C. Tarts, pudding, pies, and pastries were also consumed. Coffee was the staple beverage, although alcohol was served on Wednesday and Saturday evenings, birthdays, and holidays. Seal meat, brown bread, hot cakes, and berries were the main food of the Norwegians, a simple natural and nutritious diet. All through the winter the Nordic team built up their stock of Vitamin C and Vitamin B-complex. Their defences against scurvy and beriberi were as high as they could make them.[17]

Scott's cook, Thomas Clissold, a young mechanic, trained as a Royal Navy cook to join the *Terra Nova* expedition. He baked white bread, not brown, and used much tinned food, poor in Vitamin C. However, seal and penguin meat were eaten in pies, as well as fried, curried, or in soup, and seal liver was a breakfast specialty. It was popular amongst the men, but it was overcooked. Mutton was served once a week and beef on special occasions. The British maintained the cause of scurvy was ptomaine poisoning from tainted tinned meat.[18] Indeed, on the *Discovery* expedition, Dr Wilson recorded that the day tinned meat was served was called 'scurvy day'.[19] More attention was paid therefore to the quality of the tinned food than to nutrition.

Figure 2. Thomas Clissold baking bread surrounded by supplies (photo: Herbert Ponting, Royal Geographic Society).

Portable Food

Journey to the South Pole

After an aborted early start, Amundsen set out again 20 October, less three men, with four sledges, with thirteen dogs each, and provisions in addition to those stockpiled in the three depots planned for eight men. Amundsen's five-man team, comprised of Olav Bjaaland, Helmer Hanssen, Sverre Hassel, Oscar Wisting – all accustomed to snow and ice, proficient on skis, and expert dog-drivers – covered on average 23 miles (37 km) per day.

By contrast, the British team set off 24 October with a support party of sixteen men and four methods of transport: two experimental tracked motor vehicles, ten ponies, twenty-two dogs, and man-hauling. The British expedition was plagued with problems from the start. Only 51 miles (82 km) from base camp, the motorized sledges failed and were abandoned. Travel was difficult as the ponies sank in the snow and suffered from the cold. The first pony was shot for food on day 24, and the last one was eaten on day 39. Some of the meat was fed to the dogs, adding to their diet of Spratt's dog cakes, as well as cached for the men on the return journey. Scott joined the cavalcade on 1 November. His previous experience with dogs greatly disturbed him; Scott favoured man-hauling.

Amundsen arrived at the South Pole at 3 p.m. on 14 December 1911 with seventeen dogs and three sledges remaining. Then 'five weather-beaten, frost-bitten fists' planted the Norwegian flag, the first at the geographical South Pole.[20] Amundsen's healthy team returned from their 1500-mile (2414 km) journey on 26 January, ten days earlier than anticipated, with two sledges and eleven dogs. They had been gone for ninety-nine days. Although the men were at times very hungry, Amundsen recorded no cravings for fat or sugar, indicating that the rations' calorie content was sufficient.[21]

Meanwhile, by the middle of December, Scott's sledging party, averaging 10 miles (16 km) a day, were all beginning to suffer from food fantasies, dreaming of sumptuous banquets, and they were becoming noticeably thin. Oates was suffering from frostbite, and an upper-thigh wound from the Boer War (1901–02) was causing extreme discomfort.[22]

Sledging Rations of Scott and Amundsen

The standard sledging ration in Antarctica during the Heroic Age consisted of pemmican, biscuits, butter, cocoa, sugar, tea, and powdered milk, with various additions such as oatmeal, chocolate, cheese, and raisins.[23] These dried and preserved foods were essential for travelling light, but fresh food was crucial for long-term health, especially in preventing scurvy.

Scott's daily polar ration from the Beardmore glacier was for each man: 454 gm biscuit, 340 gm pemmican, 85 gm sugar, 57 gm butter, 20 gm tea, and 24 gm cocoa. This ration contains about 4500 calories (18,828 kJ), about 900 calories more than Scott's 1902 ration. Man-hauling sledges requires 6500 to 8000 calories per day. By contrast Amundsen's polar rations – dog driving requiring less calories – were only four in number: 400 gm biscuits, 75 gm dried milk, 125 gm chocolate, and 375 gm pemmican. However, the individual foodstuffs were more nutritious than Scott's.[24]

Pemmican, the time-honoured polar sledging ratio, originated from North America's

The Role of Food in the Race to the South Pole

Cree Indians. The pemmican that Antarctic explorers knew was finely ground dried beef with 60% added beef fats (suet) and a little seasoning, packaged in cans or square cakes. It was greasy, rich, and valued for its compressed nourishment.[25]

Scott's pemmican, made by Bovril, was a meat/fat only type. Amundsen mistrusted the commercial product. He had pemmican made to his requirements which included dried vegetables and oatmeal. Amundsen believed the pemmican and biscuits he chose made a significant contribution to their success.[26]

Scott's biscuits, made by the British company Huntley & Palmer, were apparently cooked to a secret recipe devised by Dr Wilson and the firm's chemist. Biochemists analysing these biscuits found they contained soluble milk protein, white flour which made them low in vitamin B1 (thiamine), and sodium bicarbonate, which could have further lowered the thiamine content. Thiamine deficiency can lead to beriberi, a disease not unlike scurvy. Amundsen's biscuits, made with oatmeal, sugar, and dried milk, were based on wholemeal flour and crude rolled oats with yeast as the leavening agent.[27] The oatmeal in both the pemmican and biscuits contained necessary B-Complex vitamins.

The standard sledging meal was a hot, thick stew or soup which the British explorers called 'hoosh'. It was made of crushed pemmican boiled up with water, thickened with biscuit or oatmeal, and perhaps garnished with curry powder. Any meat was also added to the stew. The Norwegians made lobscouse; a thick stew made with meat. The hoosh or lobscouse was cooked in an aluminium Nansen cooker. Amundsen, however, used a small primus stove for the polar assault.

In addition, when the final support party left to return to base, Scott made another

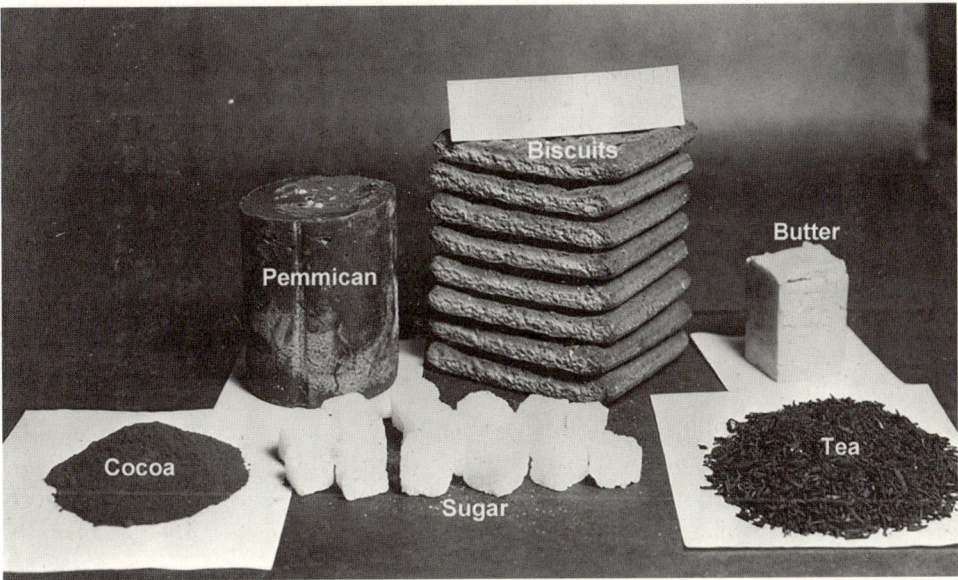

Figure 3. One day's sledging ration for one man, Antarctic British Expedition 1910–1913 (photo: Herbert Ponting, ANMM Collection).

fatal decision – to increase his polar team from four to five men. Scott had not planned to include 'Birdie' Bowers in his polar party but changed his mind just 150 miles (242 km) from the Pole. Only a few days earlier, Scott ordered the support team to depot their skis, so Bowers had to travel on foot while the others were on skis. Scott estimated they had enough food: '5½ units of food – practically over a month's allowance for five people which ought to see them through' (a unit of food was a week's supply for four men). Moreover, since the Nansen cooker contained only four mugs and spoons, cooking for five took seriously longer, which Scott noted he 'had not considered when reorganising'.[28]

The sledging party for the final assault on the South Pole comprised of Scott, Dr Wilson, Captain Lawrence E.G. Oates, Lt Henry Bowers, and Seaman Edgar Evans. At this point they were all man-hauling, four on skis and one walking. The party arrived at the Pole 17 January 1912, only to discover Amundsen's tent and the Norwegian flag he planted one month earlier.

Scott's Return Journey

Reaching the Pole after Amundsen was dispiriting; they turned wearily for home the following day. With five men man-hauling, the party was already ravaged by starvation and malnutrition, dehydration, snow blindness, exhaustion, and injury. It was getting late in the season, and the return journey became one of sheer survival. Dr Mike Stroud, nutritionist, and polar explorer, estimates that Scott's men had a deficit of 3000 calories a day. That would have meant each man lost 25 kg of body weight by the time they reached the Pole. 'You don't just lose fat; you lose muscle as well. You can't keep warm,' said Stroud. 'On the return journey the men would have been emaciated. [...] Vitamin levels – low at the start – would have dropped further. Furthermore, the last-minute decision to add a fifth man to a journey originally planned for four put extra pressure on food rations.'[29]

The first to die was Evans. He was a former Royal Navy gymnastics instructor, well-built, and weighing over 180 lbs (81.6 kg). He may have been the first affected by privations because of his size, said biochemist Robert Feeney. Big men require more calories. As well, a bad cut on his hand refused to heal. The likely cause was severe cold and lack of vitamin C. Evans became disorientated (another sign of scurvy) and finally lagged behind the group, collapsed off the trail, and died 17 February – probably from a brain haemorrhage caused by scurvy.[30]

Then Oates began to fail, particularly with frost-bitten feet and hands, and the wound on his thigh. On 16 March, Oates marched out of the tent and allegedly declared, 'I am just going outside and may be some time.' All knew he was walking to his death.[31] By this time the remaining three, Scott, Wilson, and Bowers, were also suffering in various ways, but particularly from lack of food and fuel, even with two men down. On 20 March, an extreme blizzard stopped all progress. They were within 11 miles (18 km) of One Ton Depot, which had been laid the previous summer further north than planned. If the depot had been laid correctly, the men would have had access to badly needed supplies.

The Role of Food in the Race to the South Pole

Fuel from the start was also in short supply, due to leakage from stored fuel cans sealed with leather washers. This phenomenon had been noticed by previous expeditions, but Scott took no measures to prevent it. Amundsen, by contrast, had his fuel cans soldered closed.[32] A fuel depot found 100 years later was still full.[33] Scott often complained of being thirsty, and the fuel shortage meant he was unable to melt as much drinking water as Amundsen.

On 29 March, 150 days out with supplies of food and fuel exhausted, Scott made his last diary entry and in a weakened state gave a reason for the expedition's failure. He explained how the expedition's disaster was not due 'to faulty organisation, but by bad weather and bad luck. It was no one's fault', wrote Scott. 'Every detail of our food supplies, clothing, and depots […] worked out to perfection […]. We have missed getting through by a narrow margin which was justifiably within the risk of such a journey.'[34] However, things did not work out to perfection: 'it was a miserable jumble,' according to Scott's uncensored journal entry dated 10 March 1912.[35]

Scott's last supporting party likewise suffered nutritional deficiencies. Lt Edward Evans, who refused to eat seal meat, developed scurvy, and had to be carried the last ten days of the trip; he was reported to have almost died.[36] However, all members of the supporting party had scurvy symptoms even though recent research suggests, according to Turney, Evan's returning party took more food and fuel from depots than allocated, robbing Scott of vital provisions.[37] Indeed Scott's diary entry on 10 March noted 'shortage all round in the depots'.[38]

The bodies of Bowers, Scott, and Wilson were discovered by a relief party on 12 November, nearly eight months later. It was also revealed that Scott's team was carrying 30 pounds (16 kg) of geological samples when they died. Man-hauling the extra weight would have further strained their already depleted calorie intake. The remains of Oates and Evans were never found.

Summary

There have been many analyses of Scott's expedition. The weak link was insufficient food and nutrition. This failure was not entirely Scott's fault given limited knowledge of nutrition at that time. However, a mere glance at Scott's diet reveals it was completely lacking in vitamins and low in calories. Even before setting out for the trek to the South Pole their diet was lacking; a quick comparison with the diet consumed by Amundsen's men reveals all. Scott's men ate white bread. Amundsen's team consumed brown bread fortified with wheat germ and leavened with fresh yeast, as well as buckwheat cakes, all good sources of B-Complex vitamins. Scott's team consumed overcooked meat; Amundsen's men ate fresh or frozen seal meat lightly cooked, along with berry preserves. Scott used regulation pemmican; Amundsen made his own pemmican with added vegetables and oatmeal. Scott increased his party from four to five; Amundsen decreased his party from eight to five, though his depots contained food for eight. Dehydration may also have been a factor in Scott's death.

Amundsen put on weight and brought back provisions from the polar journey as souvenirs for his suppliers. The dogs Amundsen took to the Pole substantially contributed to his success. The dogs were not only their means of efficient transport, as well as loyal friends to the men, but also provided a fresh supply of meat to both man and dog.

In pursuit of the Pole, Scott (and Wilson) did indeed take a risk knowing first hand from the *Discovery* expedition and from Shackleton's expedition that the diet lacked sufficient calories as well as nutrition. Expedition member Cherry-Garrard stated, 'in the end they starved to death'.[39] Meat and melted snow might have been their salvation. The fate of nations does indeed depend on how they are fed.

The South Pole is no longer an abstract spot on the map, or a goal of heroic struggle, but a posting for scientists and support personnel who live in permanent quarters known as the Amundsen-Scott South Pole Station. If Amundsen and Scott arrived at the South Pole today, they would find something resembling a small town. They would also find a plentiful supply of fresh vegetables. A sealed and insulated greenhouse provides hydroponic vegetables, and in the summer months fresh vegetables are flown in from New Zealand.

Notes

1 Jean-Anthelme Brillat-Savarin, *The Physiology of Taste* (London: Penguin, 1970), p. 13.
2 Roald Amundsen, *The South Pole: An Account of the Norwegian Antarctic Expedition in the "Fram," 1910–1912*, trans. by A.G. Chater (Edinburgh: Birlinn, 2002), p. 180.
3 Apsley Cherry-Garrard, *The Worst Journey in the World: With Scott in Antarctica 1910-13* (London: Picador, 1994), pp. 433-34; Chris Turney, 'Why Didn't They Ask Evans?', *Polar Record*, 53.5 (2017), 498-511 (p. 498).
4 Robert E. Feeney, *Polar Journeys: The Role of Food and Nutrition in Early Exploration* (Washington D C: University of Alaska Press and American Chemical Society, 1997), pp. 221–28.
5 Edward Wilson, *Diary Of The 'Discovery' Expedition To The Antarctic Regions 1901–04*, ed. by Ann Savour (London: Blandford Press, 1966), p. 238.
6 See Roland Huntford, *Race for the South Pole: The Expedition Diaries of Scott and Amundsen* (New York: Bloomsbury Academic, 2010); Turney, 2017 p. 489.
7 Edward Wilson, 'Medical Aspects of the "Discovery's" Voyage', *British Medical Journal*, 8 July 1905, pp. 77–80.
8 Ernest Shackleton, *The Heart of the Antarctic. Being the Story of the British Antarctic Expedition 1907–1909* (Skowhegan, ME: Kellscraft Studio, 2018), p. 5.
9 Shackleton, p. 214.
10 Frederick Cook, *Through the First Antarctic Night 1898-1899* (New York: Doubleday & McClure, 1900), p. 23; H. R. Gully, '"Polar anaemia": Cardiac Failure during the Heroic Age of Antarctic Exploration', *Polar Record*, 48.2 (2011), 157-64.
11 Jeff Rubin, 'Train Oil and Snotters: Eating Antarctic Wild Foods,' *Gastronomica*, 3.1 (Winter 2003), 37-57 (p. 38).
12 Robert Falcon Scott, *Scott's Last Expedition: The Journals* (New York: Carroll and Graf, 1996), p. 130.
13 See, *Blizzard: Race to the Pole*. Re-enactment of Scott and Amundsen's journey to the South Pole. DVD (BBC2 series, 2006).
14 Amundsen, p. 147.
15 Roland Huntford, *Scott and Amundsen* (London: Hodder and Stoughton, 1979), pp. 389–89.
16 Amundsen, p. vii.

17 Huntford, 1979, p. 388.
18 Scott, pp. 280–81.
19 Wilson, 1966, p. 287.
20 Amundsen, p. 249.
21 Amundsen, p. 209.
22 Lack of Vitamin C causes old wounds to reopen. See Paul Ward, 'Food in Antarctica – Page 2', *Cool Antarctica* <https://www.coolantarctica.com/Antarctica%20fact%20file/science/food2.php> [accessed 28 February 2023].
23 Tom Griffiths, *Slicing the Silence: Voyaging to Antarctica* (Sydney: University of New South Wales Press, 2007), p. 302.
24 Huntford, 1979, p. 581.
25 Griffiths, pp. 301–02.
26 Amundsen, p.43.
27 Griffiths, p. 348.
28 Scott, pp. 384, 386.
29 Joseph Coulson, 'Scott's Antarctic Diet: Stewed Penguin and Champagne', *BBC News*, 29 March 2012 <https://www.bbc.com/news/health-17371543> [accessed 28 February 2023].
30 Feeney, pp. 150, 158.
31 Scott, p. 430.
32 Cherry-Garrard, p. 436.
33 Chris Turney, *1912: The Year the World Discovered Antarctica* (Melbourne: Text Publishing, 2012), p. 119.
34 Scott, p. 441.
35 Huntford, 2010, pp. 385–87; Turney, 2017, p. 506.
36 Feeney, p. 159. Turney, 2017, pp. 503-05.
37 Turney, 2017, pp. 505-06.
38 Scott, p. 428.
39 Cherry-Garrard, p. 439.

The '*tacos de canasta*' (Basket Tacos) in the University: An Example of the Close Relationship between Informality and Mobility

Ayari G. Pasquier Merino and
M. Fernanda Estrada González

Urbanization has imposed multiple changes in food practices, including the type of foods consumed, transformation and preparation processes, and consumption patterns and symbolic representation systems around food.[1] The sale and consumption of food on the street is a practice that, although it has important historical antecedents, has become an icon of urban feeding.[2] The multiple manifestations of this practice in the world reflect its cultural diversity, economic significance, and importance in the daily food routine. Discussions around the topic underline its link with people's right to work, the collective right to public space, and the human right to food security.[3]

According to the International Food and Agriculture Organization, more than 2.5 billion people are fed every day on food sold on the street.[4] Its importance is largely explained by its capacity to solve a set of needs not met by other sectors, thus contributing to the functioning of the city. For vendors it represents a space for self-employment that requires little investment and training, making it an activity that makes substantial economic contributions, despite being a highly precarious labour space. In general, it is part of what is known as 'informal commerce' because it operates outside official regulations, although it usually enjoys wide social legitimacy.

For consumers, street food offers a wide range of culturally valued, low-cost dishes, available at any time of the day, that can be consumed in a few minutes, constituting an important source of energy for a great part of the low-income urban population.[5] Indeed, while it often faces significant challenges in terms of safety and nutritional quality, its availability makes it a convenient option when long working hours and long distances must be covered under precarious transportation conditions, which makes home consumption of food difficult. These characteristics have favoured a substantial increase in street food, especially in cities with rapid population growth, accentuated by internal migratory flows, and where poverty rates have increased.[6]

This topic is approached through a study carried out in Mexico City, where bicycles, tricycles, and all kinds of carts travel the streets every day to meet their customers at precise times and places. Street food offers a wide variety of foods that can take the place of the main meal or be eaten as a snack or to meet a craving. The importance of industrial products is growing and recognized as a space of wide innovation; however, traditionally prepared dishes prevail, forming an essential part of urban food cultures. The development of the present proposal is based on ethnographic work focused on the sale and consumption of a dish known as '*tacos de canasta*' that was carried out on the central campus of the National Autonomous University of Mexico during April and May 2022. The research specifically addresses the cultural meanings surrounding the sale and consumption of *tacos de canasta*, as well as the implications of the informality of their sale for distributors and consumers, seeking to identify their relationship with the mobility of supply.

State of the Art

The term 'street food' describes a wide variety of food and beverages sold in public spaces, usually on the street, although they are also present in establishments such as schools, hospitals, factories, parks, etc. Food can be prepared in advance or momentarily; consumption can occur on the spot or be taken to be consumed along a journey or elsewhere.[7]

Since the late 1990s, the study of street food as a multidimensional phenomenon has progressively expanded, mainly with the contributions of Tinker and Simopoulos and Bhat.[8] The topics covered by these studies include the economic importance of street food vending, its cultural meanings, its daily contributions to urban food, its relationship with tourism, consumer perceptions, and legislative aspects. In a review on street food literature published in 2018, Abrahale and colleagues identified more than 400 publications on the topic, about half of which had been published after 2012.[9] Three quarters of the papers analyzed in this review were conducted in Africa and Asia, and more than 85% focus on the safety issues of this type of food, showing the continuity of initial trends in the study of this topic.

Studies that inquire about social, economic, and cultural aspects have addressed different topics, including the characterization of the supply, their contributions to the income of poor households, their entrepreneurial capacities, the everyday experience of vendors, consumer behaviour, and caloric inputs for city workers.[10] Among those that explicitly focus on mobility, Acho-Chi analyzes the characteristics of mobile street food supply, and Bakić Hayden considers the implications of mobile street food offerings for consumers.[11]

Tacos de Canasta in Mexico City

Street food in Mexico City is internationally recognized, and tacos are one of the most common dishes offered.[12] *Tacos de canasta* are characterized by being a food that is easy to keep warm, easy to transport, and easy to eat. They are a dish usually consumed throughout the morning and early afternoon. They are said to have originated in the mining regions in the nineteenth century, and there are records of their consumption in the city since then, although the sale of tacos on a larger scale is intrinsically related to the popularization of the tortilla machine in the mid-1950s.[13] The current recipe is associated with the town of San Vicente Xiloxochitla, in Tlaxcala, where about 80% of the families are dedicated to its preparation. Like all tacos in Mexico, they are composed of a folded and filled corn tortilla. The most common fillings are potato, bean, adobo, and pork crackling. These are meticulously arranged inside a wicker basket, in which as many as 1000 tacos fit, then they are sprayed with boiling oil to finish cooking and wrapped in three layers that keep them warm, one from paper, the second one from cloth and the last one from blue plastic. The basket is tied to the back of a bicycle, along with a jar of hot sauce prepared with onion, cilantro, and chilli, which diners add to taste directly on the taco before eating it (Figure 1).

The tacos are usually prepared at dawn in private homes and sold by young men who pedal through the streets of the city and stop at the same places every day, at specific times, where they meet their known customers and those who pass by with a hole in their stomach or a few minutes to satiate a craving, generating ephemeral spaces of consumption (Figure 2).[14] Where and when these spaces are generated depends, also, on their capacity to establish agreements to be able to offer their products without inconvenience, many times in exchange for a quota given to a local leader who is in charge of making the agreements with the authorities. They are eaten standing up, on a piece of paper or a plate covered with a plastic bag, which is discarded after each consumption, a practice that seeks to mediate the lack of infrastructure for washing utensils with the need to maintain basic hygienic conditions that guarantee the safety of the food offered.

Although the sale of *tacos de canasta* is not in itself an illegal activity, vendors generally

Figure 1 (left). Tacos de canasta *placed in a wicker basket, and two different salsas offered by the street vendors (picture taken in Ciudad Universitaria, 2022). Figure 2 (right). Street vendor and his bicycle with the blue basket holding* tacos de canasta *and the jar of salsa (picture taken in Ciudad Universitaria, 2022).*

operate outside the laws and regulations governing labour relations, food preparation, tax contributions, and the use of public space.[15] These characteristics place it in the informal economy, a sector that in Mexico occupies 57% of the economically active population and generates around 22% of the national gross domestic product.[16] Whereas due to its nature, there is no precise data on the magnitude of street food sales, it is estimated to be equivalent to one-third of food sales, being an important activity in the local economy. In any case, it is enough to walk the streets to see its importance, both in terms of the number of people it employs and the number of people who eat a meal or snack at these stalls, including people from a variety of sectors and economic backgrounds. The specificities of supply and consumption vary according to the area and the time of day.

University students spend long days on campus, especially in a large urban area like Mexico City. Stocking up on food is essential for them, but it is not always easy. On university campuses, there is a certain variety of formal food offerings from suppliers registered with the university authorities. However, as in the rest of the city, these spaces are not exempt from informal food offerings, usually itinerant, made up of different types of dishes. *Tacos de canasta* are known by the entire population and are daily food for many students: they are considered one of the best options because of their low cost, good taste, and quick consumption. They can take the place of a meal or fulfill a craving, but there is no doubt that they have become an emblematic dish at the university, especially for the student population, as the students themselves testify:

> *Tacos de canasta* are very convenient, it's something very fast that they already have prepared and just serve us, they are very accessible between class changes. You can also customize them with what they offer, they are very tasty with lemon, onion and salsa. It's good to support vendors, I'd rather eat *tacos de canasta* than pay for a sandwich at Subway. (Student, Ciudad Universitaria)

> *Tacos de canasta* are a very good option to eat on campus, they are inexpensive, very tasty and they take away my hunger, plus we can eat them quickly because they are already prepared, not like a salad that takes longer to prepare. I eat them at least twice a week since I entered college, they are my favorite. (Dentistry student, Ciudad Universitaria)

It is also worth mentioning the association of *tacos de canasta* with more recent demands, such as the need to reduce meat consumption and support the local economy:

> I eat *tacos de canasta* at least twice a week, as a vegetarian, it's a very good alternative because many of the other options have meat products that are not balanced or do not seem to be hygienic. They are the safest option to eat something tasty, cheap and fast within the faculty, they will usually always taste good and are affordable. Besides, it's nice to be able to support small merchants and the relationship that can be built with some of them it's also enjoyable. (Student, Ciudad Universitaria)

Some vendors have been set up in the same spot for decades, sometimes with the informal backing of campus security. Others skillfully avoid security personnel or seek temporary agreements with them in case they are caught:

> *Tacos de canasta* are distinguished by their seasoning and blue baskets, other people sell *tacos* by bike or walking, but if you see a blue basket you can't miss them, it's what identifies them. It's the product for the students here, they are my main customers. I started selling *tacos* when I was 12 years old. It has always been here because at CU they sell out faster. (*Tacos de canasta* vendor on bicycle, Ciudad Universitaria)

The *canasta taco* vendors experience an accentuated labour precariousness and are exposed to various risks, related both to the informal nature of the activity, as well as to the conditions of public and road insecurity. In spite of this, they identify mobility as an advantage, so in general terms, it is an occupation appreciated by the vendors, who highlight the freedom associated with self-employment and the earnings it provides them:

> The advantages of selling by bicycle are that you can move around a lot and very fast, right now I am here, a little later in Santa Ursula, then in Xochimilco, although in street vending you run the risk that they will come and take away your basket and even your bike, you are also exposed to trucks, cars, and being mugged. (*Tacos de canasta* employee, Ciudad Universitaria)

> On the days you sell well you could earn a lot and on the days you sell poorly at least you get enough for the day. Compared to a factory where you get paid once a week and if something aches you have to wait until a fortnight or a week to buy some medicine, here that does not happen. Besides, it is a good business, I bought my house in Ajusco doing this. I only sell on campus because I already have clients that I have known for a long time here: professors, football players who are now coaches, students, tourists, on a weekend with a little luck you can finish in an hour if you get a group of twenty tourists (*Tacos de canasta* vendor on bicycle, Ciudad Universitaria).

Discussion

Tacos de canasta correspond to many of the preconceptions documented by the literature that analyzes street food: they are cheap, they constitute the daily food and work space of sectors in poverty, and they have high caloric indices and little nutritional value, in addition to being frequent vectors of gastrointestinal infections. However, this does not mean that they are no longer considered as an excellent daily food option for students and many workers. They constitute a food supply that has been adapted to the needs of different urban spaces. Their strategic packaging and the possibility of attaching them to a bicycle allows vendors to cross long distances with enough tacos to sell during the day and to have access to multiple sale points, taking advantage of different customer flows to maximize their sales, as well as being an advantage to avoid security controls that limit their sales in some areas.

The simplicity of ingredients and the use of a bicycle as a means of transportation minimize investment costs. At the same time, *tacos de canasta* respond to the taste and satiety expectations of the local food culture, making them a positively valued alternative for a wide variety of consumers. Despite their high caloric content, they are often considered a good choice in nutritional terms by the population. While safety risks, which are part of the cultural representation of street tacos, are sought to be controlled through the implementation of different strategies, both by vendors, e.g. using plastic bags to cover the plates or handle money, and by consumers, who usually have identified 'their trusted taquero', a mechanism in which repeated experience becomes the key to safe consumption.[17]

As already mentioned, much of street food operates outside official regulations, which places it in the informal sector. Although this term may seem obvious at first glance, the literature has shown that the boundaries between formality and informality are not always clear, since the actors are linked to activities in both sectors and continually move between them. Street food, like other activities characterized as the informal economy, often are small family businesses where a few members of the family nucleus work, sometimes integrating one or more assistants. However, it is important to note that the sector has developed extensively in recent decades and its operation can accommodate organizations of greater complexity.

As Tiana Bakić points out, one of the most important ways that street-food vendors have managed to secure their right to the street over the last hundred years has been through the creation of informal vendors' unions; these unions are generally highly hierarchical and led by strong and charismatic leaders, who negotiate with local authorities to secure the right to sell for their members in specific areas of the city that are under their jurisdiction.[18]

Street food sale in Mexico has pre-Columbian origins, and it was not until the nineteenth century that it began to be seen as a problem of public order and an obstacle to modernization, so they began to be displaced and criminalized by laws restricting the use of public space.[19] Regulation of street food varies greatly according to the context: in some cases there are regulations regarding the preparation and sale of food, in others, there are not, but it is widely tolerated, or if there are, they are not enforced for different reasons. The main objective of street food regulation is to promote safety, but it can also be related to bidding licenses to sell in certain spaces (giving priority to restaurants) or to avoid road congestion.[20] Development of these practices responds to a large extent to the limits of the formal economy to generate enough quality jobs for the city's inhabitants.

Among the results of the present study, the broad social acceptance of informal street food commerce by consumers stands out. The heart of this discussion is the vendors' legitimate right to work. For vendors, regulations largely imply a restriction on earning a living in a context of great need and few employment opportunities, so breaking official rules is considered a necessity. Thus, the right to work becomes one of the main arguments that legitimize these activities, both among those who work in the sector and among consumers.

The implications of informality include little or no access to infrastructure, which undermines hygienic practices in the preparation, transportation, sale, and consumption of food, as well as the working conditions of those employed in these activities, who are exposed to multiple risks as they move around the city. This same condition limits their access to certain vending spaces, from where they can be expelled by residents or those responsible for fixed businesses, in addition placing them at the mercy of the dynamics by abuse of power from official agents or de facto managers of public spaces who present themselves as mediators with the authorities in the framework of clientelistic political structures, or 'sell' security, another 'necessary evil' from the perspective of the vendors who need to comply with the conditions established in the different spaces in order to operate. As a whole, these conditions impose a high precariousness to workers in the sector, one of the manifestations acquired by the symbolic and structural violence to which street vendors are subjected.[21]

The fact that this activity takes place in public space brings to the discussion the governance of public spaces, which is particularly complex in large cities such as Mexico City. This can make vendors the target of retaliation by those in charge of public security; however, the most common situation is that a tolerance agreement is established, or informal agreements are reached through payment of fees (outside the law), either directly or with 'managers' who act as intermediaries between the representatives of the formal sector and the vendors. This situation limits local governments' planning capabilities and their ability to implement programmes to improve working conditions and hygiene in the sector – although it should be noted that local authorities have done little to document these activities and promote formalization in order to improve safety and economic development in the sector.

Mobility is a valuable characteristic: it provides flexibility and allows vendors to be in the right place at the right time to maximize sales. The fact that street *tacos de canasta* vendors have temporary locations lessens the pressure from residents, often opposed to the establishment of fixed or semi-fixed stalls on the sidewalks. At the same time, it is important to note that the mobility of food supply responds to the increasing mobility of customers in cities.[22] It also facilitates the control of the risks involved in informality, especially in terms of possible reprisals that may arise from formal authorities or de facto managers of public spaces.

Precariousness is a feature that often characterizes those who distribute food under itinerant schemes, as well as consumers who have limited economic resources and little time to feed themselves, although it should be noted that *tacos de canasta* sale in Mexico City is an increasingly well-organized activity, which has led to the formation of small family businesses that control large areas. Even so, it is generally an activity carried out by people belonging to socio-economically marginalized sectors, with little access to formal work options. Preparation of *tacos de canasta* remains in many cases a self-employment sector that requires minimal investment and little training. Preparation at home allows family members to participate in this activity without specific remuneration. In cases where people from outside the family group are employed, as is often the case with vendors, they work in the absence of formal labour agreements, which implies minimal costs for employers, but it also translates into a high degree of job insecurity for employees.

For consumers, mobility facilitates the possibility of having access to food in places where there are often not many other options, with a low cost and a high capacity to satisfy the appetite, in addition to having a broad cultural correspondence that makes them be identified as an appetizing food for a wide variety of tastes. Although their price makes them a preferred dish for sectors with scarce economic resources, people from the wealthy classes are also occasional consumers. The convenience of being able to be consumed in a short time, which adapts to the diners' routines, is also frequently highlighted.

One of the challenges frequently mentioned is safety, related to the conditions and infrastructure of the preparation areas, as well as the long days spent without refrigeration. In this regard, the sauce is generally considered to be the element of greatest risk, since it incorporates raw ingredients and is preserved in glass jars that are often exposed to the sun. It is also identified as unhealthy food, with consumers highlighting its high-fat content. Another problematic element in terms of health is street consumption, where the necessary infrastructure for the personal hygiene of vendors and consumers is lacking. However, there is a certain interest from vendors in limiting the health effects that this may cause, in order to maintain their clientele.

Conclusions

The mobility of the food supply is an adaptation to the conditions imposed by the context that makes it possible to take greater advantage of the economic activity, minimize infrastructure costs, and deal with the inconveniences of the administrative status imposed by informality. In this sense, although the taco can be considered an icon of traditional food, the street taco market is a social construction product of modernity that responds to the economic, cultural, political, and social conditions of Mexican cities.[23]

Lack of empirical correspondence of the limits between formality and informality and their amplitude and socio-economic importance in the global south has led various authors to propose the lack of relevance of the notion of 'informality' as an analytical category. In this sense, the importance of formally recognizing their contributions to the local economy, food, and welfare of cities has also been emphasized, avoiding the marginalization, criminalization, and stigmatization of which they are targets as members of the informal economy.[24] This could lead to more easily addressed health risks, especially in terms of microbiological diseases, by providing training on the subject to those who prepare and sell these foods guiding the strengthening of practices. It would also be desirable that sanitary infrastructure, important for both vendors and consumers, could be installed in the spaces where different street food vendors gather.

Notes

1. Michèle O'Deye-Finzi and Nicolas Bricas, *Nourrir les villes en Afrique subsaharienne* (Paris: Edition l'Harmattan, 1985).
2. Bruce Kraig and Colleen Taylor Sen, *Street Food around the World: An Encyclopedia of Food and Culture* (Santa Barbera, CA: ABC-CLIO, 2013), pp. 19-31.
3. Sueli Alves da Silva and others, 'Street Food on the Coast of Salvador, Bahia, Brazil: A Study form the Socioeconomic and Food Safety Perspectives', *Food Control*, 40.1 (June 2014), 78-84 < https://doi.org/10.1016/j.foodcont.2013.11.022>.
4. FAO, *Spotlight: School Children, Street Food and Micronutrient Deficiencies in Tanzania* (Rome: 2007).
5. T. Sujatha and others, 'Street Food: An Important Source of Energy for the Urban Worker', *Food and Nutrition Bulletin*, 18.4 (1997), 1-5 <https://doi.org/10.1177/156482659701800401>.
6. Primo Arámbulo and others, 'Street Food Vending in Latin America', *Bulletin of PAHO*, 28.4 (1994), 344-54.
7. Cletus Acho-Chi, 'The Mobile Street Food Service Practice in the Urban Economy of Kumba, Cameroon', *Singapore Journal of Tropical Geography*, 23.2 (December 2002): 131-48 <https://doi.org/10.1111/1467-9493.00122>.
8. Irene Tinker, *Street Foods: Urban Food and Employment in Developing Countries* (Oxford: Oxford University Press, 1997); Artemis P. Simopoulos and R. V. Bhat, 'Street Foods', *World Review of Nutrition and Dietetics*, 86 (1997), 53-99.
9. K. Abrahale and others, 'Street Food Research Worldwide: A Scoping Review', *Journal of Human Nutrition and Dietetics: The Official Journal of the British Dietetic Association*, 32 (2019), 152-74 <https://doi.org/10.1111/jhn.12604>.
10. Alves da Silva and others; Lucan and others, 'Assessing Mobile Food Vendors (a.k.a. Street Food Vendors), Methods, Challenges, and Lessons Learned for Future Food-Environment Research', *Public Health*, 127 (2013), 766-76. <https://doi.org/10.1016/j.puhe.2013.05.006>; Guillaume Iyenda, 'Street Food and Income Generation for Poor Households in Kinshasa', *Environment & Urbanization*, 13 (2001), 233-41; Annemarie Hiemstra, Koen G. Van Der Kooy, and Michael Frese, 'Entrepreneurship in the Street Food Sector of Vietnam – Assessment of Psychological Success and Failure Factors', *Journal of Small Business Management*, 44 (2006), 474-81 <https://doi.org/10.1111/j.1540-627X.2006.00183.x>; E.A Khan, 'An Investigation of Marketing Capabilities of Informal Microenterprises: A Study of Street Food Vending in Thailand', *International Journal of Sociology and Social Policy*, 37.3/4 (2017), 186-202; P.L. Malasan, 'The Untold Flavor of Street Food: Social Infrastructure as a Means of Everyday Politics for Street Vendors in Bandung, Indonesia', *Asia Pacific Viewpoint*, 60 (2019), 51-64 <https://doi.org/10.1111/apv.12217>; A. Ozcelik and O. Akova, 'The Impact of Street Food Experience on Behavioral Intention', *British Food Journal*, 123.12 (2021), 4175-93 <https://doi.org/10.1108/BFJ-06-2020-0481>; Sujatha and others.
11. Acho-Chi; Tiana Bakić Hayden, 'Street Food as Infrastructure: Consumer Mobility, Vendor Removability and Food Security in Mexico City', *Food, Culture Society*, 24.1 (2021), 98-111 <https://doi.org/10.1080/15528014.2020.1859920>.
12. 'Top 10 Cities for Street Food', *Reuters*, 20 July 2012 <https://www.reuters.com/article/uk-travel-picks-food-idINLNE86J02720120720> [accessed May 2022].
13. Domingo García Garza, 'Una etnografía económica de los tacos callejeros en México. El caso de Monterrey', in *Estudios Sociales*, 19 (2011), 33-63.
14. José Antonio Vazquez-Medina and Miriam Bertran, 'Edible Heritage: Tradition, Health, and Ephemeral Consumption Spaces in Mexican Street Food', in *Urban Foodways and Communication*, ed. by Casey Man

Kong and Marc de Ferriere Le Vayer (Lanham, MD: Roman & Littleflied, 2016), pp. 139-52.
15 Tiana Bakić Hayden, 'The Taste of Precarity: Language, Legitimacy, and Legality among Mexican Street Food Vendors', in *Street Food: Culture, Economy, Health and Governance*, ed. by Ryzia Cassia Vieira Cardoso, Michèle Companion, and Stefano Roberto Marras (London: Routledge, 2014), pp. 83-97.
16 '*Medición de la informalidad*', INEGI, 2020 <https://www.inegi.org.mx/temas/pibmed/> [accessed May 2022].
17 Victoria I. Delgado and Miriam Bertran, '*Consumo de comida callejera y riesgo de obesidad en la Ciudad de México: Una aproximación antropológica*' in *La medicina social en México IV. Alimentación, cuerpo y corporeidad*, ed. by Florencia Peña Saint Martin and Beatriz León Parra (Cdmx; ENAH ALAMES, 2011).
18 Bakić Hayden, 'The Taste of Precarity', p. 86.
19 Mario Barbosa Cruz, '*Trabajadores en las calles de la ciudad de México: subsistencia, negociación y pobreza urbana en tiemos de la reveloción*', *Historia mexicana*, 60.2 (October-December 2010), 1077-1118.
20 Kraig and Sen, pp. 19-31.
21 Bakić Hayden, 'The Taste of Precarity', p. 86.
22 Jérome Monnet and Juliette Bonnafé, *Memoria del Seminario: El ambulantaje en la Ciudad de México: Investigaciones recientes* (CDMX: México: Programa de Estudios Universitarios de la Ciudad, Universidad Nacional Autónoma de México/Centro Francés de Estudios Mexicanos y Centroamericanos, 2005); Bakić Hayden, 'The Taste of Precarity', p. 86.
23 García Garza, pp. 33-63.
24 Acho-Chi, pp. 131-48; Bakić Hayden, 'The Taste of Precarity', p. 87.

From the Steppe to Space: The Portable Power of Qurut

Simi Rezai-Ghassemi

> Oh, Lord, this is not just a stone.
> I could smell the scent of milk from them.
> — Raisa Golubeva, '*Qurt*, a Gem'[1]

Dairy is one of the main food sources of the pastoral nomads of central Asia, and has aided them in spreading their genes, language, culture, rituals, and beliefs through their conquests from China to the Black Sea and beyond.[2]

Milk from various ruminants has been consumed by people in southwest Asia for millennia.[3] Over this period, various methods of preserving milk have been developed based on the available resources and technologies of the time, most of them probably discovered by happenstance. In each case the resources available to these nomadic herdsmen and women, materials, climate, and technology have spurred this evolution and innovation.

In dry areas, the power of the sun has been utilized; in colder climates, the snow.[4] Either way, preserving food has helped mankind to prosper, offering nutrition in leaner months, a method of trade, and in some cases a reliable starter for future dairy products.[5] Drying food was one of the first methods of preservation, not only allowing the food to last a long time but also making the food smaller, lighter, and therefore portable.

There are many ways in which dairy is preserved in different cultures. In the central Asian plateau kefir and yoghurt are well documented. *Qurut*, an important food of this area and its peoples, is less well known. Preserving and fermentation are increasingly in vogue, but on the Steppe, and in nomadic life, preserved food is a necessity.

Qurut is dried yoghurt. As described in a line in the poem by Raisa Golubeva, in its traditional form *qurut* resembles a chalky whitish apple-sized rock.[6] As with most dairy products, when fresh it is milder but as it matures the flavours strengthen. The flavour of *qurut au nature* ranges from salty, tangy feta to a very strong blue cheese-like taste, especially as it ages. I like to call the flavour 'animally'. It brings soft, aromatic richness and umami when added to dishes. My mother's favourite *qurut* was made with their buffalo milk. She said it had a certain creamy sweetness.

Regardless of its form, dry *qurut* can be a snack when rehydrated a drink or, when formed into a yoghurt-like paste, made into a meal. In the more elaborate Persian

From the Steppe to Space: The Portable Power of *Qurut*

dishes such as *Oshe reshte* or *Kashke bademjan*, made with fried onions, garlic, herbs, and walnuts, *qurut* is a flavour enhancer.[7]

There is doubt about where it was first made, how it came to be, and its etymology, but there is no doubt that *qurut* is dried yoghurt, a preserved food common in the Turko-Mongolian regions. The etymology and use of the equivalent modern Iranian ingredient in Farsi, *kashk*, is better documented.[8] It's the food of my homeland and heritage: the word *qurut* in my mother tongue (Azeri) is literally an order: 'Dry it.'

Fresh spring and summer gluts of milk were transformed into various products and consumed in season.[9] Some of the yoghurt, which is unwieldy and without refrigeration and proper storage would not last the *kooch* (the seasonal nomadic movement from summer pasture to winter pasture), was left to strain away the whey. In the southern regions more than the north of the Steppe, salt was added to increase its preservability, and the stiff, salty paste was formed into round shapes, some flat, others spherical, some with a thumb mark, and left out in the arid heat of the Steppe to dry out completely. This process not only made the large quantities of milk lighter and smaller in volume, more portable, but the process also made the *qurut* easier to digest. These compact, easily transportable, calcium-rich, protein-packed balls of goodness were enjoyed in the winter and used to barter for goods and services. *Qurut* can be made into a drink or meal without many utensils or resources. This made for a nutritious meal on the hoof for the conquering Mongols, and much more recently it was part of the supplies used by cosmonauts taking off from Baikonur.[10]

Depending on the season the milk is produced, the animal, the method of making the yoghurt, the vessel it is made in, and the starter culture and length of processing, there are many types of *qurut* with different textures and flavour characteristics. It can be used in many ways: I loved reading Charles Perry's paper, 'The Horseback Kitchen of Central Asia', in which he mentions that 'qara qurut is cooked down to a very thick consistency; there are tales of automobile radiators being patched with it'.[11]

Qurut is known, named, and used differently in different geographic areas. In the north and east of central Asia, amongst the Turkic speaking countries, the name is some form of *qurut*, but in the south and west it is known as various forms of the word *kashk* in Farsi, *kishk/jameed* in Arabic, *jortan* in Armenian, and *aarull* in Mongolian.[12] Not only is it named differently, but also it is different in style and use, with each region adapting recipes to their cuisine. Whether as a powder or in balls, it is a useful ingredient both on the move and in the modern cupboard or fridge. It is common to see colourful *qurut* in the bazaars of Uzbekistan, where the addition of nuts, herbs, spices, and berries give it interesting hues. These additions are made possible because of modern preservation and storage formulas and techniques. There are also low-salt versions too. The tradition of sun-drying foods has been replaced with advanced dehydration technologies, commercially dried, sometime with additives, and also available in refrigerated pasteurized sauce form. These vacuum-packed white chalky soulless balls lack the robustness of flavour which comes with time.

In Iran we usually serve *gurut* dishes with garlic and mint. In the Yunani medicine of the Persian/Arabic tradition, food is considered medicine and dishes are usually prepared such that they are balanced with 'hot' and 'cold' ingredients. Garlic is deemed cold and mint hot (unlike in the west where mint is considered cooling). The hot food heats the humours up, and the cold brings them down. I wonder whether it is not just for the taste combination or seasonal availability of these ingredients alone that they are consumed together, or rather it may be the microbial effects of phytochemicals in the mint family (as whole) or allicin in garlic which might prohibit or reduce the effects of deterioration when included in the making of *qurut*. Either way, the ingenuity of these ancient foods is remarkable: humans have found ways not only to make dairy more bioavailable but also to partner it with ingredients which aid digestion and reduce contamination or toxicity.

On that note, in Iran they say it is important when using *qurut* that you boil it and then allow the dish to simmer for a few minutes. We rarely eat it dry as a snack as they do in Kazakhstan.[13] As with most traditionally processed food, it can go off, and in the case of *gurut* types of *C. botulinum* may be present if it hasn't been processed or stored properly. Boiling doesn't kill the spores, but it does get rid of the toxins they produce. Having said this, in a few studies in Iran it was found *gurut* was one of the least contaminated foods; however, it does occur and for this reason we don't consume *gurut* raw, although I have yet to find documentation of *qurut* poisoning.[14]

An interesting side note is that cultures who process dairy are more lactose intolerant than those who consumed milk in its simple form. This has led to some snobbery and one-upsmanship, in the former deeming the latter uncouth for consuming straight from the teat. There seems to have been a certain prestige attached to cultures who processed dairy and were consequently less able to digest raw milk![15] Instead in these cultures fresh milk played a role in sacred or ritualistic symbolism, sprinkled at weddings, behind a traveller, or on a winning horse's head.[16] All dairy, known as white products, was not only food but also had symbolic meaning and was held in great esteem, bringing health, wealth, and longevity. Even today in Iran, one of the main alms dishes offered as thanks on certain religious and high days, *Ashe reshteh*, contains *qurut*.

After researching about *qurut*, what it is called, how it is made, where it is from, and its benefits culturally, nutritionally, and for world domination and space travel, I decided to make some. It took over a week, and in Bath's damp climate in February it was quite a to-do. I used a lot of electricity to run an oven overnight to dehydrate a handful of balls, rendering it uneconomical for a small quantity – maybe I should have borrowed a dehydrator. However, the reduction in weight and size was remarkable. I was left with fifteen walnut sized, light balls.

I started by leaving a one-kilogram unopened pack of organic full fat yoghurt on the kitchen counter for a week. I then put it in a cheesecloth, covered, and left it to strain for another three days. After that it was quite stiff and reduced by a third. I put some flour on my hands and took bits that I rolled into balls. I put them on a piece of parchment paper, covered them with a light cloth, and left them for another three days. In my case, all this was done in a modern heated home. I then worried it would go off, so I put it in the oven with the light on overnight. I took them out, let them cool, and froze them. I'm not sure if I'm going to eat them, even though they looked and smelled ok.

When I was thinking back to my first memories of *qurut*, I remembered the noise of its preparation more than the flavour. There was a particular muted rhythm of the rock-hard balls being rolled around in water in a large, open, rough-textured ceramic bowl, similar to the bowl of the *suribatchi*. It had a particular slurping and clanking noise to it. My mother would sit on the floor with the bowl in front of her and roll one way then the other way. It was a labour of love, and we all loved the resulting *Ashe reshshte* and *qurut* with aubergines. Regardless of whether I eat it, mine has hardened to a rock-like ball, and I might just rehydrate it to hear that sound again.

Acknowledgements
I would like to thank Gamze Ineceli, Charles Perry, and Nader Mehravari for their encouragement and generosity.

Notes
1. Cited in Moldir Oskenbey, 'Fermented Dairy Products in Central Asia: Methods for Making Kazakh Qurt Introduction and their Health Benefits', *Crossroads*, 14 (2016), 205-18 (p. 214).
2. Susie Armitage, 'Make the Ancient Road Snack of Central Asian Nomads: Qurt Is Salty, Long-Lasting, and Packed with Protein', *Atlas Obscura*, 8 March 2021 <https://www.atlasobscura.com/articles/what-is-qurt> [accessed 18 April 2023].
3. S. Wilkin and others, 'Dairy Pastoralism Sustained Eastern Eurasian Steppe Populations for 5,000 Years', *Natural Ecology and Evolution*, 4 (2020), 346–55; Eva Rosenstock, Julia Ebert, and Alisa Scheibner, 'Cultured Milk Fermented Dairy Foods along the Southwest Asian–European Neolithic Trajectory', *Current Anthropology*, 62.S24 (2001) <https://doi.org/10.1086/714961>.
4. Brian A. Nummer, 'Historical Origins of Food Preservation', *National Center for Home Food Preservation*, May 2002 <https://nchfp.uga.edu/publications/nchfp/factsheets/food_pres_hist.html> [accessed 18 April 2023]; Harold McGee, *On Food and Cooking: The Science and Lore of the Kitchen* (New York: Scribner, 2004).
5. Helen J. Saberi, 'Travel and Food in Afghanistan', in *Food on the Move: Proceedings of the Oxford Symposium on Food and Cookery 1996*, ed. by Harlan Walker (Totnes: Prospect Books, 1997), pp. 265-73; Rosenstock, Ebert, and Scheibner.
6. Natalia Zhukovskaya, 'The Milk Food of the Mongolian Speaking Nomads of Eurasia in a Historical and Cultural Perspective', *Acta Ethnographica Hungarica*, 53.2 (2009), 307-14; Aliya Uteuova, 'Qurt: A Kazakh cheese of resilience', *BBC Travel*, 27 April 2021 <https://www.bbc.com/travel/article/20210426-qurt-a-kazakh-cheese-of-resilience> [accessed 18 April 2023].
7. Saberi.
8. Charles Perry, '*Tracta/Trahanas/Kishk*', *Petits Propos Culinaires*, 14 (1983), 58-59; Françoise Aubaile-Sallenave, '"Al-Kishk": The Past and Present of a Complex Culinary Practice', in *A Taste of Thyme: Culinary Cultures of the Middle East*, ed. by Sami Zubaida and Richard Tapper (London: Tauris Park, 2000), pp. 105-41.
9. Margaret Shaida, 'Yoghurt in Iran,' in *Milk: Beyond the Dairy – Proceedings of the Oxford Symposium on Food and Cookery 1999*, ed. by Harlan Walker (Totnes: Prospect Books, 2000), pp. 309-14; Nader

Mehravari, 'Yogurt in Persian Cookery', *Petits Propos Culinaires*, 122 (April 2022), 33-45; Najmieh Batmanglij, 'Milk and Its By-products in Ancient Persia and Modern Iran', in *Milk: Beyond the Dairy – Proceedings of the Oxford Symposium on Food and Cookery 1999*, ed. by Harlan Walker (Totnes: Prospect Books, 2000), pp. 64-73.

10 Oskenbey; Armitage.
11 Charles Perry, 'The Horseback Kitchen of Central Asia', in *Food on the Move: Proceedings of the Oxford Symposium on Food and Cookery 1996*, ed. by Harlan Walker (Totnes: Prospect Books, 1997), pp. 243-48.
12 Aubaile-Sallenave.
13 Uteuova.
14 H.R. Tavakoli and others, 'A Survey of Traditional Iranian Food Products for Contamination with Toxigenic Clostridium botulinum', *Journal of Infection and Public Health*, 2.2 (2009), 91-95.
15 Armitage.
16 Zhukovskaya.

Cooking Summer by the Beach: Rulebreaking, Modernity, and Wellness in Three 1960s Egyptian Cookbooks

Salma Serry

About a year ago, during one of my weekly visits to the used book market in Cairo, I found three of the earliest Arabic cookbooks in Egypt dedicated to summer cooking by the beach, on roadtrips, and for outdoor meals. Published in the early 1960s by popular women's magazine *Hawāa*, they appear to target the young, middle-class Egyptian housewife. What is most striking about them is the foreignness, extravagance, and novelty of most of their dishes. They stand out in sharp contrast against economic crisis, growing nationalism, and the nonchalant nature of summer. Foie gras sandwiches appear pretentious and crab platters expensive for the magazine's middle-class audience, English canary pudding looks complicated for vacation rental kitchens, and French rice salad seems to be completely incongruous with local culinary traditions.

To answer how and why such dishes made it into a mainstream consumer magazine, it is necessary to deconstruct their key elements, investigate who the target reader was, and contextualize the recipes within the historical background and social climate of the time. Building on Barbara Wheaton's approach to analyzing the reader and borrowing from the communication field, I experiment with the audience persona tool to synthesize a more intimate picture of who the readers might have been and how these recipes would have fit summer holiday lifestyles. Investigating summer vacation history, the state's socialist wellness policies, as well as the country's Eurocentric modernity, lends important contexts for the booklets' dishes, ingredients, and techniques. In the meantime, summer-themed popular culture films and literature provide invaluable insight into how summer vacations brought on an exceptionalizing rhetoric that permitted rule breaking and warranted seeming contradictions. This contextualization aims to complicate essentializing binaries (Eastern vs Western, traditional vs modern, elite vs working-class) and expand our understanding of the Egyptian society of the time and its food.

The Booklets
The three booklets are titled *Akalat Khafifa lil-Shati' wal-Rahlat wal-Varanda* (Light Dishes for the Beach, Trips, and Verandas), *Akalat lil-Rahalat* (Food for Trips), and *Tabaq Khafif, Wajba Kamila* (A Light Dish, A Whole Meal). All written by Baheya Uthman, the co-author of the seminal cookbook *Usul al- ṭahy al-Nazari wal-'Amali*

(The Art of Cooking: Theory and Practice), they were circulated as complimentary gifts with *Hawāa,* the women's magazine first published in Egypt in 1954 by the *Dar Al Hilal* house that was nationalized in 1960.¹

The three summer-themed cookbooks contain a total of seventy-six recipes, while two of them also feature general tips for preparing and transporting food on trips, for the beach, and away from home. They suggest preparing dishes on the nights before trips, leaving garnishes and sauces to just before heading out, and avoiding foods like anchovies, chocolate, or puddings that cause thirst. They recommend using portable containers, parchment paper, Nylon bags, plastic plates, and Thermos flasks. *Akalat lil-Rahalat* includes a list of daily lunch menus, each including a type of meat like fish kofta, cold boneless chicken, or roast beef; at least two types of sandwiches like tongue, egg salad, or cheese; a salad; a dessert like fruit cake, *mille-feuille,* or swiss-roll; and either lemonade, tea, or coffee. Cold dishes like aspics, sandwiches (with toasted, sliced bread – never local *baladi* bread), and desserts like ice creams are frequently listed.²

The recipes present a curious case considering the economic and political climate. *Hawāa* generally emphasized frugality in the face of harsh economic reforms and strongly encouraged local production following an aggressive nationalist surge. Yet, the recipes contain expensive items like seafood and foreign ingredients like English cheeses. *Hawāa* was also a magazine issued by a nationalized publishing house, born into a nascent socialist society that took pride in abolishing European-influenced aristocracy and its status symbols. Despite that, the dishes exude extravagance and embody a domineering European element that is hard to ignore. And most recipes needed a full kitchen to cook, bake, and refrigerate the dishes, a requirement not typically usual for temporary summer vacations away from home.

Methodology and Approach

In my methodology, I preliminarily rely on Barbara Wheaton's approach to examining cookbooks as a starting point.³ In this paper, however, I focus on the booklets' ideal reader, one of Wheaton's key steps to a systematic cookbook reading. In attempting to draw a more vivid idea of the reader, I experiment with the technique of drawing audience personas, borrowing from the field of communication. Audience persona is a marketing analysis tool that creates different character composites for different audience groups for a media outlet. This communication tool lends a creative way to reimagine history and expands our understanding of faceless readers by using research to synthesize details about them and their lives.

I also find Fabio Parasecoli's analysis of food within popular culture inspiring when applied to historical research. Juxtaposing cookbooks against films and novels from popular culture aids beautifully in creating new perceptions about who the reader might be, what the food might have looked like, what technology may have been used to cook away from home, and how food was consumed, in this case by the beach. As Parasecoli writes, 'pop culture constitutes a major repository of visual elements, ideas, practices, and discourses that influence our relationship with the body, with food consumption, and, of course, with the whole system ensuring that we get what we need on a daily basis, with all its social and political ramifications.'[4] And while it can't be claimed that cookbooks, film, and literature perfectly reflect reality, as cultural representations they might be closest to embodying everyday life in as much detail as possible.

Film and literature aid in visualizing aspects that would otherwise be hard to picture. Also, they help complete information that hardly ever makes its way into cookbooks, like aspirations associated with cooking, preparing, and consuming such foods, or moral and social meanings and dilemmas that come with living certain lifestyles. Most of all, they enable us to see the nuances in the food of everyday individual narratives beyond the generalized binaries I referred to above. This multidisciplinary application allows for more speculation and possibilities, rather than strict thesis-driven conclusions.

Background on Summer *Masyaf* and Its Food

Masyaf, a term in Arabic derived from *saif* (summer), means 'the place of summer'. It came to refer to sea-side towns by the Northern Egyptian coast where people go in summer for cooler weather, either in quick trips or as a vacation. Trips to Alexandria and its beaches had long been seen as the first-class option, the most sought after in the decades since King Fouad I (r.1922-1936) and the ousted King Farouk I (r.1936-1952) established their personal summer residences there.[5] While some areas of Alexandria did offer cheap rental cabins and apartments, more popular and affordable were other coastal towns scattered along the Mediterranean.[6]

By the 1950s, people could reach most of these *masyaf* areas by train, public bus, or car, and naturally, food was required both *en route* and at the destination. Summer guides, films, and novels tell us that most packed home-made sandwiches, cakes, nuts and seeds, and beverages in Thermos containers for the road, and they filled their bags with long-life pantry essentials like rice, lentils, ghee, jams, and canned meats and tomato paste to cook in rental homes. Trains offered light and full meals and beverages, while there were some 'rest houses' along the way that provided snacks, sandwiches, and refreshments. When it came to accommodation, *Al-Hilal* magazine in 1947 informs us that 'municipalities were keen on establishing beach cabin rentals and attentive to public wellness day and night'.[7] Some stayed in hotels, but, according to writer Amina Al-Saeed, others seemed to have preferred renting temporary vacation homes, apartments, or cabins. She specifically expressed her dissatisfaction with hotels and their food:

> Hotels are a waste to our wallets, minds, and bodies. The prices in hotels are high, their food is of low nutritional value and poor quality, their restaurants have dreadful service and very noisy [...]. If you wished to have breakfast, it will arrive by noon, cold and not fulfilling. And if you started your lunch or dinner, they would leave you unattended staring at empty plates, until the other richer hotel residents finish eating before you are even served.[8]

Summer guide books from 1959, 1961, and 1963 describe lunch, the main daily meal at Alexandria's first-class hotels and restaurants, as consisting of soup, beef or poultry, vegetables with a sauce, a dessert, and fruits, with coffee. Specialty traditional seafood stores offered grilled and fried fish and shrimps, typically for catering or pick-up. Breakfast at most first-class hotels was continental: eggs, jam, butter, toast, fruits, and pastries with coffee or tea. In comparison, first-class hotels in another town, Ras al-Barr, served traditional stewed fava beans, Egyptian cheeses, and fried eggs instead. Port-Said's best restaurants were also cheaper, with lunch options like grilled fish, roasted beef fillet, escalope, and *salade russe* for lunch. Popular eateries in most regions were kebab and skewered *kofta* specialty stores, while cheaper and more casual eateries typically offered traditional meals. These included grilled chicken, curry rice, baked meat and vegetable casseroles, baked macaroni dishes, and simple green salads.

Dining out options were plenty, including restaurants, cafes, and nightclubs. In a short story from *Al-Banat wal-Saif* (1958), Ihsan Abdel Qudoos depicts an argument over whether one of three friends in an Alexandria summer cabin should go to the city centre to get them grilled chicken for lunch or order something from their beach resort's cafeteria. 'How can you stand the cafeteria's soggy pasta and meat that taste like leather shoes!' one exclaims. And when the other insists on picking up sandwiches from the cafeteria instead, he replies, 'I'll eat anything but cafeteria sandwiches. I'll even die hungry and not have their sandwiches for lunch.'[9] They end up sending someone to buy the grilled chicken from the city instead.

Families also resorted to cooking in their temporary *masyaf* residences. Some factors facilitated cooking, including having access to one's own cooks and help, preferring traditional and cleaner home-cooked meals, and avoiding the high cost of dining out. Those who could afford to have their cooks and servers travel with them could throw alfresco lunch *uzumat* (parties revolving around food) to entertain others. A perfect example is the deputy minister and his wife in 'Finjal Ahwa wa Hallit Molokhhia' (1969), another of Abdel Qudoos's works. In this short story, the hosts entertain guests on their cabin veranda with large spreads of traditional Egyptian classics like *molokhia* (jute mallow soup), *fatta* (soaked bread pieces topped with rice and meat), *dolma* (stuffed vegetables), and grilled rabbit, all very traditional dishes associated with home-cooking and hardly found in restaurants outside of the home at the time.[10]

Furthermore, most middle-class women were restricted to humble budgets; their families could not afford to dine out every day. As we understand from one character in another story from Abdel Qudoos's 1958 anthology, 'everything in the *masyaf* [was] very expensive', whether it was a platter of grilled chicken or a humble bowl of stewed *fool medammis* (fava beans).[11] But even in most humble cabins, cooking was not difficult. Municipalities transported food supplies from nearby cities to coastal *masyaf* resorts.[12] By the 1950s, almost every Egyptian household used portable pressurized-kerosene burners, known as *waburs* in Egypt, and although the appliance itself was not meant to be used as a fixed tool inside kitchens, they were ubiquitous as a cheaper alternative to large coal-fired or electric stoves. In addition, the government subsidized kerosene gas and supported the national manufacturing of these burners, rendering this cooking technology more affordable for everyone.[13] Summer holiday films, like *Agazit Saif* (1966) and *Shate' al-Marah* (1967), show the *wabur* being used to make soups, stews, rice, and casseroles, both inside cabins and outside on the beach.

Three Personas

Presumably, the target readers of *Hawāa*, the magazine behind these guides, were mostly young, middle-class women, at least judging by the magazine's mass circulation and its content stressing frugality and self-reliance. But middle class does not reveal much beyond a broad socio-economic categorization. As I read and re-read the recipes my mind wandered, trying to draw an image of who the reader could be. There are a few illustrations and images inside the booklets, but, beyond that, there aren't many useful details. So, instead of assuming one generic faceless reader, I adopted the marketing analysis tool of audience personas that I applied in planning marketing campaigns for media agencies years ago. In communication and marketing industries, the tool is particularly helpful in understanding consumer needs and desires.

While its end goal is crafting effective messages that speak to the target consumer, the persona building and analysis process can also be a way to enhance our understanding of who the readers of these cookbooks might have been. As its guiding principle, this process relies on creating an intricate description of fictionalized characters that represent core audience groups.[14] A rich composite of the characters' social habits, aspirations, and even problems is imagined while keeping in mind variables of age, class, education, and gender.

Essentially, such depictions should be grounded in research, so instead of using marketing data, I relied on socio-historical research. A list of questions is first prepared with demographic characteristics, individual traits, beliefs and attitudes, then deeper attributes like peculiarities, idiosyncrasies, or personal struggles. These questions then get answered and analyzed to arrive at key learnings about the audience. Film and literature provided insight to aesthetic preferences, lifestyle choices, and food preferences that helped synthesize the personas. It is important, though, to be cognizant that these personas represent neither factual information nor historical absolute truth. Instead, the tool guides further historical research by allowing us to be immersed into subtleties of the individual. Additionally, it is important to be aware of one's own biases in creating the personas to avoid generalizations.

In this way, the exercise helps place the recipes in their context between the binaries explained above. By associating them with composite characters and placing them in social settings, I was able to imbue them with the potential of various cultural, social, and personal meanings, considering how food might have embodied different experiences for different readers. Such synthesis can benefit from further future assessment against other historical sources like oral history interviews, photographs, and memoirs to complete, compare, or challenge what we know about the past.

The first persona created for the booklets' readers is Mona, a creative bonne vivante housewife who would go to lengths to impress her social circle with her in-vogue lifestyle and impressive culinary taste. She could be wealthy or of a middle-income household, but her adventurous taste for trying new things, especially in summer, gave her a modern flair. Sometimes though, this taste sat at odds with the Egyptian palette, pairing ingredients like boiled beef cubes with crab meat and shredded fish all together in one salad, as one of the recipes calls for. Other times, it was out-of-the-box presentations like shrimp salad in bowls made of orange peels, according to another recipe. A recipe titled 'French rice' seems to unflinchingly blaspheme centuries-old local laws for cooking and eating rice, but Mona could have very well considered experimenting with it. Instead of the usual way of bringing rice to a boil, then letting it absorb the water or broth flavoured with meat, vegetables, or spices, this recipe requires boiling and straining it, then mixing it with raw vegetables, cold cuts, and drizzles of French salad dressing.[15]

The second reader, Amale, has a passion for socializing and throwing expensive beach cabin parties with ostentatious displays. She may have thrown these parties out of pure fun, duty as wife or daughter of a notable personality, or a competitive need to keep a shiny social image. The *Hawāa* booklets were more inspiration for her ideas than how-to guides; after all, she had a servant or several to do the shopping and cooking. Pricey foreign ingredients could have been easily accessible for her at the multiple European specialty grocers in Alexandria. One recipe titled '*Salatat al-Rahalat*' (Trips Salad), calls for luncheon meat, Cheshire cheese, and Gouda or Swiss cheese, none of which would have been found in a typical local grocery. The booklets, also, are heavy on seafood, which was not cheap. Generally, fish and seafood in Egypt were more expensive than chicken and meat, and, seasonally, their prices rocketed in *masyaf* markets.

The third persona I visualized, Laila, also enjoyed entertaining in her rental summer apartment, but on a much smaller scale with savvier spending due to budgetary constraints. Alternatives to costly ingredients were the way to go for many like Laila due to the government's policies of restricting imports and boosting confidence in national products instead. Some recipes from the booklets are aligned with this reader persona, calling for local *roumi* cheese, when Parmesan would have been the classic ingredient, and a lot of canned food. Many of the sandwich recipes list more than one type of cheese, and other dishes offer a choice between butter or 'Dutch ghee'. And perhaps, after all, if one of the visualized personas chose to splurge on a meal of seafood or imported cheese for one day, that does not necessarily mean she wouldn't be able to make up for it by having cheaper dinners for the rest of the week.

Eurocentric Modernity

For leading consumer magazines, values like unconventionality and eliteness drive the editorial process, so flashy and trendy recipes would typically attract and maintain readership. From the reader's perspective, the desire to be creative in the kitchen can rarely be underestimated, anywhere or at any point in history. Also, it was as important back then as it is today to appear well-travelled, or at least knowledgeable of global trends. That meant wearing the badge of sophistication and fashionable taste, with the concept of 'omnivorousness', as approached by sociologists Shyon Baumann and Josée Johnston, coming to mind.[16] However, for Egyptians in the middle of the century, the desire to eat foreign food as a way to collect cultural and social capital was distinctively skewed towards European dishes, owing to a pervasive Eurocentric modernity.

Modernity most certainly did not occur suddenly, nor was it exceptional to Egypt or produced by the regime in the 1950s. Modernization has been a long and steady process which had roots planted in Egypt since the middle of the nineteenth century with British imperialism and the ruling dynasty's immense fondness for French culture, fashion, and food. With time, this modernity trickled down the society's social tiers, becoming a way to acquire cultural capital.

Notwithstanding, the reason why Egyptian appetites for Eurocentric modernity reached its peak by mid-century can be explained in light of two key factors: a political agenda by the ruling regime to modernize the Egyptian woman, and a prior history that westernized domesticity. Firstly, as historian Laura Bier discusses, Nasser's regime mobilized women's employment and motherhood as ways of strengthening and advocating the socialist party's approaches towards modernity and state-building.[17] In the collective consciousness, the image of the westernized modern woman was already associated with the elite and consequently with social status, success, and progress. The new administration saw the importance of holding on to such ideals and mobilized this image to work for the 'revolution'. Women were aggressively encouraged to get an education, join the workforce, and manage households that served the nation. Food was at the heart of the latter duty.

Secondly, as early as the 1900s, when Egypt was still a British protectorate, domestic education was regarded as an extension to the colonial endeavours to 'tame, sanitize, and control the Egyptians'.[18] As early as 1910, government-run girls' schools added cookery curricula in specialized domestic education modelled after the British educational system. Baheya Uthman describes how her own background was intertwined with this education history. Like hundreds of other women around the 1930s, Uthman received state-sponsored scholarships to study domestic science in England. Upon graduation from the Training College of Domestic Subjects, Berridge House, in England, as she describes in *Usūl al- Tahy al-Nazari wa-l 'Amali*, Uthman returned to Egypt to dedicate her time and efforts to the art of cooking in service to her nation.[19]

Thus, western modernity was deeply sought-after and indoctrinated in Egyptian society, but Egyptians had agency in how they altered it to suit their tastes. It was shared with a strong inclination to Egyptian and Turko-Egyptian food, resulting in a diversified mishmash of culinary tendencies that challenged rules about what types of dishes and ingredients went together, how food was prepared, and how it was eaten.

Summer played a large part in relaxing these rules, especially when it came to spending on food.

Relaxing the Rules at the 'Anti-picnics'

'Beaches in summer have a certain magic that... liberates the spirit from the restrictions imposed on us during the rest of the seasons,' says one writer in the popular Egyptian *al-Ithnayn* magazine, in an article about Egyptian beaches from 1949. The writer expresses how the seaside, its beaches and its air, has a 'spell' that permits a certain 'enjoyment drawn from spectacle, from visual pleasure, and sexual liberation'.[20] Indeed sexual freedom is not the subject of concern here, but wherever sensual pleasure goes, food usually follows. Egypt's *masyaf* spots, as portrayed in the films and novels of the time, were a case in point: romantic dates and formal dinners, belly dancing shows and buffets, flirty beach strolls and gelato, racy veranda nights with cold beer and snacks, and chatting up attractive strangers excused by buying *friska* or soda. Food is undeniably part of the pleasure obtained from these summer trips. Ethical and sometimes even logical rhetoric were given up. And that is perhaps why the urge to splurge on food was excusable, and how the abundance of seafood, pricey sliced bread, and imported cheese in our booklets became understandable.

However, complete surrender to such hedonism might result in turning trips and vacations to some sort of suffering, or as Walter Levy coined, 'anti-picnics'. As Levy explains, 'the anti-picnic is a counterpoint to what is expected' of picnics: summer trips are supposed to be positive, joyous, and beneficial to the mind and soul.[21] Instead, in the sample of Egyptian films I examined, we find summer trips resulting in heartbreak, lust, sin, and greed, usually used just as a narrative tool to drive the plot ahead. For example, one could easily slip into shameless spending, desperately flaunting wealth, or even in some cases stealing food. The film *Agazit Saif* (1966) comically features a scene where, after spending all the family's money at the end of the trip, a man steals kabab skewers out of extreme hunger while his sister succumbs to courting a wealthy man she doesn't like because he spoils her with mouth-watering dinners.

The booklets, though, seem to acknowledge the tough economic conditions many Egyptians were experiencing at the time. They feature some recipes that salvage 'leftovers', especially of chicken and meat, in salads or sandwiches. Another recipe ingeniously presents a faux *foie gras* using minced chicken liver and mustard. So perhaps it was not out of the ordinary to 'misbehave' and treat one's self to grilled shrimp, fancy English cheeses, and maraschino cherries a couple of weeks a year in order to truly indulge in summer.

Gamal Abdul Nasser's Government and Public Wellness

By the mid 1950s, the Egyptian government was heavily promoting various programmes to support social welfare, wellness, and recreation. It also invested in building *masyaf* resorts, and camps of various scales, that catered to different social classes. These efforts were part of a larger socialist ideology that aimed to have luxuries once restricted to the aristocracy now be within the common man's reach. As a colourful poster from 1964 promoting a new chalet resort project put it:

> The municipality of Alexandria now gives the opportunity to everyone to enjoy its sea and the magic of its beaches by building luxurious fully-furnished chalets right on its shore. The chalets are built from sound and heat-proof wood and consist of two bedrooms, a living room, a veranda, and a kitchen, that is equipped with a stove, a refrigerator and a telephone.[22]

Various government efforts to promote summer vacation travel included securing its employees transportation and accommodation as well as facilitating access to food. Many governmental entities and ministries in fact had their own exclusive summer resorts, while smaller state-owned companies held lotteries with paid summer vacation trips as prizes. There, meals for the employees and their families were provided as part of a daily package at a nominal charge, if not for free.[23]

Consequently, adapting and specifying recipes and preparations for summer, beach, and vacation foods became a priority. It is no surprise, then, that these state-published booklets were, in fact, the earliest recipe books solely dedicated to summer vacations. As much as Nasser's regime might have been over-ambitious in promising the fast-growing working class the luxuries of previous decades' aristocracy, it might have at least succeeded in realistically securing an unmatched level of accessible wellbeing and interest in recreation for most, as a result blurring lines between classes and their food.[24]

Conclusion

Summer vacation foods like those invoked in the *Hawāa* culinary booklets and portrayed in popular culture, with its puzzling contradictions and playfulness, provide us with a lens to a more intimate and nuanced reading of the mid-century social history of Egyptians. The recipes link to state strategies of wellness and recreation, the politicization of Eurocentric 'modern woman' notions by the regime, and to the history of British-modelled domestic education in the country.

The audience persona tool lends a way to imagine subtleties and personal lives of past readers, while summer-themed film and literature from the time bring to life micro-narratives of aspirations, pleasure and status-seeking, selective adoption of modernity, preferences for local tastes, and occasional bending the rules for summer indulgence. Such perspectives allow us to break away from established rigid dichotomies of East vs West, modern vs traditional, and working class vs elite, that are ascribed to Egyptian society of the time. Instead they offer a grey area where contradictions, adventures, and exceptions abound. This study could be extended through comparisons with oral history accounts and further contemplation of the anxieties between reality and aspirations, as we are reminded of the intention of cookbooks and the complexities surrounding the social historiography of food.

Notes

1. Baheya Uthman, *Akalat Khafifa lil-Shati' wal-Rahlat wal-Varanda* (Cairo: Dar al-Hilal, undated); Baheya Uthman, *Akalat lil-Rahalat* (Cairo: Dar al-Hilal, undated); Baheya Uthman, *Tabaq Khafif, Wajba Kamila* (Cairo: Dar al-Hilal, undated).
2. Baheya Uthman, *Akalat lil-Rahalat* (Cairo: Dar al-Hilal, undated), p. 1.
3. Barbara Wheaton, 'Finding Real Life in Cookbooks: The Adventures of a Culinary Historian', *Humanities Research Group Working Paper 7*. (1998); Máirtín Mac Con Iomaire. 'Towards a Structured Approach to Reading Historic Cookbooks', *The Journal of Media and Culture*, 16.3 (2013) <https://doi.org/10.5204/mcj.649>.
4. Fabio Parasecoli, *Bite Me: Food in Popular Culture* (New York: Berg, 2008), p. 3.
5. Mahmoud El-Gawhary, *Ex-Royal Palaces in Egypt: From Mohamed Ali to Farouk* (Cairo: Dar al-Maarif, 1954).
6. A 1959 summer vacation guide by *Al-Jeel* magazine contains insightful detailed information ranging from hotels, cabin rental prices, transportation, food and beverage options and costs, as well as nightlife and entertainment.
7. 'Masayif al-Sharq al-Arabi', *Al-Hilal*, 1 August 1947, p. 91.
8. Amina Al-Saeed, 'Al-Masayif Al-Madriyya Limatha Yahjuruha Al-Masryoon', *Al Hilal*, 1 August 1952, p. 25.
9. Ihsan Abdel Qudoos, 'Al-Bent Al-talta', *Al-Banat wal-Saif* (Cairo: Matbu'at Akhbar Al-Youm, 1958), pp. 135-83.
10. Ihsan Abdel Qudoos, 'Finjal Ahwa wa Hallit Molokhhia', *Al-Nisaa Lahunna Asnan Bayda* (Cairo: Al-dar Al-masriyya Al-lubnaniyya, 2016), pp. 9-29.
11. Ihsan Abdel Qudoos, 'Al-Bent Ar-rabia', *Al-Banat wal-Saif* (Cairo: Matbu'at Akhbar Al-Youm, 1958), pp. 135-233.
12. Munir Nasif, 'Marsa Matrooh Masyaf Hadi' fi Qalb Al-Sahara', *Al Arabi*, November 1968, p. 77.
13. Anny Gaul, 'Kitchen Histories in Modern Africa' (unpublished doctoral thesis, Georgetown University, 2019).
14. For more on this process, see e.g., Maria A. Kopacz, 'Who is Julia? Teaching Audience Analysis Through the Concept of Audience Persona', *Communication Teacher*, 36.2 (2022), 146–52, and Joe Pulizzi, *Epic Content Marketing: How to Tell a Different Story, Break Through the Clutter, and Win More Customers by Marketing Less* (New York: McGraw-Hill, 2013).
15. This was the first time for such an avant-garde take on rice to appear in any recipe books in Egypt, since probably as far back as the fourteenth century.
16. Shyon Baumann and Josée Johnston, 'Democracy versus Distinction: A Study of Omnivorousness in Gourmet Food Writing', *American Journal of Sociology*, 1 (2007), 165–204.
17. Laura Bier, *Revolutionary Womanhood: Feminisms, Modernity, and the State in Nasser's Egypt* (Redwood City, CA: Stanford University Press, 2011).
18. Gaul, p. 87.
19. Nazira Nichola and Baheya Uthman, *Usūl al- Tahy al- Nazari wa-l 'Amali* (Cairo: Al-Nahda Al-Masriyyah, 1981).
20. Amir Boktor, 'Likull Shati' Sihruh', *Al-Ithanyn*, 1 August 1949, p. 7.
21. Walter Levy, 'The Morality of Anti-Picnics', in *Food and Morality: Proceedings of the Oxford Symposium of Food and Cookery 2007*, ed. by Susan Friedland (Totnes, Devon: Prospect Books, 2008) pp. 165-72 (p. 172).
22. Municipality of Alexandria, Alexandria summer resort commercial poster, 1964, posted in Alexandria Ayam Zaman, *Facebook* <https://www.facebook.com/alexandriazaman1/photos/pb.100069345220410.-2207520000./488307521330031/?type=3> [accessed 13 October 2022].
23. Munir Nasif, 'Marsa Matrooh Masyaf Hadi' fi Qalb Al-Sahara,' *Al Arabi*, November 1968, p. 77.
24. Laura Bier, 'The Democratization of Well Being in Nasser's Egypt', *Jadaliyya*, 28 September 2020 <https://www.jadaliyya.com/Details/41782/The-Democratization-of-Well-Being-in-Nasser%E2%80%99s-Egypt> [accessed 2 November 2022].

Pemmican: An Ideal Trail Food

C. Thomas Shay

'The pemmican, which is so useful, and in fact almost essential, to the traveller'
— George Back, Arctic Explorer[1]

Meat has always been part of human cuisines.[2] Indeed, the term 'hunter-gatherer' applies to all past societies before the emergence of food production, ten to twelve thousand years ago.[3] When one or more animals were killed, everyone feasted on freshly-roasted flesh, while the excess was probably dried to preserve it for later use as a kind of 'pantry staple'. Sometimes this dried meat would be mixed with fat and bone marrow, creating a tasty concoction that in North America came to be known as pemmican, a word taken from the Cree language.

The mixing of dried meat and fat is ancient and may have begun in the Lower Palaeolithic, some 400,000 years ago.[4] The practice is worldwide, stretching from the sun-drenched plateaus of Central Asia to the grassy plains of North America, from Alaska's frozen tundra to Patagonia's lofty meadows and from the jerky-like *bakkwa* of China and Malaysia to the dried *kilishi* made by Hausa cooks in Nigeria and the *kavurmeh* (or *kawurmeh*) of central Africa.[5] There are many variations of this iconic portable food among indigenous North American cultures, and it has also been used by polar explorers and the military. It is still popular among some groups today.

Salt and spices both help preserve and flavour the various recipes but, in practical terms, drying is what keeps the meat from being spoilt by bacteria.[6] For example, a beef steak is about 70 percent water and has a limited 'shelf life' while dried beef, such as jerky, has less than 20 percent water and can be stored without refrigeration for a very long time. Being lighter, dried meat is also much easier for nomadic people or travellers to carry.[7]

While meat offers protein, added bone marrow and fat provide additional energy. Marrow is the spongy tissue in large leg bones where both white and red blood cells are formed. In nutritional terms, marrow contains a high-quality fat plus valuable vitamins and minerals. For example, one tablespoon (14 grams) of raw caribou bone marrow provides 14 calories (kcal) or 59 kilojoules of protein and 110 calories or 460 kilojoules of fat.[8] This fat thus supplies about eight times more nutritional energy than protein, a crucial consideration under sometimes trying physical circumstances.

Eighteenth-century armies also took advantage of the energy found in animal marrow. Judge John Joseph Henry's account of the failed attempt by Americans to capture Quebec City in December 1775 notes: 'We feasted till noon, and in the intermediate moments, culled the entrails for the fat: we even broke the bones, and extracted the marrow, under the full persuasion, that food of an oily nature, is one of strongest mainstays of human life.'[9]

Pemmican in Indigenous North America

The word pemmican is derived from the Cree, *pimîhkâ*, meaning 'manufactured grease'.[10] Ingredients and recipes vary across the continent, depending upon the animals and plants available. For instance, some Arctic Canadian Inuit use seal meat.[11] For the Dene of subarctic Canada, it might be caribou, moose, rabbit, or fish.[12] Maryann Sam of the James Bay Cree, located at the southern end of Hudson Bay, asserts that powdered fish makes the best pemmican. The powder can be mixed like a batter with melted lard, or the fat of caribou, goose, or moose.[13]

Among the Minnesota Anishinaabe (Ojibwa) to the south, deer, moose, and bear were dried and eaten as jerky. Timothy Roufs relates the words of Gabe-bines, who tells us that he much preferred bear jerky:

> I think they called it 'jerk steak' in English because it was so dry you had to jerk it to take a bite off and eat it. My mother made jerk steak from deer and bear, but bear dried meat was the best meat we ever tasted.[14]

Along the Rainey River between Minnesota and northwestern Ontario, sturgeon was favoured.[15]

Pemmican across the Great Plains

Archaeological evidence of making bone grease, a key ingredient in the food, suggests pemmican was made on the Great Plains between five and six thousand years ago.[16] A description of the bone grease-making is offered by Julia McDonald, a Vuntut Gwich'in resident of Old Crow Flats in the Canadian Yukon:

> After the meat has been cut off, the bones are left for one day, which allows them to dry a little. If the bones were left for two or three days, the bone grease made from them would taste too strong to be pleasant. [...] The bones [...] are smashed into little pieces, 'as big as finger nails,' with the back of an axe. [...] The broken bones are then put in a kettle with a little cold water and placed on the fire.[17]

Pemmican has been made by southern tribes such as the Comanche (Nermernuh)[18] and across the plains to the north, by the Blackfoot of Alberta and Montana. Ethnographer Clark Wissler wrote of the Piegan Blackfoot group Piikáni:

> While the Blackfoot had no cereal from which such bread substance could be made, they found a substitute in a compound of berries and flesh generally known as pemmican. For this, the best cuts of buffalo were dried in the

usual manner. Then they were pounded on a stone until fine. [...] Just before pounding, the pieces of dried meat were held over the fire to make them soft and oily. Marrow and other fats were heated and mixed with the pounded meats, after which crushed wild cherries were worked into the mess.[19]

Reducing the meat to a powder and smashing the bones into small pieces must have demanded many hours of work. Food historian Rachel Laudan points out that 'grinding and pounding are some of the heaviest tasks humans have ever undertaken'.[20]

Pemmican Ingredients and Recipes
Recipes depend upon the available ingredients: type of meat, fat source, preferred spices, and a sweetener. One issue common to all, however, is what should be the ratio between meat and fat? Among the Anishinaabe (Chippewa) of Turtle Mountain, North Dakota, the ratio is five parts meat to four parts fat. The pemmican recipe prepared by the Sisseton-Wahpeton Oyate (Dakota) and analyzed at the University of South Dakota consisted of 7.4 parts fat to 12.7 parts protein. A popular website prefers 1:1, except in a hot climate where it suggests one part fat to two parts meat.[21]

Choice of sweetener is another major issue. Ripe berries and fruits may serve but, where it has become available, sugar is used. Among the Lakota, whose name for pemmican is *wasna*, four fruits were routinely added: juneberry (also known as saskatoon), wild plum, sand cherry, and chokecherry. Other wild fruits have been used elsewhere.[22]

Energy from Pemmican
Carrying firewood back to camp, dragging a sled through the snow, or paddling a canoe for several hours all demand energy. An active person, whether working or simply on the trail, needs copious amounts of it, expending upwards of 3,000 to 4,000 calories (12,600 to 16,750 kilojoules) a day.[23] Where should this energy come from? Pemmican supplies calories from both fat and protein. Fat yields 9 calories (37.7 kilojoules) per gram, protein only 4 calories (16.7 kilojoules). Exercise physiologist Loren Cordain estimates that the typical pemmican recipe contained one part dried meat and one part fat. This works out to 73.7% of the energy from fat, and 26.3% from protein.[24] Cordain considers this nearly ideal. A calculation for the Turtle Mountain Anishinaabe recipe mentioned above produces 74% fat energy with 26% protein energy. The Sisseton-Wahpeton Oyate recipe yields only 57% fat. Proportions were probably altered, depending upon the pemmican's intended use. Notably, health problems result if the proportions stray too far from Cordain's ideal.[25]

Pemmican in Spiritual Life
Like many foods, pemmican became part of the people's spiritual life as Lois Frank notes in the introduction to *Foods of the Southwest Indian Nations*: 'The acts of hunting, growing, gathering, cooking, and eating take on a spiritual aspect akin to

prayer.'²⁶ This sacredness extends to the added ingredients such as the berries. In the past, pemmican featured in religious rituals including the Lakota Ghost Keeper mortuary ceremony, the Blackfoot Horns Society, and as a substitute for bread in the Christian sacrament of Holy Communion within the Red River region.²⁷ It also found its way into Indigenous stories.²⁸

The Fur Trade and Pemmican

By the early seventeenth century, most fur-bearing mammals in England and much of northern Europe had disappeared or become scarce because of over-trapping and habitat loss. Meanwhile, the demand for fur was on the rise as beaver hats came into fashion.²⁹ In response, traders increasingly turned to North America where, a century or so earlier, explorers had described a wilderness teaming with beaver and other wildlife. Trade in furs began on the Atlantic coast in the 1500s and gradually moved inland, reaching the St. Lawrence River in what is now Québec by the early 1600s.³⁰

As Europeans demanded more and more fur for hats, coats, jackets, capes, linings, muffs, boots, stoles, shawls, gloves, slippers, trimming, and other apparel, the pressure on the major trading companies to increase their supply led them to expand into the boreal forests of northern Alberta and British Columbia. Feeding their many traders and canoe men (voyageurs) became a serious problem since game in the north was scarce. Pemmican became the solution.

The demand for pemmican in northwest Canada grew until its production and distribution became a major part of the trading efforts. During the 1860s, the Hudson's Bay Company needed over 90,000 kg (220,000 lbs) a year just to feed its boatmen.³¹ Much of this pemmican came from the prairies of southern Manitoba and Saskatchewan, produced by the Mètis (people of mixed European-First Nations ancestry).³² In his celebrated book, *Pemmican Empire*, George Colpitts detailed this episode of fur trade history.³³

From Food to Fuel

Did this energy-rich trail food truly provide the calories needed by those intrepid voyageurs who paddled large canoes, heavy with furs and trade goods, over long distances?³⁴ The voyageurs reportedly ate regular meals containing carbohydrates from maize, wild rice, and ship's biscuit. Roasted meat and peas provided protein. Meat and maize would also provide some minerals and electrolytes. Pork fat used in cooking added to the mix. Wild fruits, wild rice, and flour furnished vitamin, minerals, and fibre. Pemmican offered both fat and protein.

Was this food intake enough to fully fuel their paddling and carrying? Did they even have time to prepare and consume this much food? While they may have practiced 'carb loading' as some endurance athletes do, I can't find evidence of this.³⁵ They may not have even needed to burn as many calories/joules after the first few days of hard paddling because, over time, human bodies can adapt to arduous tasks and use fewer calories than expected.³⁶ Tapering off of energy expenditure also applies to contemporary paddlers of the Canadian

Voyageur Brigade Society.³⁷ Even the gut flora of endurance athletes have been shown to alter in a beneficial way.³⁸

In a search for answers, I examined the lifestyle of these men and learned that human energy needs can be met in ways beyond pure caloric intake. Psychological adjustment and some routine activities probably helped everyone rise to the gruelling physical demands. For example:

- Pipe breaks were taken every fifteen to twenty miles. Tobacco is supposedly an appetite suppressant though its effects may be only temporary.³⁹ Beyond that, these breaks could have been a time of de-stressing that enabled the men to rest just enough for the next strenuous bout of paddling. Injuries and portages also offered time to rest.⁴⁰
- Songs helped mark the paddling cadence and boosted morale.⁴¹
- Following versus leading. The lead boat was cutting new water, but the boats that followed benefited from something called the 'wash riding' effect. This means they didn't need to work quite as hard to cover the same distance.⁴²
- Teamwork. Being together for weeks must have spawned teamwork among the crew as well as a desire by some to lead that team. Individual canoes may have challenged others in the brigade to do their best just to keep up. As P.E. Vernon wrote about the sport of sculling, 'For a boat to go well and win its races there must be developed both a close physiological and psychological interconnection between members of the crew.'⁴³ Something similar would have also helped the voyageurs.

Even so, at times, I think so-called 'caloric deficits' were more common than not. In addition, various injuries – similar to those suffered by Atlantic rowers today – were not uncommon and these would have increased nutritional needs that, perhaps, could not be met in the field.⁴⁴

Polar Explorers and the Military Use Pemmican

Polar lands, especially the Arctic, have attracted European explorers for centuries. Beginning with John Cabot in 1497, dozens of expeditions sought the fabled northwest passage to the Orient. Others ventured to Antarctica. Though spurred by commerce and national sovereignty, these adventurers followed their passion and challenged themselves. The hardship and danger involved is almost unthinkable.

Compact and nutritious, pemmican became a mainstay of many expeditions. A list of explorers who have packed it on their journeys reads like a who's who – Perry, Amundsen, Scott, Shackleton, Nansen.⁴⁵ South Pole explorer Robert Scott summed up the experience of many: 'There can be little question, therefore, that polar sledging runs an easy first as a hunger-producing employment.'⁴⁶ That hunger was often satisfied by chewing on this potent mixture.⁴⁷ In a slight twist, the Danish army fed both men and dogs in their sledge patrols across Greenland what was called pemmican.

It consisted of soya protein, pea flour, milk protein, vitamins, and minerals.[48]

It has been said that armies travel on their stomachs, and logistics demand lightweight and nutritious field rations. As a result, pemmican and jerked beef were common in military campaigns for several decades after the American Civil War.[49] However, experiences during World War II were mixed as variations on the Indigenous product were tried. In 1942, a test of dehydrated and compressed meat bars (essentially cold hamburgers) was abandoned when the food did not keep well.[50] After the war, another test fed an infantry platoon nothing but tea and pemmican. This trial was also abandoned when the men became 'listless with drawn faces and sunken eyes'.[51] The men recovered after eating pemmican mixed with carbohydrates.[52]

Pemmican Today

The traditional version of pemmican made with bison meat is still being produced and eaten by Indigenous groups across the plains. A leading commercial favourite is the Tanka Bar. The company's mix of bison meat and cranberries is a story of rocketing success followed by almost complete failure followed by a promising comeback.

Started in 2006 on the Pine Ridge Reservation in South Dakota, Tanka introduced the first meat and fruit bison bar with a plan to create a product that would help ease the soaring local unemployment rate and also restore the bison's presence among the Oglala Lakota people at Pine Ridge. Over time, Tanka became a commercial success. Then a giant food company launched its own bison bar in 2018. The national publicity generated by that company increased demand for Tanka's product, but they could not scale up fast enough. Tanka almost went under as dozens of nimble competitors took over the market. In 2020, however, Tanka entered into collaboration with a network of independent family farmers and ranchers that enabled it to secure a supply of bison meat and operating capital that helped them address their production issues. Its future again looks bright as the top meat bar supplier.[53]

Without a doubt, pemmican is one of the oldest and most widely eaten portable foods on the planet. While maintaining its historic roots and recipes, it has also evolved culturally to become a familiar part of today's world of portable food.

Postscript

> A wonderful food is the pemmican
> It's been made since the world began
> Just a morsel of meat
> can feed one for a week
> I'm darned if I know how the hellican!

Acknowledgements

I appreciate the help and advice of Carolyn Podruchny, Margaret Pearce, Herman

Pemmican: An Ideal Trail Food

Pontzer, Brian Smith, Mark Lund, Ted Bentley, Trevor Connor, Mark Willems, George Colpitts, Jack Brink, John Speth, Douglas Bamforth, and Linea Sundstrom, and the assistance of Beth Page.

Notes

1. George Back, *Narrative of the Arctic Land Expedition to the Mouth of the Great Fish River and Along the Shores of the Arctic Ocean, in the Years 1833, 1834, and 1835* (London: John Murray, 1836), vol. 92, p. 501.
2. C.S. Larsen, 'Animal Source Foods and Human Health During Evolution', *The Journal of Nutrition*, 133.11 Sup. 2 (2003), 3893S–3897S <https://doi.org/10.1093/jn/133.11.3893S>.
3. Greger Larson and others, 'Current Perspectives and the Future of Domestication Studies', *Proceedings of the National Academy of Sciences*, 111.17 (2014), 6139–46 <https://doi.org/10.1073/pnas.1323964111>
4. R.J. Blasco and others, 'Bone Marrow Storage and Delayed Consumption at Middle Pleistocene Qesem Cave, Israel (420 to 200 ka)', *Science Advances*, 5.10 (2019) <https://doi.org/10.1126/sciadv.aav9822>.
5. This jerked meat, when dried, is broken into small pieces and stored in pots full of clarified and melted butter. Richard F. Burton, *The Lake Regions of Central Africa: A Picture of Exploration*, 2 vols (London: Longman, Green, Longman, and Roberts, 1860), II, p. 285 <https://burtoniana.org> [accessed 20 May 2022]; Edward N. Wentworth, 'Dried Meat: Early Man's Travel Ration', *Agricultural History*, 30.1 (1956), 2–10.
6. A. Casaburi and others, 'Bacterial Populations and the Volatilome Associated to Meat Spoilage', *Food Microbiology*, 45 (2015), 83–102.
7. D.G. Lim and others, 'Effects of Different Drying Methods on Quality Traits of Hanwoo Beef Jerky from Low-Valued Cuts during Storage', *Food Science of Animal Resources*, 32.5 (2012), 531–39; U.S. Department of Agriculture, Agricultural Research Service, 'Beef, ground, 80% lean meat / 20% fat, raw', FoodData Central <https://fdc.nal.usda.gov> [accessed 7 February 2022].
8. Rachel Link, 'Bone Marrow: Nutrition, Benefits, and Food Sources', *Healthline Media* (2019) <https://www.healthline.com/nutrition/bone-marrow> [accessed 7 February 2022]; J.W. Brink, 'Fat Content in Leg Bones of Bison bison, and Applications to Archaeology', *Journal of Archaeological Science*, 24.3 (1997), 259–74 <https://doi.org/10.1006/jasc.1996.0109>.
9. J.J. Henry, *An Accurate and Interesting Account of the Hardships and Sufferings of that Band of Heroes, Who Traversed the Wilderness in the Campaign against Quebec in 1775* (Lancaster PA: William Greer, 1812), p. 46.
10. *Online Cree Dictionary* <https://www.creedictionary.com> and *The Canadian Encyclopedia* <thecanadianencyclopedia.ca/en/article/pemmican> [accessed 29 April 2022].
11. Liam Frink and Celeste Giordano, 'Women and Subsistence Food Technology: The Arctic Seal Poke Storage System', *Food and Foodways*, 23.4 (2015), 251–72 <https://doi.org/10.1080/07409710.2015.1099906>.
12. H.J. Brumbach and R. Jarvenpa, 'Ethnoarchaeology of Subsistence Space and Gender: A Subarctic Dene Case', *American Antiquity*, 62.3 (1997), 414–36 <https://doi.org/10.2307/282163>.
13. Maryann Sam, 'Pemmican', in *Traditional Indian Recipes From Fort George, Quebec* ([n.p.], 1967), p. 27, referenced by S. Hicks, 'Eating History: An Experiential Examination of Pemmican', <http://activehistory.ca/2019/07/eating-history-an-experiential-examination-of-pemmican/> [accessed 17 March 2022].
14. 'Chapter 9 Bears', in *When Everybody Called Me Gabe-bines, Forever-Flying-Bird: Teachings from Paul Buffalo*, ed. by T.G. Roufs <https://www.d.umn.edu/cla/faculty/troufs/Buffalo/pbwww.html> [accessed 17 May 2022].
15. T.E. Holzkamm, V.P. Lytwyn, and L.G. Waisberg, 'Rainy River Sturgeon: An Ojibway Resource in the Fur Trade Economy', *The Canadian Geographer/Le Géographe canadien*, 32.3 (1988), 194–205.
16. D. Bamforth, 'Origin Stories, Archaeological Evidence, and Post-Clovis Paleoindian Bison Hunting on the Great Plains', *American Antiquity*, 76.1 (2011), 24–40 <https://doi.org/10.7183/0002-7316.76.1.24>; Jonathan Douglas Baker, 'Prehistoric Bone Grease Production in Wisconsin's Driftless Area: A Review

of the Evidence and Its Implications' (unpublished master's thesis, University of Tennessee, 2009) <https://trace.tennessee.edu/utk_gradthes/508> [accessed 22 May 2022]; A. Janzen and others, 'Smaller Fragment Size Facilitates Energy-efficient Bone Grease Production', *Journal of Archaeological Science*, 49 (2014), 518–23; and Jack Brink, personal communication, 9 February 2022. Eugène Morin has shown, through careful study of bone fragments plus experiments in fracturing bones of red deer (*Cervus elaphus*), the practice of bone grease production in ancient sites: see E. Morin and M.C. Soulier, 'New Criteria for the Archaeological Identification of Bone Grease Processing', *American Antiquity*, 82.1 (2017), 96122; E. Morin, 'Rethinking the Emergence of Bone Grease Procurement', *Journal of Anthropological Archaeology*, 59 (2020), 101178; and E. Morin, 'Revisiting Bone Grease Rendering in Highly Fragmented Assemblages', *American Antiquity*, 85.3 (2020), 535–53.

17 D. Leechman, 'Bone Grease', *American Antiquity*, 16.4 (1951), 355 <https://doi.org/10.2307/276988>.

18 T. Kemper, 'Life Among the Comanches', *Westview*, 14.3 (1995), Article 10 <https://dc.swosu.edu/westview/vol14/iss3/10> [accessed 29 April 2022].

19 Clark Wissler, *Material Culture of the Blackfoot Indian* (New York: Anthropological Papers of the American Museum of Natural History, 1910), pp. 22–23.

20 Rachel Laudan, *Cuisine and Empire: Cooking in World History* (Oakland: University of California Press, 2015), pp. 31–33; and Rachel Laudan, 'Pounding and Grinding' <http://www.rachellaudan.com> [accessed 17 May 2022].

21 Kade M. Ferris, 'Pemmican: The Indigenous Super Food', 1 December 2019, Turtle Mountain Chippewa Heritage Center <http://www.chippewaheritage.com> [accessed 30 April 2022]; S.P. Stluka and M.L. Gengler, 'Nutritional Composition of Selected Traditional Native American Foods', *The Journal of Undergraduate Research*, 2.1 (2004), Article 2 <http://openprairie.sdstate.edu/jur/vol2/iss1/2> [accessed 30 April 2022]. For recipes, see Alderleaf Wilderness College, 'Four Pemmican Recipes' <https://www.wildernesscollege.com/pemmican-recipes.html> [accessed 16 May 2022].

22 T.M. Ngapo and others, 'Pemmican, an Endurance Food: Past and Present', *Meat Science*, 178 (2021), 108526 <https://doi.org/10.1016/j.meatsci.2021.108526>.

23 Energy expenditure may be up to 7000 Calories (29,300 kilojoules). R. Passmore and J.V. Durnin, 'Human Energy Expenditure', *Physiological Reviews*, 35.4 (1955), 801–40; J.H. O'Keefe and others, 'Organic Fitness: Physical Activity Consistent with Our Hunter-Gatherer Heritage', *The Physician and Sportsmedicine*, 38.4 (2010), 11–18; Captain Calculator, <https://captaincalculator.com> [accessed 25 April 2022]; G. Colpitts, *Pemmican Empire: Food, Trade, and the Last Bison Hunts in the North American Plains, 1780–1882* (Cambridge University Press, 2014), p. 33.

24 Loren Cordain, 'Pemmican: A Plains Indians Staple Food that Prevented Protein Poisoning', *The Paleo Diet*, 17 June 2016 <https://thepaleodiet.com/pemmican-a-plains-indians-staple-food-that-prevented-protein-poisoning> [accessed 21 May 2022].

25 J.D. Speth and K.A. Spielmann, 'Energy Source, Protein Metabolism, and Hunter-Gatherer Subsistence Strategies', *Journal of Anthropological Archaeology*, 2.1 (1983), 1–31. A person who gets all, or nearly all, of their energy from fat will generate ketones, resulting in ketosis. Ketosis does not, in itself, mean a person is starving or otherwise not meeting their caloric needs although ketosis will also occur during starvation (Herman Pontzer, personal communication, 25 October 2022). For weight loss, some experts recommend ketosis-type diets (B. O'Neill, and P. Raggi, 'The Ketogenic Diet: Pros and Cons', *Atherosclerosis*, 292 (2020), 119–26; W. Dafoe and G.T. Gyenes, 'Comments on "The Ketogenic Diet: Pros and Cons",' *Atherosclerosis*, 296 (2020), 1).

26 Lois Frank, *Foods of the Southwest Indian Nations: Traditional and Contemporary Native American Recipes* (Berkeley CA: Ten Speed Press, 2002), Introduction.

27 B. Hungry Wolf, *The Ways of My Grandmothers* (New York: William Morrow & Company, 1980), pp. 183–89 and B. Hungry Wolf, 'Life in Harmony with Nature', in *Women of the First Nations: Power, Wisdom, and Strength*, ed. by C. Miller and P. Chuchryk (Winnipeg: University of Manitoba Press, 1996), pp. 77–82. For ritual use among the Lakota, see William K. Powers and Marla M.N. Powers, 'Metaphysical Aspects of an Oglala Food System' in *Food in the Social Order: Studies of Food and*

Festivities in Three American Communities, ed. by Mary Douglas (Russell Sage Foundation, 1984), pp. 40–96. For the Blackfoot, see Dave Melting Tallow and Joe Gambler, 'Interview with Joe Gambler' (University of Regina: Indian History Film Project, 1966) <http://hdl.handle.net/10294/62> [accessed 17 May 2022]. For Holy Communion, see Linda W. Slaughter, 'Leaves from Northwestern History', in *Collections of the State Historical Society of North Dakota* (Bismarck: Tribune, State Printers and Binders, 1906), vol. I, pp. 200–23 (p. 223).

28 F. Ballinger, 'Coyote, He/She Was Going There: Sex and Gender in Native American Trickster Stories', *Studies in American Indian Literatures*, 12.4 (2000), 15–43; M.W. Beckwith, 'Mythology of the Oglala Dakota', *The Journal of American Folklore*, 43.170 (1930), 339–442 <https://doi.org/10.2307/535138>.

29 J.F. Crean, 'Hats and the Fur Trade', *The Canadian Journal of Economics and Political Science/Revue canadienne d'Economique et de Science politique*, 28.3 (1962), 373–86.

30 H.A. Innis, *The Fur Trade in Canada: An Introduction to Canadian Economic History* (1930; repr. Toronto: University of Toronto Press, 1999).

31 W.A. Dobak, 'Killing the Canadian Buffalo, 1821–1881', *The Western Historical Quarterly*, 27.1 (1996), 33–52 (p. 44).

32 W.B. Merriam, 'The Role of Pemmican in the Canadian Northwest Fur Trade', *Yearbook of the Association of Pacific Coast Geographers*, 17.1 (1955), 34–38; V. Benoit, 'French Presence in the Red River Valley Part I: A History of the Métis to 1870', in *L'Heritage Tranquille: The Quiet Heritage, Proceedings from a Conference on the Contributions of the French to the Upper Midwest, November 9, 1985*, ed. by C. A. Glasrud (Moorehead, MN: Concordia College, 1987), pp. 116–33.

33 The North West Company merged with the more powerful Hudson's Bay Company in 1821. Colpitts, *Pemmican Empire*; A. Carlos, 'The Causes and Origins of the North American Fur Trade Rivalry: 1804–1810', *The Journal of Economic History*, 41.4 (1981), 777–94; Adam R. Hodge, 'In Want of Nourishment for to Keep Them Alive: Climate Fluctuations, Bison Scarcity, and the Smallpox Epidemic of 1780–82 on the Northern Great Plains', *Environmental History*, 17.2 (2012), 365–403; R. J. Perry, 'The Fur Trade and the Status of Women in the Western Subarctic', *Ethnohistory*, 26.4 (1979), 363–75.

34 Trips averaged 28 or 42 days on the 1700 km journey between Lachine, Quebec and Fort William with 38 energy-intensive portages between Lachine and Fort William. One journey took 79 days (Margaret W. Pearce, John Macdonell, and Charles M. Gates, 'The intricacy of these turns and windings: a voyageur's map', Journey Cake, Marshall, MI (2005), OCLC Number 58983930).

35 N. Clark, 'Carbo-loading: Tips for Endurance Athletes', *Palaestra*, 23.1 (2007), 44–46.

36 Hermon Pontzer, personal communication, 8 August 2022; C. Thurber and others, 'Extreme Events Reveal an Alimentary Limit on Sustained Maximal Human Energy Expenditure', *Science Advances*, 5.6 (2019) <https://doi.org/10.1126/sciadv.aaw0341>.

37 Ted Bentley, personal communication, 12 August 2022; Brian Smith, personal communication, 9 August 2022; Mark Lund, personal communication, 13 August 2022.

38 D.M. Keohane and others, 'Four Men in a Boat: Ultra-endurance Exercise Alters the Gut Microbiome', *Journal of Science and Medicine in Sport*, 22.9 (2019), 1059–64; G. Miranda-Comas and others, 'Implications of the Gut Microbiome in Sports', *Sports Health*, p.19417381211060006.

39 K.A. Perkins and others, 'Acute Effects of Tobacco Smoking on Hunger and Eating in Male and Female Smokers', *Appetite*, 22.2 (1994), 149–58 <https://doi: 10.1006/appe.1994.1014>; M. Pilhatsch and others. 'Nicotine Administration in Healthy Non-Smokers Reduces Appetite But Does Not Alter Plasma Ghrelin', *Human Psychopharmacology: Clinical and Experimental*, 29.4 (2014), 384–387; T. Hu, Z. Yang, and M.D. Li, 'Pharmacological Effects and Regulatory Mechanisms of Tobacco Smoking Effects on Food Intake and Weight Control', *Journal of Neuroimmune Pharmacology*, 13.4 (2018), 453–466.

40 Carolyn Podruchny, personal communication, 23 July 2022.

41 'Songs of the Voyageur', Les Productions Rivard <http://rendezvousvoyageurs.ca/en/world/leisure/songs.html> [accessed 10 October 2022].

42 G.L. Gray, 'Oxygen Consumption During Kayak Paddling' (unpublished master's thesis, University of British Columbia, 1992) <https://doi.org/10.14288/1.0077099>.

43 P.E Vernon, 'The Psychology of Rowing', *British Journal of Psychology*, 18.3 (1928), 317–331 (p. 330).
44 C. Podruchny, 'Werewolves and Windigos: Narratives of Cannibal Monsters in French-Canadian Voyageur Oral Tradition', *Ethnohistory*, 51.4 (2004), 677–700. For specific injuries, see W.J. Galsworthy, J.A. Carr, and R.P. Hearn, 'Common Health Issues Encountered by Ultraendurance Ocean Rowers', *Wilderness & Environmental Medicine*, 33.1 (2022), 97–101.
45 Ngapo and others.
46 R.F. Scott, *The Voyage of the Discovery*, 2 vols. (New York: Charles Scribner's Sons, 1907), I, p. 323.
47 Ngapo and others.
48 L. Vanggaard, *The Effects of Exhaustive Military Activities in Man. The Performance of Small Isolated Military Units in Extreme Environmental Conditions* (Royal Danish Navy Gentofte: Danish Armed Forces Health Services, 2001), pp. 9–10.
49 Franz A. Koehler, *Special Rations for the Armed Forces,* 1946-53 QMC Historical Studies, Series II, No. 6 (Washington DC: Historical Branch, Office of the Quartermaster General, 1958).
50 D. Fitzgerald, 'World War II and the Quest for Time-Insensitive Foods', *Osiris*, 35.1 (2020), 291–309 (p. 295).
51 J.S. Edwards, E.W. Askew, and N. King, 'Rations in Cold Arctic Environments: Recent American Military Experiences', *Wilderness & Environmental Medicine*, 6.4 (1995), 407–22, (p. 412).
52 Perhaps the three-day trial was not long enough for the men to adjust to the pemmican diet (John Speth, personal communication, 23 October 2022).
53 M. Noble, 'One Year after Native-owned Tanka Bar Had Lost Nearly Everything, the Buffalo Are on Their Way Back', *The Counter,* 24 Jan 2020 <https://thecounter.org/tanka-bar-niman-ranch-bison-grassfed/> [accessed 14 May 2022].

'The Clever Dining Car Conductor': Creating a Luxury Dining Experience on the Move in Britain, 1879-1948

Chloe Shields

This paper focuses on the working lives of dining car conductors and the role that they played in creating a luxury dining experience for people on the move in Britain between 1879 and 1948. The setting of the dining car aligns itself with the symposium theme of 'Portable Foods: Foods Away from the Table' or at least food away from a stationary table to a table in a dining car in the industrial space of the railways, travelling at roughly sixty miles per hour. The paper will focus on dining car conductors and how they were portrayed by railway companies and passengers, and what these depictions imply about railway travelling in this period. The services of the Great Western Railway will be the primary focus of this paper, analyzing a company dining car worker's instruction manual from March 1920 to illustrate the railway company's expectations of the dining car service. I will use a wide variety of additional sources, including railway magazines, menus, and newspaper 'letters to the editor' to investigate the role of the dining car conductor in creating luxury dining experiences on the move.

Prior to the introduction of the dining car in Britain in 1879, both railway catering and general passenger comforts and safety were heavily criticized by the travelling public. During the nineteenth century, as the railways spread across Britain and passenger services increased, there was growing demand from the public for railway companies to improve both the safety and the comfort of railway travel. In 1871, an anonymous article in *The Graphic* titled the 'Pace That Kills' discussed the haste which ensued during rail travel from arriving at the platform to buying and eating a meal. The author stated that 'It must surely be that heart-disease and nervous complaints are largely engendered by what we go through when we travel.'[1] Travellers were anxious about the safety of the railways as well as how the railways increased the speed in which they lived their lives. The author went further to suggest ways in which railway companies in Britain could improve their services taking inspiration from abroad, such as the luncheon baskets in France and the restaurant car in the United States.[2]

Travellers turned to the railway companies for improvements. However, when the companies fell short of expectations, railway travellers used the resources at their disposal to make their voices heard. Many, for instance, wrote 'letters to the editor' to complain about the poor railway standards. These letters were written to the editors of newspapers by the public, quite often to complain about aspects of daily life, and they were then published in the weekly newspapers. Railway catering was not immune to these criticisms in newspapers; both articles and letters to the editor complained about railway refreshments, particularly the poor quality of the food on offer, the inflated prices, and the short refreshment stops.

One of the most complained about stations was Swindon Junction on the Great Western Railway. When their first refreshment room was built, the owners of the building struck a deal that meant every passenger train that passed through the station had to stop for ten minutes for passengers to visit the refreshment room.[3] Passengers complained about the brevity of these stops and the quality of the food on offer. The most infamous complaint was written by Charles Dickens in *The Boy at Mugby* (1866) where he described the refreshment room workers as uncaring and related how they served appalling, stale plates of 'sawdust sandwiches'.[4] His tale was often quoted by others when writing to the editor of *The Times* to complain about the quality and prices of railway fares. These complaints were not just about bad coffee and stale sandwiches; they also reflected the bad reputation of the railway companies. The poor food they served was seen as another example of railway companies' apathy towards their passengers. For railway travellers, the poor meals reflected their anxieties about how the railways were affecting their everyday lives.

However, as the nineteenth century progressed, there was a surge of developments in railway technologies that improved passenger travel including the invention of the dining car. The first dining car service was introduced to Britain in 1879. It was the creation of an American inventor and entrepreneur called George Mortimer Pullman, and had run in the United States since 1868.[5] The first dining car in Britain was called the 'Prince of Wales' and was reserved for first class travellers. Introduced by the Great Northern Railway, its first official run was on 1 November 1879 from Leeds to London. The dining car resembled the luxury and comfort of a high-end restaurant or a gentleman's club. The carriage was split into several sections. There was a dining room, a smoking room, and a small kitchen on one end 'with all culinary appliances' including a coal burning range, a pantry, and a wine and spirit store. There were ten seats upholstered in crimson velvet with tables in both the dining room and the smoking room. Beside each table there were electric bells which rang outside the steward's room. The dining room interior had 'a light and charming effect' created by a veneer of white oak panels on the ceiling with painted flowers, and the floor laid with 'Brussels carpet' with 'ottomans to match'.[6]

The carriage was luxuriously decorated, and the food was a vast improvement on the stale buns that had previously been on offer. Instead, passengers were invited to enjoy a leisurely hot meal in the dining car, cooked by the cook in the kitchen and served to them by the dining car conductor and attendants. The *Wakefield Express* informed that the dining car served both a 'substantial lunch consisting of mutton cutlets and green peas, cold meat and mashed potatoes, cheese and celery' and a 'first-class dinner of six courses, consisting of soup, fish, roast joint, with vegetables and sweets, and café to follow'. The meal was seen as a success, and they went further to note the stability of the meal. Despite the speed in which they were travelling no wine spilled on the floor, no plates toppled into passengers' laps, and no decanters slid off the table. The reporter even stated that 'there was an entire absence of the feeling of rapid travelling'.[7]

This last point, whilst an exaggeration, highlights a shift in public thinking about rail travel. The introduction of the dining car and other comforts on the railway showed changes in both how railway companies approached passenger rail travel and public perception. Amy Richter described how in the United States, Pullman used 'beautiful interiors, comfortable furnishings, and well-designed amenities' to recreate domestic space on the train.[8] Similarly in Britain, railway companies hoped to use the dining car, their food offerings, and their employees to make railway travelling a less industrial experience by imitating typical luxury gastronomic settings such as restaurants, gentleman's clubs, or the home. The experience of eating in the railway dining car distracted travellers from the experience of rail travel. Passengers were not fooled into thinking they were sat in a stationary restaurant; however, they were removed as much as possible from the industrial movement of the train. Unlike previous dining experiences where passengers were forced to rush to eat what food they could buy at the platform in the allotted time that the train stopped, the passengers in the dining car could eat at a slower, more leisurely pace. The Great Northern Railway dining car was widely praised, and by the early twentieth century most of the company's competitors had invested in their own. The railway companies' investments in these new technologies meant that, rather than railway companies being criticized about their lack of safety and passenger care, they were praised for their investments in improving comforts for railway passengers. Essential to these comforts were the passenger-facing railway workers, such as the dining car conductors, who made these experiences possible. Railway companies and the travelling public alike portrayed their role as vital to creating these experiences.

Dining cars in the late-nineteenth to the mid-twentieth century varied in size and design: some having a separate kitchen car and multiple dining cars, others having one car split with both a kitchen and seating area. As technology throughout the period developed, however, the types of workers and the hierarchy of their employment was consistent throughout. Whilst there were varying numbers of workers, the hierarchical structure maintained a standard. The most junior members of staff were page boys, then pantry and kitchen assistants, dining car attendants, assistant cooks, cooks, and then the dining car conductor. Dining car workers in Britain were predominantly men; nevertheless during the first and the second world war more women were hired to replace men who had joined the war effort.[9] The dining car conductors managed the dining cars and were responsible for the success of the operation, supervising the other dining car workers and ensuring the comfort of those travelling.

The responsibilities of dining car conductors included every aspect of the passenger dining experience, and railway companies expected a high standard of service to be kept. The Great Western Railway published a manual for their restaurant car workers in 1920 which included one hundred and two instructions for their workers, sixty-nine of which were specifically dedicated to the dining car conductor. The prescriptive nature of the manual perhaps suggests more about the railway companies' expectations of their dining car conductors than it does about the workers themselves. For example, number eighty-one of the responsibilities of the dining car conductor was to collaborate with the cook to decide on the menu, and the conductor was responsible for quickly and accurately relaying the passenger food orders to the cook to prevent waste. However, the instructions went further to say that the conductor was to supervise the cook and that 'All food served to passengers must be of ample portions and "follows" must be served on request. Passengers must not be stinted in any way'.[10] This perhaps suggests, as the conductor was expected to collaborate with the cook on the menu and to make sure that the cook was performing his duties properly, the railway companies expected the dining car conductor to understand what food should be served on the menu and the processes involved in cooking the meals properly. However, what is most telling is that the railway company did not want the passenger to feel stinted. It suggests how the company valued the passenger's opinion of their dining experience. Providing plentiful portions and the opportunity for a second helping suggests that the railway company aspired for the dining car to be a space where passengers are left satisfied by their experience and the conductor is responsible for maintaining these standards.

Alongside the quantity of the food served, the manual also instructed the conductor to maintain certain standards when it came to how the food and drinks were served and prepared. 'China Tea', coffee, and cocoa were to be served at all times when required by passengers; it was imperative that silver teapots were used unless a passenger asked for a china teapot. Cheese was to be handed round whole, placed on 'a flat dish covered with a d'oyley' and 'Fruit Tarts must always be cooked in pie-dishes and served hot. Stewed fruit and baked crust cooked separately must not be served as fruit tart'.[11] This last instruction was given to both the cooks and the conductors, most likely because the conductors were to supervise the food being prepared. These instructions suggest how the Great Western Railway expected their dining car workers and in particular the conductors to maintain a meticulous standard. By serving and making foods in the same way they would be made in the kitchens of a restaurant or a gentleman's club, the railway company could retain the illusion that the dining car provided an equivalent experience of authentically made food, despite travelling at sixty miles per hour. The dining car conductor's role in maintaining this high-quality service meant that passengers could rely on having a luxury dining experience.

One task that conductors performed was to seat passengers in the dining car, resembling the service in a restaurant or a gentleman's club. However, unlike in a restaurant or a gentleman's club, passengers could be seated at a table with strangers due to the limited space of the dining car. Originally there were separate dining cars for first and third class, but as the twentieth century progressed, regular dining cars were introduced which seated all classes of passenger with a ticket. The unexpected encounter with a stranger on a train was a source of anxiety and anticipation and became a common theme in railway writing of the early nineteenth and twentieth century. Peter Ritchie Calder, a journalist, wrote a non-fiction book titled *Roving Commission* in 1937. The book documented a trip he took around Britain where he worked in different occupations and recounted his experiences. As part of his travels, he worked as a dining car attendant working on the 'Flying Scotsman'. He discussed many aspects of working as a dining car attendant, including playing cupid where he matched a woman he named 'Diana of the Dining Car' with a handsome gentleman. By seating both passengers at the same table, Calder enabled the passengers to overcome the indecency of two strangers, particularly of different sexes, sharing a meal together. He stated how, by placing them together and allowing them to make introductions to one another, 'the proprieties had been satisfied'.[12] Whilst this is a glamourized account of a dining car encounter, it does highlight the role of the workers in seating passengers and maintaining decorum.

Part of the dining car conductor's role was to specifically seat people in the dining car. In doing so, they prevented any disputes between passengers. The Great Western Railway went slightly further within their instruction manual by stating that whilst 'there is no distinction of class as regards seats in the regular cars, discretion must be used in keeping first and third-class passengers separate, as far as practicable'.[13] These stipulations in the instruction manual portrayed how part of the conductor's role was to enforce class boundaries, and in doing so maintaining a luxury experience for first class passengers who would no longer have to engage with the lower classes as they might when at a refreshment stop at a provincial station.

The dining car conductor was seen as an essential part of the dining car experience, particularly by the Great Western Railway. Their role helped to create a comfortable and luxurious experience for railway passengers. Railway companies were not alone in highlighting the importance of the dining car conductor, an article in *The Windsor* in 1905, included a section titled 'The Clever Dining Car Conductor' in which the merits of these workers were discussed:

> when they are up to their work, [conductors] are amongst the best canvassers employed by the railway companies. There are hundreds of businessmen travelling constantly on the main routes of commerce, to whom a comfortable meal on the train is a great solatium after a harassing day. The good dining-car attendant shows the same solicitude as a good valet or club servant. He knows every regular customer by face, if not by name, studies his peculiarities, anticipates his wants, and stands between him and many of the minor worries of the journey.[14]

This indicates how the travelling public viewed dining car conductors. The model worker went beyond seating them and serving their meals; they were seen as equivalents to valets or club servants, who remembered their regular customers' faces if not names, anticipating their wants. The best dining car conductors were expected to provide the same familiarity that a person might experience when frequenting their regular haunts such as a gentleman's club. By providing this familiarity, dining car conductors imitated their counterparts in leisure and domestic service, and in doing so created a sense of comfort and luxury for the traveller attending to 'many of the minor worries of the journey'. The author states that by providing these additional services, the dining car conductor became the railway company's 'best canvasser' by attracting people to travel with them. This highlights a shift in travelling habits, where comfort and improved railway services such as the dining car had become a deciding factor for railway passengers. This shift emphasizes the important role that the dining car conductor played not just in creating these luxury experiences for passengers, but also in representing and promoting the railway company.

In the late-nineteenth to the mid-twentieth century, the dining car was a space in which contemporary railway companies and passengers attempted to replicate stationary eating experiences whilst on the go. Through their expectations of the dining car conductor and his role, contemporaries aimed to recreate the same foods, table settings, and rituals that they would experience in a restaurant or a gentleman's club. Whilst the passenger could sit leisurely in their seat, the dining car conductor could recreate the comforts of leisurely eating experiences, from remembering their names and how they like their tea, to serving them an authentically made tart and offering them cheese to follow 'served whole on a flat tray with a d'oyley'. The role and expectations of the dining car conductor convey the sheer determination of both the railway companies and passengers to protect the eating experience from the industrial speed and movement of the train.

Notes

1. 'Pace That Kills', *The Graphic,* 30 December 1871, p. 11.
2. 'Pace That Kills', p. 11.
3. Richmond, National Archives of the UK, Lease from Joseph Drown Rigby & Charles Rigby (builders) to Samuel Young Griffith (hotel keeper) of refreshment rooms at Swindon Station, 24 December 1844, RAIL 252/174.
4. Charles Dickens, *The Boy at Mugby* (London: Chapham & Hall), p. 75.
5. Chris de Winter Hebron, *Dining at Speed: A Celebration of 125 Years of Railway Catering* (Kettering, Northants: Silverlink Publishing Ltd, 2004), p. 128.
6. 'The Pullman Refreshment Saloon', *Wakefield Express,* 1 November 1879, p. 2.
7. 'The Pullman Refreshment Saloon'.
8. Amy Richter, *Home on the Rails: Women, The Railroad, and the Rise of Public Domesticity* (Chapel Hill: University of North Carolina Press, 2005), p. 60.
9. Susan Major, *Female Railway Workers in World War II* (Barnsley, South Yorkshire: Pen & Sword Books, 2018).
10. Richmond, National Archives of the UK, Great Western Railway, Hotels, Refreshment Rooms, and Restaurant Cars Department Booklet, March 1920, RAIL 1135/57.
11. RAIL 1135/57.
12. Peter Ritchie Calder, *Roving Commission* (London: Methuen & Co., 1935), pp. 89-90.
13. RAIL 1135/57.
14. 'The Clever Dining Conductor', *Barnsley Chronicle,* 6 May 1905, p. 6.

Connected Food: Preserving Traditional Food Practices via Portable Foods

Sevgi Mutlu Sirakova

Preparations for tomorrow's return to Germany have reached the final stage. Ever since settling there, a familiar flurry of activity occurs whenever it is time to leave Turkey: saying last goodbyes to relatives, checking all the documents for border crossings, and, importantly, making sure that all the food – prepared over the last few weeks – is safely packed into the car. The drive will take two days, with a precious cargo of food products that will bring Turkish flavours to the table in Germany throughout the year.[1]

I spent my childhood in a small village in the Rhodope Mountains of Bulgaria, where many foodstuffs such as pickles, *tarhana* (a dried fermented soup base), and cheese were prepared during the summer to be consumed in winter. Later, I migrated with my family to Bursa, a large city in Turkey.[2] Throughout the years following our migration, my family kept transporting traditional foodstuffs from villages to the city. During this time, I familiarized myself with portable foodstuffs and learned different preservation techniques to produce them.

These experiences sparked my academic interest in exploring the socio-ecological relationships in which these portable foodstuffs are nested. In my early research, I conducted multi-sited ethnographic fieldwork in Turkey and Bulgaria with a focus on fermented foodstuffs (Figure 1).[3] During my preliminary work, I noticed that these informal food transfers happen not only within the borders of the countries but also between different destinations within Europe. Germany is the preferred destination among Turks and Bulgarians. Currently, my ongoing doctoral research focuses on the informal food transfers of the Turkish-speaking community in Germany, which occur after visits to homes and relatives in Bulgaria and Turkey.[4]

These cross-border food circulations are more common than I had anticipated, but they remain neglected in the academic literature. In a globalized world where more and more people are migrating, it is more crucial than ever to understand such circulations and the relationships they represent. This article seeks to broaden the discussions around food and migration, using portable foodstuffs as the entry point of inquiry. Here, the focus is on the food directly carried by migrants who visit their home villages seasonally (rather than commercially exported food or food parcels sent via post).

This paper is structured in three parts. The first part provides context for these foodways with a brief overview of the current literature. The second part focuses on modes of preservation by introducing *tarhana* as a prominent traveller in these journeys. Considering the socio-ecological relations between underrepresented migrant people and foodstuffs, I propose the term 'connected food' to describe portable food that links the consumer directly to its production, creating intimacy in the process. Drawing on initial insights from my ongoing research, the third part reflects upon the current trends and changes in these foodways. In so doing, I put these foodstuffs into the broader context of ongoing socio-ecological changes and pose questions for future research.

Figure 1. Examples of the transferred foodstuffs from my fieldwork. Pictures are arranged in rows: a) local meat products, canned meat (kavurma), sun dried sausage (sucuk) and intestines; b) dry soup base (tarhana, a sun-dried fermented foodstuff made of curd, tomato, spices, and flour); c) home-made cheese-making processes; d) preparation of pine syrup; e) dried wild porcini mushrooms or manatarka (Boletus edulis). (All photos by the author.)

Preserving Traditional Food Practices via Portable Foods

Foodways of Migration

There were 50–60 jars remaining on the floor, some of which came from the market. The items were carefully placed in bags and then into the car's trunk, on and under the seats, and in the footwells. Although the car was tightly packed, the number of products still waiting to be loaded seemed to only increase as the day progressed. 'Take these green olives; look, we can put the rest of these jams by your feet,' said the mother of the household. Over 30 jars of jam, ten jars of tomato paste, various pickled vegetables, and green olives in five-litre plastic containers were already packed. Deciding which items to bring took almost the entire day. The most valuable home-made tarhana, dried spices, dried apples, syrups, and pickles – some of which came from the garden – were placed into the car as priority passengers. Linden, rosehip teas, and dried porcini mushrooms from the mountains were still waiting their turn to be crammed into any remaining nooks and crannies.

In this article, I adopt the term 'informal food transfers' to reflect both the informality and mobility of these foodways, though they have also been called various other terms: non-market food flows, non-market food transactions, or informal rural-urban food transfers, to name a few.[5] The mounting literature in this field documents 'the importance of ongoing informal rural-urban links and their role in providing food for urban residents'.[6] Studies on African cities show that while only 'a minority of urban households produce any food, a much larger number depend on informal food transfers from their rural homes'.[7] Although these studies use various approaches and methods to investigate foodways, their common point is that they recognize the transfers to be essential for the well-being of many families.[8] Hence 'urban-rural reciprocity', as Bruce Frayne points out, is not 'only a one-way movement of people and resources from the urban to the rural areas, but also a transfer of food from rural to urban households'.[9]

The edited volume *Food Parcels in International Migration* is a rare contribution to the cross-border aspects of these food transfers. The book presents detailed ethnographic descriptions of receiving and sending food parcels internationally via cargo. The prevalence of these 'food transfers has long been acknowledged but has only rarely been the focus of serious scholarly attention' write Mata-Codesal and Abranches. Accordingly, they suggest that food parcels are a powerful tool for understanding the 'intimate connections across the complex spaces of international migration'.[10]

Rather than focusing on food parcels, my research is concerned with the portable foodstuffs that are directly circulated by people who move back and forth to visit their home countries seasonally. The primary focus is on the home-made foods that are prepared for winter (*kışlık erzak* in Turkish), such as the preservation of fruits and vegetables while they are plentiful in summer and autumn. These foodstuffs are prepared using locally-grown ingredients, often from personal gardens, but also including wild foods (which are highly regarded ingredients considered to be the most 'natural').[11]

When people migrate to other areas to find work or education, these portable foods accompany their migration. They are either prepared by relatives or directly by the migrants, who are often multi-sited and participate in the production process while visiting their rural homes. Summertime is a particularly special period when many migrants return to their villages, reunite with their relatives, and collectively produce foods.

Though many of the traditional preservation methods were originally developed to preserve food over the winter, such methods now make it possible to transport these foods over long distances.[12] During my field visits, I recorded several methods of preservation, which can be divided into four broad categories: fermentation (both animal- and plant-based), drying (sometimes including curing with salt and spices), canning (which involves boiling the cooked or semi-cooked foodstuffs), and freezing (Figure 2). In practice, these preservation techniques are often used in combination to prepare portable foodstuffs. *Tarhana* is a good example that reveals how challenging it is to separate these categories from each other.

Figure 2. Main preservation methods: a) Fermentation (from left to right): pickling; yoghurt; wine); b) Drying: dry wild herbs (to make tea or to be used in cooking); cured meats (dry intestines and home-made sausage, sucuk); tarhana (dry, fermented soup base); c) Canning: canned cooked meats (kavurma); canned wild berry syrups; shelves full of a variety of canned foods; and d) Freezing: home-made butter made from yoghurt; a variety of frozen foods; minced local meats. (All photos by the author.)

Nowadays, *tarhana* is one of the fermented foodstuffs that many Turkish and Bulgarian people long for when they are abroad. It is an instant soup base that is consumed as comfort food. Its preparation is based on sourdough fermentation facilitated by lactic acid bacteria. Flour and yoghurt are the main ingredients, but the production process and additional ingredients vary widely from region to region. It is said that there are more than fifty different recipes for *tarhana* in Turkey. Despite the availability of market versions, *tarhana* is among the most carried portable foodstuffs. This is likely due to the diversity of preparation methods and therefore flavours; everyone longs for their own mother's version of it.

Figure 3. Preparation phases of tarhana. From left to right: fermentation, resting, drying, and cooked for consumption. (All photos by the author.)

Tarhana dates back to ancient times and is still prepared in areas ranging from the Balkans to the Middle East.[13] Its main ingredient, yoghurt, is the oldest dairy product, and the geographical range of *tarhana* overlaps with the places where agriculture emerged. *Tarhana* is likely one of the oldest portable foodstuffs, which therefore deserves a special place in this year's symposium.

Traditionally, it is prepared in large amounts in late summer (when the vegetables are seasonal and plentiful), through the collective work (*imece* in Turkish) of more than one woman. I observed the preparation of 5 kg of *tarhana* (sufficient for a family of five for a year) in Bursa, Turkey (Figure 3).

The ingredients for *tarhana* are 2 kg of peppers, 0.5 kg of tomatoes, 1–2 onions, around 3 kg flour (as much as the mixture takes), 0.5 kg yoghurt, and various spices (these vary widely by region and taste choices, but generally include chilli, black pepper, thyme, and mint). The recipe I observed for *tarhana* involves four distinct processes.

Preparation: since the distinctive taste and texture of *tarhana* is produced by sourdough fermentation, a starter culture (*maya*) is necessary, provided by yoghurt and *tarhana* from the previous year. Starter can be saved from the previous year's *tarhana* in two ways: either by putting a handful of the *tarhana* dough (before drying) in the freezer for the following year, or by putting the dry *tarhana* into water a few days before processing and leaving it in a cool (but not refrigerated) place to sour. In the meantime, yoghurt is left to ferment at room temperature for a few days. It is recommended that the yoghurt to be used in *tarhana* should be quite sour. The peppers, tomatoes, and onions are cut into small pieces and cooked in steam with little water. These cooked vegetables are pressed through a colander to puree them. Flour is then added until the correct consistency of dough is achieved. Then, soured yoghurt and the *tarhana* starter are mixed with this dough and kneaded.

Resting: the dough mixture is left to ferment for around a week, covered with a moist cotton cover to prevent it drying out. The rising dough is mixed daily by hand to control the process. If necessary, a little more flour is added. When the smell and texture are satisfactory, the dough is divided into small balls, which are laid out on a clean cotton sheet and left to dry.

Drying: *tarhana* is left to dry for 5–7 days in a light, airy place that does not receive direct sunlight. The pieces are turned daily and broken into smaller pieces by hand. Once they are well dried, these small pieces are crushed and granulated by hand. Although electrical mixers are sometimes used in this last step, the women I talked to said that dry *tarhana* should be treated carefully to prevent it from crumbling too much and that it is more delicious when rubbed by hand. The powdered *tarhana* is stored in glass containers for later consumption. It can be easily made into a soup as comfort food throughout the winter.

Preparing a soup: dry *tarhana* is placed in warm water to dissolve for at least half an hour (for each serving of *tarhana* soup, one tablespoon of dry *tarhana* suffices). If desired, you may add *kavurma* (meat sautéed in its own fat and preserved for later use), onions, or tomato paste; these ingredients are lightly fried in oil and then water is added. The rehydrated *tarhana* is slowly added to this mixture and cooked by adding more water. Spices such as chilli, thyme, and mint can be added if desired. The soup may be served with grated *kashar* (medium-hard cheese) or crumbled white cheese (feta).

Connected Food

The Ark of Taste is an online catalogue maintained by the Slow Food Foundation for Biodiversity that aims at drawing attention to the traditional foods that 'might disappear within a few generations'.[14] Interestingly, in this database, seven of the seventy-nine products given for Turkey are types of *tarhana* (Beyşehir Tarhana, Çerez Tarhana, Gömbe Tarhana, Koca Tarhana, Sakizi Tarhana, Sour Tarhana, Yelten Tarhana).[15] They include recipes with resin (*sakiz*), with sour cherries, and without tomato. The recipes represent *tarhana* practices from different regions that are at risk of extinction.

The Slow Food website states that 'agricultural biodiversity and small-scale, family-based food production systems are in danger due to industrialization, genetic erosion, changing consumption patterns, climate change, the abandonment of rural areas, migration, and conflict'.[16] Broadly true though that is, informal food transfers like those I study show how migration does not necessarily curtail the production of traditional food. Portable foods become a way to keep traditional food practices alive, as the knowledge and skills required to produce them are passed on. The role that these widespread foodways play in maintaining traditional food practices deserves further consideration.

In its 'Manifesto for Quality', Slow Food calls upon the practice and dissemination of a 'broader concept of food quality' that recognizes and counters the harm done by the industrialization and homogenization of food. Quality food, they say, must be 'good, clean and fair': *good* means 'quality, flavorsome and healthy food'; *clean* stands for 'production that does not harm the environment'; and *fair* requires 'accessible prices for consumers and fair conditions and pay for producers'. Everyone can contribute to good, clean, and fair food 'through their choices and individual behavior'.[17] However, I suggest that an essential prerequisite is missing here, which has to do with what I call 'connected food'.

Inspired by informal food transfers, connected food is food that has been prepared by oneself or somebody one knows, so that the consumer participates in – or is connected to – the intricate story of its production. In this article, I have given examples of food prepared in one's hometown with or by one's family before being transported to other destinations. In such cases, one becomes an embodied participant in the foodways and socio-ecological relations crucial to the food one eats. One is intimately connected to the culture and environment upon which the food relies.

These portable foodstuffs carry not just flavour and nutrition, but also memories and tastes of home. They therefore keep families socially, emotionally, and microbially connected to their homeland. Even more crucially, skills and knowledges are being preserved and transferred alongside the foods themselves, thus passing on – and sometimes modifying – the cultural traditions essential to their production. These foodways therefore provide a way to understand the maintenance, reproduction, and evolution of the skills and knowledges of traditional food practices.

Changing Foodscapes

Where is the motherland, where is the longing? Foods are home to us.[18]

A flavour carried from one's homeland contains many stories in itself: both intimate memories of the place left behind and aspirations for the new home. Transported food is therefore a fruitful object of study for thinking about changing socio-ecological aspects of production, circulation, and consumption. In my ongoing research, I am conducting semi-structured interviews to understand why these foods are carried and the factors influencing their transfers.[19]

In this last section, I will share some preliminary results from my ongoing research. Aside from the durability and portability of the transported foodstuffs, my interviewees emphasized three main reasons for justifying their food transfers – all of which correspond to the idea of connected food, which connects consumers to different geographies, temporalities, and imaginations.

The first reason is unique taste. Collectively, my informants emphasized the specific taste of these foodstuffs as the main reason for transporting them. For them, the taste of the home-made products prepared with local ingredients using traditional methods cannot be substituted with consumer-ready products. When tasting a specific transported food, they always linked it to their homeland and childhood memories. It is clear that the taste of these foodstuffs serves as a cultural reminder of the hometown and the past, which is worth further consideration on its own. The second reason is that these foods are deemed healthier and more 'natural' than commercially-available foods. For them, these products represent 'real organic' foods. This is closely connected with the third reason, which is the trust and confidence people have in these foods. Intimate awareness of the ingredients, preparation methods, and those doing the preparation creates confidence in such foods. When I asked why these food flows are essential, people would proudly say: 'because I know where my food comes from and how it is produced'.

That given, there is a significant decrease in the amount and variety of food transported compared to twenty or thirty years ago. Although foodstuffs are still transported, my participants kept telling me they are bringing less and less every year. There seem to be many ongoing changes that may have led to this development. Below, I summarize the contributing factors frequently mentioned in my interviews.

Relations with the countryside: those who maintain their social ties with rural areas, visit their relatives, and stay in the countryside during their visits bring products more.[20] Those who have lost ties to rural places are more likely to transport commercially-produced foods.

Mode of transportation: the mode of transportation affects the type and quantity of what can be brought. Since aircraft use has increased in recent years, some products cannot be brought in the large quantities that were possible in the past by car. It is important to note that a significant amount of food is still brought by car.

Availability of products: when Turkish grocery stores opened up in the 1970s and '80s in Germany, families started to buy some products from them. For example, tomato paste, pickles, local spices, and olives are now available in Turkish grocery stores. One participant shared the observation that 'instead of preparing them in their villages, people started to take the easy way out and buy them'.

Generational differences: those born and raised in Germany are less interested in these products, even though they became familiar with them through their parents. As one of my participants summarized, 'those who grow up with the food culture are looking for that taste and still bringing it'.

Border controls: in the 1990s, certain weight restrictions were imposed on animal and plant products transported by non-EU passengers, which also affected the amount of food that travellers brought. One of my informants observed that when the customs controls on the Austria–Germany border increased due to human smuggling in the 2000s, the passengers' food was also affected. After that, 'cars began to be searched thoroughly, illegal food products were increasingly found, and heavy fines were imposed on them'.

Economic and social status: at least among my participants, the cheaper cost of these food products does not play a decisive role in the choice to bring them to Germany. The social ties with the countryside are more important than the costs involved. However, my participants told me that with migrants who marry foreigners, the demand for and consumption of these products decreases significantly.

Changes in villages: the decrease in traditional food production and global decline of small-scale family agriculture is also reflected in these foodways. Many of my interviewees say that these foods are not produced in the villages any more, reflecting wider urbanization trends and increasing dependency on large-scale agriculture and industrial food production.

One participant, who has been living in Germany for more than forty years and regularly carries foodstuffs from Antakya, shared that 'maybe we halved the food we brought'. She explained why:

> Now very few families are left in the countryside who can farm and plant. Young people went to the cities for jobs, and no one is left to look after the animals and cultivate the fields. Also, we are getting lazier and prefer to buy them from the market here in Germany, instead of preparing them in Turkey. These foods require a lot of time and labour. But people are still bringing them, especially those who go to the countryside and have relatives; they still produce and carry traditional foodstuffs. For us also, home-made *yeşil zeytin* [green olives], *nar ekşisi* [pomegranate syrup], *salça* [tomato paste], and spices are still a must that we continue to prepare and bring even when we take a flight.

Portable Food

Although they are changing, these informal food flows continue to provide many people access to healthy, traditional foods. The sociological and ecological dimensions of connectedness are obvious, but perhaps there is also a spatio-temporal connectedness at play. While preserving food is helpful to provide food for the barren winter months, it also allows for transportation across large distances. Portable food can connect people with traditional techniques – and the places in which they are practised – through the time-bending properties of preservation techniques.

As it was getting closer to midnight, choosing what to bring was becoming more challenging. 'Let's get up early in the morning and have a final look,' said the son, who was going to drive the car. Since three people were making the trip to Germany, their luggage could be put on the back seats. 'We can leave you here to make room for even more things,' the son said, tiredly joking with his mother. Nobody had the energy to even laugh. Before going to bed, the mother gave a last reminder: 'There is butter and cheese in the freezer; we must not forget them in the morning.' As everyone fell asleep, their minds continued to race, worrying about forgetting something important, as they ran over the products that were already packed and still needed to be packed (Figure 4).

Figure 4. *Portable foodstuffs, waiting for their turn to embark.*

Acknowledgements

This article enormously benefited from the Rachel Carson Center works-in-progress sessions between 2019-2022 and Oxford Food Symposium Panels in 2022. I want to express my warmest thanks to all the participants who commented and improved this ongoing research. I also extend my gratitude to my informants, who generously shared their time, experiences, and valuable responses with me. Finally, I am immensely grateful to the Andrea von Braun Foundation for its generous two-year funding, which has allowed me to continue my research.

Notes

1. Italicized vignettes used in this article are adapted from my field notes to capture the preparations for the travel from Turkey to Germany.
2. The Turkish ethnic group in Bulgaria represents about 9% of the population (National Statistical Institute of the Republic of Bulgaria, '2011 Population Census – Main Results', *National Statistical Institute* <https://www.nsi.bg/census2011/PDOCS2/Census2011final_en.pdf> [accessed 16 October 2022]. Their history goes back to the late fourteenth century, when Bulgaria became part of the Ottoman Empire, and nomads from Anatolia settled across the Balkans (see Halil İnalcik, *The Ottoman Empire: 1300–1600* (Hachette UK, 2013), p. 11. After the collapse of the socialist regime in 1989, a substantial migration wave of this community occurred from Bulgaria to Turkey (see Ayşe Parla, 'Longing, Belonging and Locations of Homeland among Turkish Immigrants from Bulgaria', *Southeast European and Black Sea Studies*, 6.4 (2006), 543–57 (p. 545)), including my family and some of my relatives.
3. George E. Marcus, 'Ethnography in/of the World System: The Emergence of Multi-sited Ethnography', *Annual Review of Anthropology*, 24.1 (1995), 95–117. At this stage, I employed sensorial and multispecies approaches to explore how people perceive and engage with multispecies entanglements in fermentation. For sensory ethnographic approaches, see Sarah Pink, *Doing Sensory Ethnography* (London: Sage, 2009). For multispecies ethnography, see S. Eben Kirksey and Stefan Helmreich, 'The Emergence of Multispecies Ethnography', *Cultural Anthropology*, 25.4 (2010), 545–76.
4. Distances between Germany and Turkey (2000–4000 km) or Bulgaria (1500–2200 km) can be travelled by car, which makes these transfers different from overseas or longer migration routes.
5. On informality and mobility, see: Bruce Frayne, 'Migration and Urban Survival Strategies in Windhoek, Namibia', 306 *Geoforum*, 35.4 (2004), 489–505; Jonathan Crush and Mary Caesar, 'Food Remittances: Rural–Urban Linkages and Food Security in Africa' (London: International Institute for Environment and Development, 2017), report, i–39 (p. 6). On other terms, see: María Rivera and others, 'Assessing the Role of Small Farms in Regional Food Systems in Europe: Evidence from a Comparative Study', *Global Food Security*, 26 (2020), 100417; Chiho Kamiyama and others, 'Non-market Food Provisioning Services via Homegardens and Communal Sharing in Satoyama Socio-ecological Production Landscapes on Japan's Noto Peninsula', *Ecosystem Services*, 17 (2016), 185–96 (p. 191); Bruce Frayne, 'Pathways of Food: Mobility and Food Transfers in Southern African Cities', *International Development Planning Review*, 32.3/4 (2010), 291–310.
6. Jonathan Crush and Mary Caesar, 'City without Choice: Urban Food Insecurity in Msunduzi, South Africa', *Urban Forum*, 25.2 (2014), 165–175 (p. 173).
7. Jonathan Crush, Bruce Frayne, and Wade Pendleton, 'The Crisis of Food Insecurity in African Cities', *Journal of Hunger & Environmental Nutrition*, 7.2–3 (2012), 271–92 (p. 286).
8. Crush and Caesar, 'Food Remittances'; Godfrey Tawodzera, 'Vulnerability and Resilience in Crisis: Urban Household Food Insecurity in Harare, Zimbabwe' (unpublished doctoral dissertation, University of Cape Town, 2010).
9. Frayne, 'Pathways of Food', p. 489.
10. Diana Mata-Codesal and Maria Abranches, *Food Parcels in International Migration* ([n.p.]: Palgrave

11 Industrial products bought from the market are also transported. However, the amount of these ready-made products has been changing significantly under the influence of various factors that I will discuss in more detail in the last part.
12 Jyoti Prakash Tamang and Kasipathy Kailasapathy, eds. *Fermented Foods and Beverages of the World* (Boca Raton, FL: CRC press, 2010).
13 Soultana Maria Valamoti, 'Ground Cereal Food Preparations from Greece: the Prehistory and Modern Survival of Traditional Mediterranean "Fast Foods"', *Archaeological and Anthropological Sciences*, 3.1 (2011) 19–39.
14 'Ark of Taste', *Slow Food Foundation for Biodiversity* <https://www.fondazioneslowfood.com/en/what-we-do/the-ark-of-taste/> [accessed 30 May 2022].
15 'Ark of Taste products in Turkey', *Slow Food Foundation for Biodiversity* <https://www.fondazioneslowfood.com/en/nazioni-arca/turkey-en/> [accessed 30 May 2022].
16 'Where Heritage Meets Biodiversity', *Slow Food USA* <https://slowfoodusa.org/ark-of-taste/> [accessed 30 May 2022].
17 'Good, Clean and Fair: The Slow Food Manifesto for Quality', *Slow Food* <https://www.slowfood.com/wp-content/uploads/2015/07/Manifesto_Quality_ENG.pdf> [accessed 4 August 2021].
18 Murathan Mungan, 'Dönmek', performed by Yeni Türkü, *Dünyanın Kapıları*, 1987. I adapted these lines from Munga's Turkish poem: '*Neresi sıla bize, neresi hasret / Yollar bize memleket*' (Where is our motherland, where is the longing? Roads are home to us).
19 To my knowledge, no studies have so far investigated how much and which foods are transferred in these cross-border journeys.
20 The origin of these products depends on where these families are visiting and whether they still have networks with rural areas or not. If they come from the countryside and stay there and meet with their relatives during their visits, they are more likely to bring more home-made products than ready ones. But if they lack such networks with rural areas and stay only in the big cities during their visits, they are more likely to bring only ready products they bought from the market. In both cases, each car travelling from Bulgaria and Turkey to Germany carries a significant amount of food.

Portable Poetic Commensality: Reflections on the Use of Food Language in Ancient Israel's Pilgrimage Songs (Psalms 120–134)

Michelle A. Stinson

First aired in 1974, McDonald's television commercial for its Big Mac hamburger featured a jingle that has stuck with me for my entire life: 'Two all-beef patties, special sauce, lettuce, cheese, pickles, onions, on a sesame-seed bun.' I have fond childhood memories of singing this song while travelling in our family station wagon, riding on school buses, and cycling through the neighbourhood with friends. Somewhat ironic is the fact that I have no actual recollection of ever *eating* a Big Mac as a child.

I don't know if MacDonald's could have imagined back then that their Big Mac song would become such a lasting hit. While working on this essay, I conducted an informal experiment to confirm my own suspicions. I would start to sing the opening words of the Big Mac jingle ('two all-beef patties, special sauce…') and wait to see if those around me would join the chorus. Depending on their age-demographic (there seems to be a sweet spot for those of us who were youngsters back in the 1970s), many of those I encountered would quickly add their voice to this moment of musical 'commensality'. In this case, instead of commensality's traditional definition of 'the act of eating together', this experience involved 'the act of singing together … about eating'.

In what follows, I will reflect on the idea of 'portable poetic commensality'. In particular, I focus on how songs about food – sung on a journey as a communal chorus – can sustain a group emotionally and even spiritually, just as actual food provisions packed for a journey would nourish a company of travellers physically. This essay considers the collection of fifteen songs found within the Hebrew Bible that bear the heading a 'Song of Ascents' (i.e. Psalms 120–134). Tradition has it that these psalms may have been sung by ancient Israelite pilgrims as they journeyed to Jerusalem with the bounty of their harvests for the three annual festivals of Passover, *Shavuot*/Weeks, and *Sukkot*/Tabernacles. Within this group of songs, 'food language' – words, images, scenarios drawn from the arenas of food production, distribution, preparation, consumption, and disposal – pervades the collection (e.g. Pss 124, 126, 127, 128, 129, 131, 132, 133).[1] In this essay, I will explore three themes that arise from these food-infused songs: the portability of poetry, food language as sustenance, and the experience of conviviality within a communal chorus.

The 'Songs of Ascents' as a Pilgrimage Songbook

The fifteen–psalm collection known as the 'Songs of Ascents' (Pss 120–134) is found in the concluding section of the Hebrew Psalter (i.e. Book V, Psalms 107–150). These psalms draw their name from their unique heading (שיר המעלות), a combination of the Hebrew word for 'song' (שיר) and a word meaning 'the steps, stairs, ascents' (המעלות).[2] While the noun מעלות can refer to the steps of a throne or an altar, the verbal form עלה 'to go up, ascend' is most often used to describe 'going up' to a sanctuary site (e.g. to Bethel, Gen 35:1; Judg 20:18; 1 Sam 10:3) or, more specifically, to Zion or Jerusalem (e.g. Ps 122:4; cf. Ps 24:3; 1 Kgs 12:28; Isa 2:3; Mic 4:2; Jer 31:6; Zech 14:17–19).[3] Although there is no consensus opinion, a significant number of biblical scholars hold to the view that these songs would have been used by pilgrims during their journeys to Jerusalem for Israel's three annual festivals.[4]

The three main festivals that marked ancient Israel's cultic calendar were Passover, *Shavuot*/Weeks, and *Sukkot*/Tabernacles (Deut 16:16–17; cf. Exod 23:14–17; 34:18–23). In the Hebrew Bible, these festivals were associated with various agricultural seasons. Passover coincided with the spring barley harvest; *Shavuot* marked the completion of the wheat harvest.[5] *Sukkot* occurred in the autumn, following the processing of grapes and summer crops, and marked the end of agricultural work for that growing year.[6] It should not be surprising that Israel's cultic calendar would be linked to the agricultural year, since, as scholar Ellen Davis contends, 'throughout the Iron Age and into the Persian Period at least, the vast majority of Israelites – eighty-five percent or more – were farmers'.[7]

Psalms of Ascents as 'Portable Poetic Commensality'

The remainder of this essay considers three thematic trajectories that connect these psalms to the 2022 Oxford Food Symposium theme of 'Portable Food: Food Away from the Table'. These trajectories include the portability of poetry, food language as sustenance, and the experience of conviviality within a communal chorus.

The Portability of Poetry

Three distinct characteristics of this collection of pilgrimage songs set it apart within the Psalter: the brevity of the songs, the use of a step or terraced-line technique (versus the more typical use of parallelism), and its use of formulaic expressions and key words.

Psalm 131 serves as a helpful example since it exhibits all three literary characteristics:

> 1 O LORD, my heart is not lifted up,
>   my eyes are not raised too high;
>   I do not occupy myself with things
>   too great and too marvelous for me.
> 2 But I have calmed and quieted my soul,
>   like a weaned child with its mother;

>   my soul is like the weaned child that is with me.
> 3 O Israel, hope in the LORD
>   from this time on and forevermore.[8]

Brevity. Psalm 131 is composed of just three 'verses', making it a 'bite-sized' song.[9] It could be sung quietly to oneself to sustain a weary walker or even begun by one pilgrim and then joined by others in one's group of travellers. The brevity of these psalms is in fact one of the hallmarks of the collection. Most of the psalms are between three to nine verses in length.[10] As Zenger observes: 'The brevity of these psalms could be an indication that they could easily be learned by heart.'[11] And thus, a characteristic that greatly increased their 'portability'.

Step/Terraced Lines. Many of the psalms – including Psalm 131 – employ a 'step technique' (*anadiplosis*) where a word or phrase is used at or near the end of a line/cola and then is repeated at the beginning of the next line/cola. Within the Songs of Ascents, this 'doubling back' appears within individual psalms, as well as occurring across adjacent psalms. In Psalm 131, the repetition of the phrase 'like a weaned child' (כגמל) serves to link the final two segments of v. 2 together. One finds numerous examples of repetition through the resumption of words or phrases across the collection (e.g. Pss 121; 122:2–4; 123:2–4; 124:1–5; 125:2, 3; 126:2, 3; 129:1, 2).[12]

Theme Words and Formulaic Expressions. It should not be surprising that songs composed for a journey to a central worship site would draw upon the rhetorical richness of key words and phrases taken from the storehouse of ancient Israel's theological traditions. Liebreich observes that the collection as a whole seems to draw language from the great Aaronic blessing of Numbers 6:22–24:

> The LORD bless you and keep you;
> the LORD make his face to shine upon you, and be gracious to you;
> the LORD lift up his countenance upon you, and give you peace.

Key terms from this benediction such as 'bless' (ברך; 128:4, 5; 129:8; 132:15; 133:3; 134:3), 'keep' (שמר; 121:3, 7; 127:1; 130:3, 6; 132:12), 'gracious' (חנן; 123:2, 3), 'peace' (שלום; 120:6, 7; 122:6–8; 125:5; 128:6) are woven throughout the Collection. Formulaic or liturgical expressions also appear with great regularity. One finds an example of this is Psalm 131 with the expression 'from this time on and forevermore' which closes out Psalm 131 (cf. Pss 121:8; 125:2).[14] Within the 'Songs of Ascents', repetition of various kinds appears on the level of lines, within individual psalms, and across the collection as a whole. This use of repetition is, as Zenger notes, 'quite remarkable' in light of the brevity of the psalms.[15]

In order to be accessible to a potentially diverse group of pilgrims, these compositions would need to be songs that could be learned easily and passed down generationally.

The young and the old would need to be able to remember the words, thus their brevity. The whole community would need to be drawn into the chorus, possibly explaining their reliance of repetition. The songs also needed to stand the test of time, thus their frequent use of language from the history and traditions of Israel. I would argue that all of these features helped to make these psalms a 'snack-sized' treat that could be brought out of one's memory and enjoyed at a moment's notice. For poetry becomes 'portable' only when it gets embedded in your mind and can be carried in your heart, wherever your feet may take you.

Songs as Sustenance: Food Language as Emotional and Spiritual Nourishment

As noted earlier, the 'Songs of Ascents' rely heavily on the language of food, most often from the arenas of food production and food consumption. Scenes of agriculture appear with great regularity, whether this is the act of ploughing (Psalm 129:3), the sowing of seed (Psalm 126:6), the harvesting of crops (Psalm 126:5, 129:7), or the binding of sheaves (Psalm 129:7). In these psalms, the psalm-singers give thanks for the 'fruit' (פרי) of one's labours (Psalm 128:2), as well as celebrating the 'fruit' (פרי) of the womb, i.e. the gift of children (Psalm 127:3, 132:11).[16]

Many of these psalms consider the hoped-for conclusion of agricultural labour – food consumption. Language of the Mediterranean triad of olives, grapes, and grains sprout up from the soil of these texts. Thriving grape vines and olive plants (Psalm 128:3) as well as olive oil (Psalm 133:2) appear within the collection. References to bread – a dietary stable in the ancient world – occurs in Psalm 127:2 as the 'bread of painful toil' (לחם העצבים) and in Psalm 132:15 as the divinely-promised provision, that the poor would be 'satisfied with bread' (שבע לחם).

Water – required for the preservation of physical life – also springs forth in this collection at various junctures. Seasonal rains appear as both a threat to life (Psalm 124:4, 5) and the longed-for renewal of the ground (Psalm 126:4). The image of abundant moisture, as will be seen in Psalm 133's reference to the morning dew flowing down a mountainside, offers a picture of refreshment and vitality in an arid landscape (v. 3).

These psalms reflect the lives of small farmers and their families, gleaning much of their imagery and language from field and orchard, table and hearth.[17] While Psalm 131 offers an intimate domestic picture of emotional contentment and peace as a 'weaned child' no longer experiencing the frantic search for sustenance at its mother's breast, Psalm 126 takes us outdoors into the fields and hillsides of Israel in this six verse song:

> 1 When the LORD restored the fortunes of Zion,
>   we were like those who dream.
> 2 Then our mouth was filled with laughter,
>   and our tongue with shouts of joy;
>   then it was said among the nations,
>   'The LORD has done great things for them.'

3 The LORD has done great things for us,
    and we rejoiced.
4 Restore our fortunes, O LORD,
    like the watercourses in the Negeb.
5 May those who sow in tears
    reap with shouts of joy.
6 Those who go out weeping,
    bearing the seed for sowing,
    shall come home with shouts of joy,
    carrying their sheaves.

Psalm 126 – a communal prayer – holds together God's past faithfulness (vv. 1–3), Israel's present need for restoration (v. 4), and a hope for a future renewal (vv. 5–6). Like in Psalm 131, this song employs the step-technique of *anadiplosis*, here being used to emphasize God's faithfulness in the past with the repeated phrase 'The LORD has done great things for them/us' (vv. 2–3). The remainder of the psalm draws on images of hoped-for restoration. Verse 4 compares their expectation to the longing for the late autumn rains that would turn dry desert riverbeds into flowing streams (v. 4). The final verses (vv. 5–6) paint a picture of hope based in the sureties of the agricultural year, that seed sown in the autumn will bring forth a harvest in springtime.[18]

For pilgrims who travelled to the festivals, these images drawn from fields and hillsides reflected the scenery that they would have passed while traveling to Jerusalem. Travellers from the south would be walking through the arid lands of the Negeb. Those journeying from the north and central regions would spend their journeys walking alongside recently harvested fields and orchards and vines. The images mentioned in these psalms would have come alive before them and very likely would have drawn to their lips the words of a particular song.

Conviviality within the Experience of Communal Chorus

The final theme considered in this essay is the idea of conviviality within a communal chorus. The biblical instruction for these annual festivals depict them as broadly inclusive and joy-filled events: 'Rejoice before the LORD your God – you and your sons and your daughters, your male and female slaves, the Levites resident in your towns, as well as the strangers, the orphans, and the widows who are among you – at the place that the LORD your God will choose as a dwelling for his name' (Deut 16:11; cf. v. 14). The communal nature of these pilgrimage psalms becomes apparent through a quick survey of the collection. Some psalms exhibit a collective voice (e.g. Pss 124, 126), some identify the community of Israel as the chorus (e.g. Pss 124, 129, 130, 131), while others move between a singular and plural voice (e.g. Pss 121, 122, 123, 128, 134).[19] Examples of what appears to be a community 'call and response' occurs in Psalms 124:1–2 and 129:1–2. And within the collection, references to a variety of groups mentioned in

Deut 16:11 appear within the songs, for example children (e.g. Pss 127:4, 128:3, 131:2) and servants (Ps 123:2), as well as brothers (Pss 122:8, 133:1).

Central to the contemporary Slow Food movement is the idea of 'conviviality', the experience of 'taking pleasure in the processes of cooking, eating, and sharing meals with others'.[20] I would like to propose that a communal chorus can be a version of 'conviviality', when this experience involves creating music, singing, and sharing in song with others, especially in those contexts where the chorus is singing about food.

Psalm 133, the penultimate psalm of this collection, celebrates the beauty of 'life together' – the literal definition of 'conviviality':

> 1 How very good and pleasant it is
>   when kindred live together in unity!
> 2 It is like the precious oil on the head,
>   running down upon the beard,
>   on the beard of Aaron,
>   running down over the collar of his robes.
> 3 It is like the dew of Hermon,
>   which falls on the mountains of Zion.
>   For there the LORD ordained his blessing,
>   life forevermore.

Similar to the previous psalms, the brevity of Psalm 133 makes it easy to memorize and call to mind on a journey. This short psalm of conviviality is saturated with the life-sustaining elements of olive oil and morning dew. Behind the expressed abundance of these two ingredients lies an implicit celebration of another successful olive harvest and gratitude for the divinely-bestowed blessing of moisture for a new day. And central to this song is the sharing of life together, as the 'kindred' of Israel, brothers and sisters together on pilgrimage.

Conclusion

Bodenhamer has observed that 'throughout the history of interpretation, the Songs of Ascents have been associated with movement or a motif of journey'.[21] The agrarian focus of these pilgrimage songs is not surprising if the chorus was made up of farmers heading to Jerusalem with the bounty of their harvests. One finds a beautiful honesty in these songs that reflect a shared agrarian livelihood with its concern for fertility and fields. Regarding this Pilgrimage Collection, Zenger observes: 'Here one sees individuals in their anxieties and hopes; the crises and joys of families; the farmers' daily life of plowing, sowing, and harvesting; the villages or small towns – and above all the many people who gather in Jerusalem and there experience themselves, at the pilgrimage feasts, as a broad family of "brothers" (siblings) blessed by YHWH.'[22] What we find in Israel's Pilgrimage Songbook is that words shared together can become as nourishing and sustaining as food enjoyed in community.

Portable Poetic Commensality

Notes

1. In his study *Cooking, Cuisine and Class: A Study in Comparative Sociology,* Jack Goody introduces a classification system that identifies five categories for considering food – its production (including various processes and aspects of agriculture and husbandry), distribution (involving storage, transport, and exchange of foodstuffs), preparation (processing and cooking of food), consumption (serving, eating, and clearing away of food), and its disposal (disposing of the leftovers from the meal) (*Cooking, Cuisine, and Class* (Cambridge: Cambridge University Press, 1982), pp. 44–49). For an extended discussion of these dimensions at work in the Psalms, see Michelle A. Stinson, '"A Table in the Wilderness?": The Rhetorical Function of Food Language in Psalm 78' (unpublished doctoral thesis, University of Bristol, 2017).
2. This collection of songs is unique within the Psalter in that all the 'Songs of Ascents' are placed together. As Grossman notes, 'it is only the Songs of Ascents which appear consecutively-no Song of Ascents appears in the MT outside of this small collection. None of the other groupings of psalms with like superscriptions are drawn completely together' (*Centripetal and Centrifugal Structures in Biblical Poetry* (Atlanta: Scholars Press, 1989), p. 15). In light of this fact, it is quite likely that these psalms existed as a freestanding collection before being incorporated into the Psalter.
3. As Nancy DeClaissé-Walford aptly observes: 'Since Jerusalem sits on a hill, no matter where one comes from, one always "goes up" to Jerusalem' (Nancy L. DeClaissé-Walford, Rolf A. Jacobson, and Beth Laneel Tanner, *The Book of Psalms,* NICOT (Grand Rapids: Eerdmans, 2014), p. 887).
4. As James Mays notes: 'The most likely and widely held theory about the superscription is that "ascents" refers to the journey made by pilgrims to the three annual festivals observed in Jerusalem (Duet 16:16)' (*Psalms*, IBS (Louisville: John Knox, 1994), p. 385). Others holding this view include Frank-Lothar Hossfeld and Erich Zenger (*Psalms 3: A Commentary on Psalm 101–150*, Hermeneia, trans. by Linda M. Maloney (Minneapolis: Fortress Press, 2011), p. 287) and Hans-Joachim Kraus (*Psalms 1–59: A Continental Commentary*, trans. by Hilton C. Oswald (Minneapolis: Fortress Press, 1993), p. 24). Kraus, in fact, calls the Collection the 'songbook for pilgrimages' (p. 24). See Zenger for a discussion of additional proposed settings for these psalms (pp. 288–94).
5. Philip J. King and Lawrence E. Stager, *Life in Biblical Israel* (Louisville, KY: Westminster John Knox, 2001), p. 86.
6. Biblical instructions for the specific timeframes for these festivals differ slightly across the corpus of the Hebrew Bible. Simeon Chaval notes: 'In the specific context of the festivals, H.L. Ginsberg argued that the farmer's distance from the single temple made it imperative in D [Deuteronomy] that the Harvest Festival close the harvest season (16:9–12) rather than inaugurate it, as in Exod 23:16a. By the same reasoning the Ingathering Festival must take place not after the gathering in of the harvested crops, as in Exod 23:16b, but only after their having been processed for the coming winter (Deut 16:13–15)' ('The Second Passover, Pilgrimage, and the Centralized Cult', *Harvard Theological Review*, 102.1 (2009), 1–24 (p. 16); citing H.L. Ginsberg, *The Israelian Heritage of Judaism* (New York: The Jewish Theological Seminary of America, 1982), pp. 58–60).
7. Ellen F. Davis, 'Propriety and Trespass: The Drama of Eating', *Ex Audito*, 23 (2007), 74–86 (p. 74).
8. All biblical quotations are from the *New Revised Standard Version* unless otherwise noted.
9. A verse or 'line' of Hebrew poetry is typically made up of two to three 'cola' (segments). A two-part 'line' is called a 'bicolon', a three-part line a 'tricolon'.
10. The exception is Psalm 132 which extends to eighteen verses.
11. Hossfeld and Zenger, p. 295. Zenger goes on to note: 'It is possible that they were songs sung to catchy melodies. Their redactional description as שיר might lead us to that conclusion' (p. 295).
12. For a detailed list of repetitions, see DeClaissé-Walford, Jacobson, and Tanner, p. 888–90.
13. Leon J. Liebreich, 'The Songs of Ascents and the Priestly Blessing', *Journal of Biblical Literature*, 74 (1955), 33–36.
14. Other formulaic expressions that appear are YHWH as 'creator of heaven and earth (Pss 121:2; 124:8; 134:3) and the prayer, 'peace be upon Israel' (Pss 125:5; 128:6).
15. Hossfeld and Zenger, p. 295.

16 Howard Eilberg-Schwartz notes: 'The symbolic connection between agricultural and human yields is given linguistic and literary expression in Israelite writings. In Hebrew, as in English, a single term (*zera*) is applied to agricultural and human "seed." The Hebrew stem that means "be fruitful" (*peru*) derives from the same stem as the word for "fruit" (*peri*). Firstborn children are referred to as a person's first yield (Ps. 105:35; Deut. 18:4; 21:7)' (*The Savage in Judaism: An Anthropology of Israelite Religion and Ancient Judaism* (Bloomington: Indiana University Press, 1990), p. 158).

17 Loren Crow was one of the first scholars to propose an underlying 'agrarian provenance' for many of these texts. Crow contends: 'The nucleus group is composed of songs whose place is in the workaday life of farmers and other laborers' (*The Songs of Ascents (Psalms 120–134): Their Place in Israelite History and Religion*, SBL Dissertation Series 145 (Atlanta, GA: Scholars Press, 1996), p. 156). According to Crow, this Collection consist of two layers. The oldest layer, which is the core of the Collection, derives from a northern-Israelite agrarian provenance. The Jerusalemite redactional layer deliberately gives the older songs a new purpose, namely to persuade northern Israelites, possibly of the Persian period, to go on pilgrimage to the Jerusalem temple (p. 157).

18 Just a few psalms later, Psalm 129 offers a strikingly different view of life for those in the group who have toiled as tenant-farmers, with no legal right to the lands they tended. In this tenuous existence, oppression and even abuse were an all-to-familiar reality. Like in Psalm 126, we find the language of food production, but here the imagery of seedtime and harvest are used as an avenue to process past trauma in the presence of one's community. The psalm opens with a recollection of this trauma:

>   1 'Often have they attacked me from my youth'
>   – let Israel now say –
>   2 'often have they attacked me from my youth,
>   yet they have not prevailed against me.
>   3 The plowers plowed on my back;
>   they made their furrows long.'
>   4 The LORD is righteous;
>   he has cut the cords of the wicked.

As we know from the experience of a global pandemic, communal and collective grief expressed through lament is a powerful practice for sustaining hope for a better future.

19 Regarding the collective nature of these psalms, Mays remarks: 'The name Israel is used in a frequency untypical of the psalms, nine times for the company the songs concern. A number of them alternate between individual and corporate style, so that the song is suitable for individuals as part of a company and a company made up of persons who assemble as individuals (121; 122; 123; 129; 130; 131)' (p. 386).

20 This description of Slow Food by Richard McCarthy is quoted in Rhonda Phillips and Chris Wharton, *Growing Livelihoods: Local Food Systems and Community Development* (New York: Routledge, 2015), p. 39.

21 K.W. Bodenhamer, 'Dwelling Together: Psalms 133 and the Psalms of Ascents.' *Review & Expositor*, 116.2 (2019), 219–224 (p. 220).

22 Hossfeld and Zenger, p. 297.

Travelling Silver for Those Not to the Manor Born: Old Sheffield Plate and Electroplated Silver in Travel Equipage and Cutlery, from 1730 to the Belle Epoque

Carolyn Tillie

Introduction

In 1743 a gentleman walked into Sheffield's Cutlers Company with a broken knife. Thomas Boulsover (1705-1788), the British craftsman tasked with repairing the utensil, accidentally overheated the knife, almost melting the metal. Looking closely at the damage, he could easily discern two separate layers, one copper and one silver. As he continued the repair, he observed that once fused together, the copper and silver behaved as one cohesive material. Thus, Old Sheffield Plate (OSP) was accidentally discovered. With this fortuitous invention of a new manufacturing technique, a burgeoning industry of affordable luxury metal goods was created for a growing middle class with upper-class aspirations.

This paper will review the growth in popularity of travel equipage and cutlery as a result of advances in British metalworking trades and techniques, including the creation of silver-plated equipage for leisure, travel, and the military, during the era that began with the invention of OSP and extended into the twentieth century.

Sterling vs. Plate and the Industrial Revolution

English silver has a reputation as the finest in the world due in large part to the unique system of traditional hallmarking as defined by the Guild of Goldsmiths, and to the Protestant Reformation in France which brought the cream of Huguenot silversmiths to London as refugees at the end of the seventeenth century. The beginning of the eighteenth century witnessed the Industrial Revolution, coupled with the growing availability of personal silver and silverplate articles from cutlery to service ware. With the introduction of tea in the mid-1600s, which became the national beverage, these events provided an impetus for the ordinary person to acquire a silver teapot or full tea set, both to use and to showcase one's affluence and good taste.[1]

Although the method and subsequent growth of OSP were revolutionary, silver plating was not. Close plating, in existence since medieval times, was a craft wherein a thin layer of silver was applied to an already-manufactured article made of a base metal, usually iron or steel. However, the process was not without its pitfalls. If a close-plated item became too hot, its silver would disappear. If the item became too wet, the silver could peel off entirely. Due to these shortcomings, close plating was largely relegated to small items such as buckles or minor cutlery.[2] During the Middle Ages, there was a tendency toward unethical practices in early plating attempts – gilding brass to look like gold or tinning steel to appear as silver. In 1300 Edward I (1239-1307) decreed that 'Guardians of the Craft' should go from 'shop to shop' to verify the metal content and stamp a leopard's head to confirm each item's sterling or gold standard.[3]

The manufacturing process of Old Sheffield Plate was innovative in that the base metal – usually copper due to its malleability – was sandwiched between thin sheets of sterling silver. It was then heated in a coke-fired furnace into a solid ingot which would fuse the silver to the copper. Once cooled, the ingot would be rolled out into sheets that could be then fashioned into anything from cutlery to candlesticks to teapots. No matter how thinly the sheet was rolled out, the fused copper always retained its silver covering.[4] Within twenty years of his discovery, Thomas Boulsover had moved on to other inventions: casting steel and saws and providing sheets of fused plate to other craftsmen. These other businessmen expanded upon Boulsover's innovation with the manufacture of sauce boats, salvers, tankards, and cutlery. Sheffield plate – or Old Sheffield Plate as the fused plate was now called – opened a new market for ready-made products by answering the demand for highly ornamented objects, seen at the time as modern luxuries.[5]

For the first few decades after Boulsover's invention, OSP remained mainly in the hands of members of the Cutlers' Guild that had been founded in Sheffield during the reign of Edward III (1327-1377).[6] These creations were limited to small items: buttons, buckles, wine labels, and snuff boxes. Joseph Hancock (birthdate unknown-1791), an

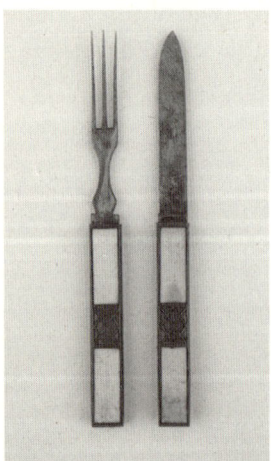 

Figure 1 (left). George Patten, Traveling fork and knife set, c. 1750-1800, steel, pewter, and ivory, England: Sheffield (Colonial Williamsburg, Virginia). Figure 2 (right). Maker unidentified, Nested traveling equipage, c. 1810, gilt-lined fused silverplate, England: Birmingham or Sheffield (Colonial Williamsburg, Virginia).

apprentice to Boulsover and credited as the father of OSP manufacture, was the first to make fused plate using powered machinery: water-driven rolling mills.[7] By the 1760s, silversmiths were recruited to the industry, bringing craftsmanship and skills that elevated the craft to an artistic level it might not otherwise have achieved.[8]

It was commonplace for individuals to travel with their own sets of cutlery until the late 1700s and early 1800s, through the beginning of the Industrial Revolution. The earliest examples depict bespoke travel cases in ivory, shagreen, or sterling, mostly for knives, but as they became more common forks and spoons were included.[9] As evidenced in their construction and design, these were intended to be worn on a belt or to be seen. They were often highly ornamental status symbols with the utilitarian portion of the cutlery in steel, with handles crafted in sterling, gold gilt, ivory, jade, or agate. A notable example is this simple but effective design dated circa-1770 wherein the dyed ivory handles were devised to form the sheath for its companion piece, becoming a solid bar when packed for transport (Figure 1). So popular was this design that a similar version was included as part of the campaign equipment of George Washington (1732-1799), and similar examples of this brilliantly engineered fork and knife set would reappear again throughout the next three centuries, notably among the military.

Sterling silver as cutlery was confined mainly to the landed gentry or the aristocracy who could afford to engage a silversmith in the manufacture of flatware, epergnes, and full dining and serving sets.[10] According to Sarah Coffin in *Feeding Desire: Design and the Tools of the Table 1500-2005*, 'the custom [of carrying one's own cutlery] did not die out when hosts started to provide flatware for their guests. Travelling sets continued to be used when visiting inns [. . .] and for picnics, hunts, and other outdoor occasions well into the twentieth century.'[11]

Silversmiths quickly envisioned fused plate as a substitute for sterling, finding that their ancient smithing techniques of raising, engraving, fabricating, soldering, chasing, and repoussé worked as well – if not better – on the fused plate. Items made in OSP could be one-third the price of pure silver ones and appear visually the same. For example, in 1796, a fused plate chocolate pot and stand sold for £3 and 3 shillings,[12] a mere quarter of the cost of its sterling counterpart. Gordon Crosskey, in his book *Old Sheffield Plate*, stated that, 'the ultimate success of the industry was the universal patronage of the aristocracy. That in an age of social emulation, conspicuous consumption was of inestimable value'.[13] One such example of the sumptuous offerings is this 1810 nested travelling set (Figure 2). Made of gilt-lined fused OSP, this set would have been marketed to the well-to-do. The large bowl-shaped container snugly contains a spouted creamer in which a small, liftable handle is hidden, a sugar bowl, a waste bowl, and finally a small hinged-topped mustard pot which, at one time, probably had a cobalt glass insert. Elegantly designed, the four-banded horizontal rims on each piece create a cohesive, radiating pattern when all pieces are nestled together. This type of set was suited for family outings and may, at one point, have had a custom-built carrying case.

In its infancy, OSP was the darling of the gentry with many early examples bedecked with family crests. Nothing brings home more clearly the wholesale adoption of OSP by the aristocracy than the innumerable crests and coats of arms found on older pieces.[14] This would begin to change with the rise of the factory system during the latter half of the eighteenth century when more pieces could be produced faster and cheaper. Machines developed for the textile industry were readily adapted to the silver trades: steam-powered mills (to roll out the fused sheet as was once done by hand), iron and steel dies (stamping *en masse*), and an assembly-line division of labour: cutters, hammermen, die sinkers, polishers, and burnishers. The Old Sheffield Plate industry quickly burgeoned in just over three decades since Boulsover's invention. The Clerk of the Sheffield Cutlers' Company, in a report to the 1773 Committee, documented an estimated 468 persons involved in the Sheffield silver trades, a twenty-fold increase from its inception.[15]

The skill and craftsmanship of fused plate travel items manufactured during the first half of the nineteenth century flourished, with designs becoming more elegant and refined. For a hunt luncheon, rather than bringing the fine silver outdoors, a circa 1800-1820 travelling tea service would meet the occasion with practicality and style (Figure 3). The teapot's lid is fully removable so that the accompanying sugar bowl, cream pitcher, and waste bowl all fit perfectly inside for portability. The open-scrolled silver handle for the cream pitcher screws on, while the removable wooden handle for the teapot is designed with flanged sockets to engage the fittings on the side of the pot, locking it into place. These are yet more examples of the cleverness and economy of design used for travel items. For the travelling gentleman, a 'Bachelor's Set', dated 1830, included a tooled leather, velvet-lined carrying case to cradle an OSP beaker, into which cushions a block securely holding a knife, spoon, and fork, all with detachable handles for ease of transport (Figure 4).

 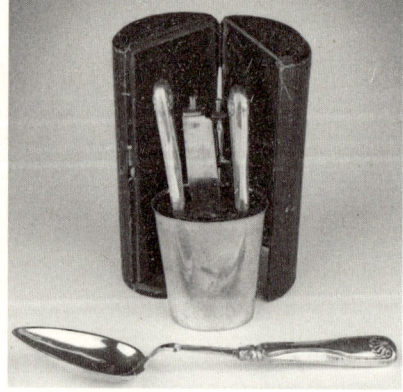

Figure 3 (left). Maker unidentified, Travelling tea service: teapot, sugar bowl, cream pitcher, and wastebowl, c. 1800-1820, gilt-lined fused silverplate, England: Birmingham or Sheffield (Colonial Williamsburg, Virginia). Figure 4 (right). Maker unidentified, Bachelor's travel canteen, c. 1830, block, leather, velvet, fused silverplate cup, knife blade, fork tines, spoon bowl, and handles, England: Birmingham or Sheffield (Colonial Williamsburg, Virginia).

As more of these innovative items became available, an even bigger change was on the horizon – one that would start to bring silvered items to the growing middle class. In 1800, Alessandro Volta (1745-1827) invented the first electric battery, and within two years his close friend, fellow Italian Luigi Brugnatelli (1761-1818), laid the groundwork for the demise of fused plating with his invention of electroplating using silver fulminate.[16] It would take another forty years of experimentation until George Elkington (1801-1865) of Misters Elkington & Company of Birmingham turned electroplating into a practical technique.[17]

In contrast to OSP, where the sheet is clad with silver and then the item is fabricated, with electroplating the teapot, spoon, or creamer is made from an economical metal and then placed in a bath containing a dissolved salt of silver such as silver nitrate.[18] The metal object is connected to an electrical current, and dissolved silver is fused to the inexpensive metal item. The layer of actual fine silver that has been electroplated is extremely thin in comparison to fused plate, enabling silver clad luxury articles to be made at a fraction of the cost: a price the middle class could afford. The increasing popularity of electroplated travel equipage and other metal goods surged, effectively spelling the demise of Old Sheffield Plate by 1840.[19]

Consumer Revolution and Leisure

With the Industrial Revolution came a consumer boom in England, fostering newfound leisure time and a growing desire for luxury products. The popularity of indulgent merchandise reached epic proportions: by the end of the eighteenth century, a greater proportion of the population than heretofore in human history was able to purchase non-essentials.[20] The rich led the way in the consumer revolution as they engaged in lavish spending on a wide range of fashionable items, bought not for necessity but as superfluous extravagances.[21] The middle ranks of society imitated the aristocracy in their desire to spend and, in turn, so did the rest of society as far as it was able.[22] Kenneth Quickenden and Arthur J. Kover in their detailed analysis of luxury marketing in late eighteenth-century Britain, state, 'The nascent form of what we now call marketing took place in Britain at that time. With it, the culture of consumption took its form, particularly among those who had previously no access to many manufactured goods, and certainly not luxuries.'[23]

Among the working class, refinement, elegance, and taste were aspirational attributes, and new physical consumer goods found a large new base of customers.[24] Fashionable cookbooks of the day encouraged readers to include silver service. The 1827 tome, *Domestic Economy and Cookery, For Rich and Poor*, exhorted, 'Families who cannot afford plate should economize till they can obtain at least four or five covered dishes, a sufficient quantity of forks, and two sauce boats of silver, made perfectly plain, for every day's use. [There] are very nice dishes now made of prince's metal, excellent for family use.'[25] As a result, demand in the economy was stimulated to an unprecedented degree, especially for those middle-class aspirants to social distinction.²⁶ Until the second half of the eighteenth century, London retained its supremacy of silversmiths and shops, when expansion of the Sheffield and nearby Birmingham manufacturers began to offer a challenge. In the acquisition of silver, Helen Clifford of the Victoria and Albert Museum posits:

> the relationship of the 'base' to the 'precious' was not one of simple emulation. [...] Sometimes the designs that originated with the makers of plated wares had to be followed by the London silversmiths because they set the fashion. The 'ready made' Sheffield plate that appeared in the shops, and which could be bought direct from the Sheffield and Birmingham manufacturers offered an alternative, as much as an imitation, to commissioned silver. Sheffield plate had its own place in the home as an indicator of novelty and variety, one that did not merely reflect the status of silver, but that had achieved its own position of regard.²⁷

The success of portable travel items is evident in the Sheffield firm of Joseph Rodgers and Sons who advertised in 1833 over forty categories of goods, adding the description 'of all sizes' or 'of all kinds' including fish knives, asparagus tongs, and portable travelling knives and forks with cork-screw.²⁸

Use by the Military, The Advent of Gentlemanly Soldiering

The growth of travel equipage literally went further afield than the elegant country estate hunt luncheons, weekend picnics in the Cotswolds, and train rides to the Brighton shore. British Imperial expansionism during the eighteenth and nineteenth centuries required commissioned soldiers to go on campaign in India, Africa, the East Indies, and other exotic locales. Starting in 1683, during the reign of Charles II (1630-1685), an officer in the British Army could purchase his commission to bypass the wait for meritorious promotion. Until this practice was abolished as part of the Cardwell Reforms in 1871, it ensured a social exclusivity to the officer class. Buying one's commission also generated an inherent snobbery among the wealthy against the more studious officers who had earned their positions. The Crown neither clothed nor fed an officer, though, since 1811, it had made a small allowance for wine.²⁹ In addition to paying for his commission, the new officer's out-of-pocket expenses included purchasing his own uniform, bed, and sundries.³⁰

It was assumed that when setting out on a military campaign, the officers would enjoy the same standard of living as they had in England. This, of course, included their elaborate table settings, tea sets, and utensils. An extensive industry of portable campaign furnishings and accoutrements arose with transportable military items excelling in function, while maintaining a beautiful aesthetic. The only real difference between fine household furniture and its campaign counterpart was that the latter could quickly be folded up, packed away in boxes, transported, and – without the use of nails, tacks, or tools – reassembled in 'some corner of another foreign field that was forever elegantly furnished England'.³¹

Travelling Silver for Those Not to the Manor Born

Military outfitters, J.W. Allen of 37 West Strand, in their *Illustrated Catalogue of Portable Furniture* from 1860, advertised a complete cavalry officer's 'barrack room' that included a portable bed with mattress, pillows, and 'name-printed' linens for £20, a recumbent chair in maroon or green leather with leg rest and travelling chest for £11, a valise with vanity (name-engraved, of course) for £3, 15s and 6p, mahogany wash table and cupboard for £12, 10s. The all-important dining equipage is described as:

> *A strong oak Canteen containing Breakfast Service for Three Persons. Electro-plated fittings. 26 ½ inches long, 16 inches wide, 15 inches deep. £24.10s. complete. Containing: 1 large dish, 2 10-inch dishes, 6 breakfast plates, 3 cups and saucers, 1 butter cooler, 3 egg cups, 1 slop basin in white china with gold edges. 2 pint decanters, 1 vinegar cruet, 1 mustard pot, 1 sugar basin, 1 cream jug, 2 salt cellars. 1 candle box, 1 sugar canister, 1 coffee canister, 1 spice box, 1 Lucifer box. Tea caddy, toast fork, corkscrew, spirit flask and cutlery.*[32]

Just what type of furnishings, equipment, and personal effects that were selected and how they were conveyed on a military campaign depended upon the kind and amount of transportation available, the nature and duration of service, and weather. During the hundred years that OSP was a viable commodity, this circa-1840 travelling pot with cover and stove would be an expected item in an officer's provision kit (Figure 5). Measuring almost 13 cm high and wide, the lidded pot – with its retractable handles that hug the circumference when closed – holds the burner insert and three supports, ready to be fired up with the addition of neutral spirits and a small wick.

For members of the officer class, maintaining decorum abroad socially as well as militarily became an extension of observing English standards of gentlemanly behaviour.[33] To facilitate acquiring these necessities of English civility, the Army & Navy Co-Operative Society Ltd. was established in 1871, becoming the one-stop shop for any officer and his family needing to provision: from groceries to guns, from portmanteau to plate.

The compact knife-and-fork set previously cited became commonplace for soldiers in the Boer and Crimean wars, the former ivory handles replaced with simple wood or pewter (see Figure 1). What had previously been sold as an elegant Bachelor's Set, with its tooled leather case, was reinvented as an officer's mess kit with the inclusion of a cylinder for salt, pepper, and mustard, and a fork that now included a cork-pull (Figures 5 and 6). Based on the number of kits researched that are monogrammed, one can assume these may have been purchased as gifts for a newly commissioned officer by a family member. It is gratifying to know that besides having the necessary cutlery with which to dine, those serving as officers in the military also were able to carry seasonings for their meals as well as a wine opener. The luxury hampers of Fortnum & Mason converted for use by the miliary were shipped to The Crimea in 1856, to bring the pleasure of the picnic and the comforts of home to those on campaign and would soon see expansion with the inclusion of plated cannisters and cutlery.

Portable Food

Railroads and Picnics

The popularity of travel equipage gained more enthusiasts with the invention of the locomotive in the early nineteenth century.[34] The Stockton & Darlington Railway line opened in 1825 and was considered the first passenger train, but travellers did not become important railway customers until the 1840s. Railway travel became more popular, but remained prohibitively expensive for many. That eased somewhat with the Gladstone Act of 1844 that introduced the Parliamentary Train, which enabled third-class travellers to pay a penny a mile.[35] The concept of travelling by rail for holidays took off with new railway lines from Liverpool and Manchester to Blackpool. Seaside spas started as playgrounds for the wealthy, but the penny-a-mile railway coupled with summer factory closings made possible the modern weekend trip and summer holidays for all who embraced music, dancing, and the sea air. The bourgeoisie were starting to blur the lines of social distinction with their own acquisition of decorative silverware, where only the trained eye could distinguish between sterling and the more affordable plate.

Wakes Weeks, originally a secular holiday, created vacation opportunities for workers when factories in different towns would take turns closing for a week for maintenance from June through September, ushering in mass tourism.[36] Seaside locations were so inundated with tourists that it led to the development of specialty tour firms like Thomas Cook who would charter trains and organize packaged excursions.[37] Salaries and wages had noticeably increased, and passengers brought with them travel equipage,

Figure 5 (left). Maker unidentified, Traveling pot with cover and stove, c. 1840, fused silverplate pot, cover, burner, burner insert, pot supports, England: Birmingham or Sheffield (Colonial Williamsburg, Virginia). Figure 6 (right). William Hutton & Sons, Travel mess kit, c. 1850-1880, monogrammed leather box, block, fused silverplate beaker, condiment holder for salt, pepper, and mustard, ivory-handled fused silverplate spoon, fork, knife, corkscrew, England: Sheffield (collection of the author).

meals, and refreshments. Packed in their travel hampers and baskets would be the meal: a hand-raised pork pie, ploughman's lunch, or pudding, all eaten with their cutlery and washed down with liquid refreshment from their flask. These baskets or canteens soon included spirit kettles for brewing tea away from the stove, along with all the necessary accoutrements for proper teatime: sugar bowl, cream pitcher, slop bowl, cups and saucers, and sugar tongs.

In her book *Household Management,* Isabella Beeton was very specific on 'things not to be forgotten at a picnic: a stick of horseradish, a bottle of mint-sauce well corked, a bottle of salad dressing, a bottle of vinegar, made mustard, pepper, salt, good oil, and pounded sugar. If it can be managed, take a little ice'. Importantly, decorum must not be sacrificed; she entreated, 'It is scarcely necessary to say that plates, tumblers, wine-glasses, knives, forks, and spoons, must not be forgotten; as also teacups and saucers, 3 or 4 teapots, some lump sugar, and milk, if this last-named article cannot be obtained in the neighbourhood. Take 3 corkscrews.'[38] She advised including both pounded sugar and lump sugar, the latter which would necessitate sugar tongs. While the Romantics set the precedent of outdoor dining during the Georgian era, it was the Victorians and their ability to create inexpensive equipage in electro-plate that propelled the rise of the fully-equipped picnic basket.[39]

G.W. Scott & Sons – wicker basket makers since 1661 – began to partner with metalsmiths in 1897, and by 1914 was listed as celebrated manufacturers and original inventors of fitted leather luncheon and tea cases.[40] For the well-heeled, Fortnum & Mason's classically elegant wicker basket would come pre-filled with delicacies such as game pies, fruits, cheeses, fruit cake, and beverages.[41] Military families ready to follow Mrs Beeton's recipe to prepare their own luncheon need look no further than to the ample offerings of the Army and Navy Stores, Ltd. for their own 'Jaycol' basket of woven English willow, filled with all the silvered accoutrements ready to be stocked for an afternoon tea: plated biscuit box, two 'Thermos' flasks, milk bottle, cups, saucers, plates, sugar and condiment canisters, and cutlery (Figure 7).

Historically, the development of picnicking coincided with the shift from pastoral to urban living, the decline of villages and the rise of modern cities, and changes in work conditions that were the

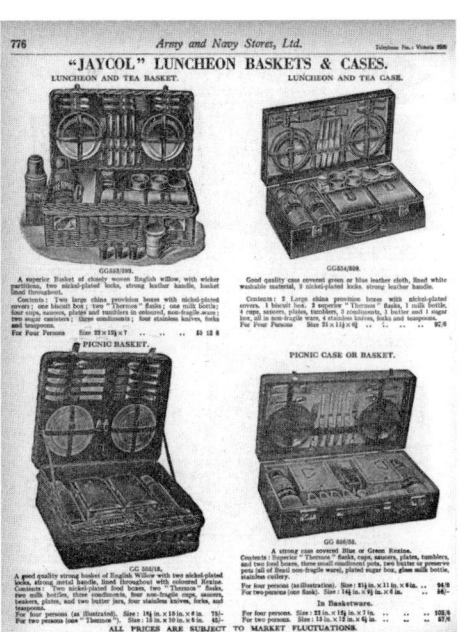

Figure 7. Army and Navy Stores, Ltd. catalogue page, c. 1910

result of improved technology, industrialization, and modes of travel. In 1801, twenty percent of the population lived in towns, while a hundred years later that population skyrocketed to eighty percent.[42] Importantly, as people began working indoors in cities, there was a resurgence in the need for recreation: to leave the city for the country, a lawn, or a grove of trees.[43] In the attempt to recreate indoor dining in an outdoor environment, the entirety of the setting – from tablecloth to meal to utensils – needed to be portable. Happily, the concurrent trends of silver-plating technology, urbanization, and railway travel converged to foster an explosion of craftsmanship and creativity that gave the not-so-wealthy an opportunity to picnic and adventure in style.

Acknowledgments

Thanks to: Janine E. Skerry, Senior Curator of Metals, Colonial Williamsburg, Virginia; Rich Cincotta, Southborough, Massachusettes; Maria-Lorraine Binchet, Yountville California; and Dr Andrew Calman, San Francisco, California

Notes

1. Arthur Reade, *Tea and Tea Drinking* (London: Sampson Low, Marston, Searle, & Rivington, 1884), p. 6; Maxine Berg, *Luxury and Pleasure in Eighteenth-Century Britain* (Oxford: Oxford University Press, 2007), loc. 618 of 4959 Kindle ebook.
2. R.A. Robertson, *Old Sheffield Plate* (London: Ernest Benn Limited, 1957), p. 26.
3. The Goldsmiths' Company Assay Office, *History of Hallmarking* <https://www.assayofficelondon.co.uk/about-us/history-of-hallmarking> [accessed 18 March 2023].
4. Frederick Bradbury, *History of Old Sheffield Plate* (Sheffield: J.W. Northend Limited, 1968), pp. 12, 14.
5. Maxine Berg, 'From Imitation to Invention: Creating Commodities in Eighteenth-Century Britain', *Economic History Review*, LV.1 (2002), 1-30 (p. 9).
6. R.E. Leader, 'Sheffield Cutlery and the Poll-Tax of 1379,' *Journal of the British Archaeological Association*, 10:3 (25 September 2017), 226-33 (p. 227).
7. Bradbury, p. 28.
8. Robertson, p. 54.
9. Sarah D. Coffin, *Feeding Desire – Design and Tools of the Table, 1500-2005* (New York: Assouline, 2006), p. 52.
10. Bradbury, p. 396.
11. Coffin, p. 52.
12. Berg, loc. 2177 of 4959.
13. Gordon Crosskey, *Old Sheffield Plate* (Sheffield: Treffry Publishing, 2013), p. 256.
14. Bradbury, pp. 87, 88.
15. Molly Duerdin, *Sheffield Silver 1773-1973* (Sheffield: Sheffield City Museums, 1973), p. 7.
16. J.C. Garcia and T.D. Burleigh, 'ECS Classics: The Beginnings of Gold Electroplating', *The Electrochemical Society Interface*, 22.2 (January 2013), 36-38 (p. 36).
17. Megan Elizabeth Gleason, 'From Vulgarity to the Current Fashion: The Impact of Electroplating on Victorian Industry, Marketing, and Design' (unpublished master's thesis, University of Glasgow, 2001), p. 18.
18. The inexpensive metal was usually a silver-colored alloy such as Britannia metal (92% tin, 6% antimony, and 2% copper) or German silver (60% copper, 20% nickel, and 20% zinc).
19. Robertson, p.19.
20. Steven King and Geoffrey Timmins, *Making Sense of the Industrial Revolution* (Manchester: Manchester University Press, 2001), p. 151.

21. Berg, loc. 302 of 4959.
22. Neil McKendrick, J. Brewer, and J. Plumb (eds), *The Birth of a Consumer Society* (London: Longman, 1982), p. 6.
23. Kenneth Quickenden and Arthur J. Kover, 'Did Boulton Sell Silver Plate to the Middle Class? A Quantitative Study of Luxury Marketing in Late Eighteenth-Century Britain,' *Journal of Macromarketing*, 27.1 (2007), 51-54 (p. 51).
24. Berg, loc. 328 of 4959.
25. Named after Prince Rupert of the Rhine, Duke of Cumberland (1619-1682) who was an amateur metallurgist, 'prince's metal' was also referred to as 'Bristol Brass' – an alloy of three parts copper to one part zinc, combined with charcoal.
26. Peter Borsay, *A History of Leisure: The British Experience since 1500* (London: Palgrave, 2006), p. 90.
27. Helen Clifford, 'Concepts of Invention, Identity and Imitation in the London and Provincial Metal-Working Trades, 1750-1800', *Journal of Design History*, 12.3 (1999), 241-55 (pp. 241, 253).
28. King and Timmins, pp. 155-56.
29. H. Moyse-Bartlett, 'The British Army in 1850', *Journal of the Society for Army Historical Research*, 52.212 (Winter 1974), 221-37 (p. 228).
30. Robert Burnham and Ron McGuigan, *The British Army Against Napoleon – Facts, Lists and Trivia: 1805-1815* (Yorkshire: Frontline Books, 2010), p. 147.
31. Nicholas A. Brawer, *British Campaign Furniture: Elegance Under Canvas, 1740-1914* (New York: Harry N. Abrams, 2001), p. 19.
32. Brawer, p. 62.
33. Brawer, p. 56.
34. John Urry and Jonas Larsen, *The Tourist Gaze* (London: SAGE Publications, Ltd., 2011), p. 17.
35. Bill Cormack, *A History of Holidays 1812-1990* (London: Routledge/Thoemmes Press, 1998), p. 23.
36. Lord Amulree, 'Industrial Holidays', *Journal of the Royal Society of Arts*, 87.4486 (11 November 1938), 3-14 (p. 6).
37. Urry and Larsen, p. 24.
38. Isabella Beeton, *Mrs. Beeton's Book of Household Management* (1861, repr. New York: Skyhorse Publishing, 2015). p. 241 of 306, Kindle ebook.
39. Diana Noyce, 'The Rise of the Picnic Hamper: Its Pleasurable and Macabre uses in Nineteenth-Century Britain,' in *Food & Material Culture: Proceedings of the Oxford Food Symposium on Food and Cookery 2013*, ed. by Mark McWilliams (London: Prospect Books, 2014), pp. 224-33 (p. 224).
40. *London Trades Directory*, 1914, p. 1403.
41. Noyce, p. 228.
42. Urry and Larsen, p. 18.
43. Walter Levy, *The Picnic: A History* (Plymouth: Rowman & Littlefield, 2014), p. 6.

Food for Walking: On the *Camino Real de Tierra Adentro*

Jaime Iram Vargas Barrientos

During Mexico's colonial period, the Spanish monarchy pushed the development of resource exploitation centres throughout New Spain. This American colony expanded northward with a rugged climate and geography, inhabited by indigenous communities that opted for confrontation in response to attempts at military and religious subordination. The discovery of silver mines of Zacatecas was one of the most important enrichments of the Spanish Crown and marked the most populated and farthest north horizon until the military and missionary exploration of the novo-Hispanic north. During the following centuries, the command for spiritual pacification and the ambition for new deposits impelled the exploration of the territories, creating, in turn, spaces of settlement as a strategy for the evangelization of Indians as well as defence against those who remained rebellious and bellicose.

These population units were also self-sufficient production centres that generated provisions for the communities dedicated to the increasingly remote and isolated mining centres of northern New Spain. The exploitation of indigenous labour in the extraction of silver and the establishment of villas and communities required the settlement of groups from the centre and south, since the demographic density of the northern Indians was insufficient and unstable.

In this way, the population was organized according to the European model of urbanity in the northern frontier communities, integrating Spaniards, who were Castilians and Basques; Indians from the centre and south of New Spain; and Africans brought as a labour force for the work in *haciendas*.

Faced with the obstacles of settlement, exploration, and exploitation of the resources of the north, the map of New Spain expanded until its farthest frontier was Santa Fe, a town that crowned a vast territory besieged by rebel groups, a harsh climate, and a mostly flat territory with irregularities that hindered communication between economic centres from north to south.

From Pre-Hispanic Ways to Royal Roads

Prior to the encounter with the Spaniards, there was commercial communication between Mesoamerican and Arid American populations. Even between distant and culturally differentiated social units, there was an exchange of goods through trade routes traced by natural or geographically accessible roads.¹ The ancestral knowledge of these groups was taken advantage of by the Spaniards in their crusade through unknown lands. After the fall of Tenochtitlan, expeditions to the north established Indian villages and Spanish republics using the roads travelled by the original groups.

The discovery of silver sites pushed the transit of people more and more towards the north of New Spain and other cardinal points of public interest, consolidating the most travelled routes and watched over by the monarchy as *Caminos Reales* (Royal Roads).² These roads were part of the project of integrating the American territory into the Spanish Empire, so the roads were taken care of to promote economic development and protect the security of passers-by.³ In New Spain, there were four such roads that departed from the ancient Tenochtitlan to Veracruz, Acapulco, Guatemala, and Santa Fe.⁴

The Royal Road to Zacatecas, used frequently for transferring metals, was popularly known as the *Camino de la Plata*, at least until its extension to Santa Fe, where it was distinguished as the *Camino a Santa Fe* or *Camino de Tierra Adentro,* since it passed through the current Mexican states of Durango and Chihuahua and the US territories of Texas and New Mexico.

Muleteers and Travellers

In addition to the flow of gold and silver, the movement of books, paintings, techniques, utensils, food, art, and knowledge through the royal roads made it possible to amalgamate a common culture among the territories it crossed. Behind this movement of goods there were traders, merchants and muleteers, as well as individuals with family, economic, and diplomatic interests, which also contributed to the reciprocal exchange between the economic centres.⁵

The merchants of the ancient trails of the Mesoamerican political units were a guild known within the Mexica empire as *poshtecas*, who had traced the mercantile routes of ancient Tenochtitlan.⁶ In their journeys they were accompanied by the *tamemes,* or loaders, who were the vehicles used to transport the merchandise; even after the introduction of mules and horses, they continued to be used as an option to move loads. After the Spanish colonization, the figure of the muleteer migrated along with the European road system, that is to say, the one who brought beasts on the roads to move goods.⁷

Within the job, there were different functions and obligations according to the hierarchy between wagon owners and loaders, as well as the time spent on their journeys, as there were those who travelled routes of several days and others of only one morning. There was also racial difference, as Spaniards, mulattos, and Indians in different proportions were at the head of the transportation organization.⁸

These particularities are important when analyzing the food needs of this group, as they show that their circumstances influenced the travel experience, where the economic position, the complexity of their work, and the culture of the muleteer came into play.

Walk and Eat on the Road

Riding on the trails was not always safe, because during the nineteenth century assailants, Apache attacks, weather conditions, and the broken and irregular geography of the Mexican territory constantly worried travellers and caused them to delay or pause their itineraries.[9]

In the case of transporting merchandise, drivers needed to know certain rules within the wagon road system that would allow for safe traffic between riders. The *jornadas* were known as a full day's journey from one point to another and consisted of two kinds of crossing: the first by means of shortcuts and known short roads and the second by the roads on horseback with a light load.[10] A single *jornada* could represent up to ten leagues or 45 km (28 miles).[11]

Whatever the case of those who walked on the roads, it was necessary to know strategic places to rest, take shelter, and replenish provisions. Since the sixteenth century in New Spain, the first ordinances were decreed for the establishment of inns, places that offered lodging and food to strangers and their horses.[12] There were also houses and *ventas* (shops) of private individuals who offered meals at the roadside.[13]

In northern New Spain, *presidios* (forts), Indian hospitals, and missions along the road offered lodging from very early times, especially where inns were distant or did not exist. The other option known among travellers was to stay at the parish priest's house, which was 'as cheap as any roadside inn but with better food'.[14]

In the case of passengers on the post office lines, the transportation service included points along the route for food or lodging during the night, and on the daytime route they had fifteen to thirty minutes to have breakfast, lunch, and snacks along the way.[15] This short time allowance suggests that these stops were exclusively to stop the carriage and take out food that travellers had brought with them.

Although there were options for resting and eating on the roads, this did not mean that they were sufficient on the routes that branched off from the main roads, nor that these were the most affordable options in all cases. Complaints of excessive food prices, poor quality, or little variety were noted in different travel diaries, which detracted from the promise of food available from these businesses.[16]

A dialogue that reflects food shortages in places like *fondas* is between Branz Mayer and an innkeeper in Mexico in the middle of the nineteenth century:

> We asked for breakfast, but the answer was the slow movement of the long forefinger from right to left and a 'No hay!'
> 'Any eggs?'
> 'No hay' [there's not]
> 'Any tortillas?'
> 'No hay'
> 'Any chile?'
> 'No hay'

'Any water?'
'No hay'
'What have you got then?'
'Nothing!'[17]

Faced with such circumstances, carrying food was a permanent option that even generated a category for the kinds of preparations intended for this purpose.

Incorruptible Food

Food preservation is a constant concern of mankind, so techniques have been developed to prolong its edibility by modifying its organoleptic properties as well as its physical characteristics according to the tools and processes used. Some methods consisted of avoiding the proliferation of pathogenic agents by reducing water content through dehydration or prolonged exposure to the sun or a heat source. Other ways were by preventing the contact of surfaces with oxygen, light, and moisture; coating foods with impermeable substances; or suspending them in saturated solutions of salt, sugar, vinegar, or fat.[18]

Most of these preparations can be considered within the term 'subsistence cooking' or 'resistance cooking', since they encompass culinary practices that rescue and take advantage of food wastage as well as processes that delay degradation.

Achieving the incorruptibility of supplies has not only been a function of family spending in times of scarcity and crisis; the conditions of the transverse roads of the novo-Hispanic north challenged the resources of the wagon drivers and expeditionaries, who in long journeys and hostile scenarios required food that did not suffer acidification in prolonged exposure to heat, was easy to transport, had high caloric content, and was affordable at any time of the year and sufficient for a communal meal. The distance between Mexico and Parral by wagon alone in the seventeenth century could be up to four months, which implied a series of challenges for travellers.[19] The *itacates* (supplies) of the walkers must therefore have been loaded with processed foods, taking a practical taste for salty, sour, or sweet preparations, which were already common in the food culture of the northern societies through which the *Camino Real de Tierra Adentro* passed.

The travel chronicles of the nineteenth century are a window into the expeditionaries' habits and protocols along the *Caminos Reales* and transversal trails. They all mention the constant concern for food and water, which was not always solved in the provisioning spaces, so the only safe thing to do was to undertake the passage with supplies to be consumed along the way.

According to Josiah Gregg's chronicle, the muleteers 'always camp out, being provided with their cooks and stock of provisions, which they carry with them'.[20] A trekker's food consisted of cheeses, tortillas, dried meat, preserves, dried fruit, bread, coffee, and, if possible, wine and cold meats, all for a three-day trip where there were no supply centres.[21]

The leather saddlebags were loaded with water and canned food sometimes imported from Europe such as olives stuffed with anchovies, fish in oil, and sardines that could be obtained in some stores in the cities.[22] For those who left Mexican homes as traders or muleteers, some canned fruit, dried meat, cheese, or any 'food for walking' must have accompanied them.

Food for Walking

The cookery recipe books are an interesting source of information on the food intended exclusively as provisions, and they also show that in the nineteenth century there was a continuity in the publication of these recipes that could well be considered as a sign of their importance to travellers.

The recipe books can be seen as instructive texts that were directed to a certain part of the society with economic affordability, so that the foods selected for their composition were often outside the reach of the bulk of the population, especially at a time when the conditions of most of the citizens were different from those who could read and write recipes.

In addition, the printing of Mexican recipe books in the nineteenth century favoured foreign preparations reproduced by families of Spanish descent and adjusted in French publishing houses settled in Mexico, who used European models in the making of national cookbooks.[23] Thus, the preparations mentioned *para caminar* (for walking) in Mexican cookbooks are in some cases adaptations of foreign recipes to local resources.[24]

The term 'for walking' indicates a specific function for mobilization, although, depending on the case, the preparation could have been consumed at home without the intention of being a provision for travellers. 'For walking' seems to be more a suggestion of the authors of the recipe books, who saw in these foods a viable option for travel because of their resistance to rottenness. The term 'resistance food' is also mentioned in culinary books in the nineteenth century, indicating food whose ability not to decompose was sufficient for a short trip or a day in the fields.[25]

Although most of the preserved foods or those belonging to subsistence cooking, such as dried meats, were known to be provisions for long trips, the recipes 'for walking' can be divided into two subcategories: concentrates and cold meats.[26]

Concentrates had the advantage of taking up little space and flavouring a large quantity of water when dissolved in it.[27] Fruit syrups for the preparation of soft drinks on the road had this function as a concentrated solution of sugar and fruit pulp. Another one was coffee syrup, which also shortened the time of preparation.[28]

Other recipes within this subcategory are pills, an equally practical resource for transportation due to their size.[29] The recipe books mention two of these preparations designed for the activity of the walker; those of lemon that were made by reducing the citrus juice along with sugar as a tablet, which could be dosed on the way to prepare a fresh drink, and the pills of broth, that according to their description were 'extremely comfortable for those who go on the way', since, being very well covered, they were incorruptible for several years and only needed to be rehydrated in hot water to make a substantial soup.[30]

In this regard, a note from the weekly publication *Semanario Económico de México* of 1810, mentions:

> Walkers and travellers usually find nothing to eat on the deserted roads inland, and even in our most frequented inns they usually find a cup of broth for a sick or delicate person. Tortillas, *mole de vaca,* and beans, is the most common thing in the shops, and this many times cannot be passed except by force of a very active hunger. To meet this need, a portable broth is made, which is preserved for more than a year [...].[31]

The broth tablets could accompany the traveller stored in a glass jar or pottery, solving the impossibility of carrying soups that turned sour in the course of a single day in hot climates. To avoid the corruption of the broth over short distances, one suggestion was to add a piece of iron to the vessel, which by its properties attracted the acidifying agents and prevented fermentation of the soup.[32] The 1844 *Mexico as It Was and as It Is* refers to the luggage for a trip where 'two mules have been hired and loaded with a good store of supplies such as hams, corned-beef, portable soups, sausages, sardines, and wine, and these are put under the charge of an *arriero*'.[33]

The impossibility of carrying fresh meats for the road was solved by subjecting the proteins to cooking in vinegar and spices. The knowledge of seasonings and certain herbs as antiseptics was already applied in different cuisines around the world, especially in hot climate zones where bacteria proliferated on food surfaces.[34] The use of acetic acid could also prolong the life of a protein for up to a year, although the use of oil kept the pieces 'fresh' for longer.[35]

As for the second subcategory, nineteenth-century recipe books repeatedly mention *piernas de carnero a la benazon para el buen camino*, a castellanization of the French word *venasion*, a preparation in which goat legs were cooked for a long time in strong vinegar, cloves, cumin, pepper, cinnamon, thyme, and laurel, which made the joint incorruptible for up to thirty days, enough time for a journey; when consumed without the need to reheat, it freed the traveller from searching or carrying fuel.[36] The recipe does not mention its packaging, although it is most likely that for portability, as with other prepared meats, a covered *cuñete* was used, which was a kind of barrel with a lid that was fastened with metal hoops.[37]

Cooking an animal 'for walking' was one of the female tasks to ensure food was available for the members who wished to migrate.[38] The muleteer activity was regularly a family enterprise, so it would not be surprising to find an internal organization to solve the journey's necessities.[39]

In other circumstances, according to the information in nineteenth-century newspapers in the Mexican capital, cold meats were obtained by order from specialized businesses such as *fondas* (inns).[40]

Another food was *jamones para el camino* (hams for the road), a piece of ham cooked in wine and a mixture of spices that was kept on the fire until the remaining liquid evaporated. The cold cuts and the pressed hams followed the same cooking with aromatics and vinegar, perhaps differing in the species of the cooked cattle and the way they were transported, because as their name mentions, the pressed hams were wrapped in cotton blankets and compressed with weight to release their cooking liquids and make them more compact.[41]

Butifarras are the only sausage explicitly mentioned in the recipe books for use on the road, but, as mentioned in the *Novísimo arte de cocina, salchichón, salchicha, chorizones, longanizas, butifarras y morcón*, they could be used to garnish stews or eaten 'alone, particularly for walking'.[42]

Hunting was also part of the activities of the expeditionaries who took advantage of the natural resources. There are several references in the travel literature to the hunting of birds and deer during the journey, including fishing and the ways of processing these foods. In the case of fish, according to the cook books, a piece of soft bread soaked in sugarcane distillate was placed in their mouths before wrapping in straw and blankets. This formula ensured the freshness of the fish up to twelve days on the road.[43]

Mesoamerican Traditional Food for Walking

For the muleteers' parties, beasts were also destined for their consumption, although the availability of pasture, the space in the vehicle, or the difficulties of moving the animal did not always make this an option.[44]

One of the traditional preparations still common along the route of the *Camino Real* is the *barbacoa*, a Mesoamerican technique that consists of cooking the carcass of any animal underground. This makes sense with the system of transhumance of wool cattle that was traced from the capital and New Mexico to the textile centres of New Spain.[45]

This preparation could represent a lot of food for one person, but for the journey on the northern roads it was better to undertake the march in retinue, since a large party of walkers could better defend themselves against the hosts of rebellious Indians, which then would mean communal meals along the frontier.[46] In the transport of *recuas* (transport animals), there should be at least two muleteers in the group, one in front of the animals, and one behind, watching that they do not leave the road.[47]

On the routes it was possible to spend the night by the side of the road or at points known to the muleteers to rest and take food; make a fire to heat tortillas, coffee, or chocolate. The explorer Gustav Tempsky mentioned that on the road chocolate 'is the food that is best prepared for a day's journey, since it is not serious for the stomach and resists their longest cravings'.[48] This could be easily prepared in hot water or also a mix of *cacao*, cinnamon, sugar and corn in pods that could be transported and mixed with a liquid.[49]

Through ethnographic exercises with the rural population of the current state of Durango, it has been possible to detect foods in family traditions that would have fulfilled the function of providing energy to the travellers. Corn, as a staple food of the

Mexican people, has been used in countless preparations fulfilling the function of dish, utensil, and also as an envelope, which has protected its contents from contact with contaminated surfaces.

Some other foods that we could consider to 'walk' from the Mesoamerican tradition include those dry foods that, without humidity, resist microbial and fungal decomposition. In that sense, Mexican dry foods such as *pinoles, tostadas*, tortillas, and tamales have been ideal for transportation.[50]

Pinole is pulverized roasted corn used as an energizer by the ancient northern Indians. It was also important as a provision in the evangelization project of Archbishop Tamarón y Romeral, who travelled the roads throughout Nueva Vizcaya (north of México) in the eighteenth century. *Pinole* is used to make *kipii*, an indigenous *O'dam* ceremonial candy that is eaten as a snack during journeys.

Following this idea, other popular preparations in northern Mexico based on corn could also be considered useful for their portability and consumption in the places of the ancient north. The *gordita de horno*, or *gorditas de comal*, are known for their traditional use as provisions in the procession from one point to another in different northern states of the ancient *Camino Real de Tierra Adentro*.

In the mid-nineteenth century, the book *Mexicanos pintados por si mismos* mentions *gordas* as a typical dish for muleteers:

> Seated in front of a good fire, where the *atajador* cooked the very tasty *gordas*, which he said more than once had caused a rational creature to come out into the world before its time. [...] The two muleteers lamented not having for dinner that night more than the fried *gordas*, very good indeed for the palate that is not condemned to taste them daily [...].[51]

The *gordita de horno* consists of a corn dough beaten with lard and stuffed with beans, cumin, and red chili, cooked in an earth oven with mesquite wood. This preparation is found in handwritten nineteenth-century recipe books from Durango and survives in today's culinary tradition as a food for travelling.[52]

The term *gordita* is used in the same way as a tortilla cooked on a *comal* and filled with different stew. These are mentioned in Durango from the early eighteenth century. The advantages of wrapping a semi-humid stew in baked dough are the retention of heat and its stability during the hustle and bustle of transportation.

Currently, throughout the Mexican Republic there are dishes known for their ability to withstand both acidification and ravages of transport, many of them coined with the term '*arriero*', for its common use in feeding them, including *tostadas arrieras* (toasted tortilla), *burro arriero* (bean and chili wrap), *nopales de arriero* (prickly pear), *guiso arriero* (stew), and others.

In the north of Mexico, as in Durango, this food has been reappropriated for some special ritual dates, as in the case of Lent, where communities make dehydrated preparations up to six months in advance, such as *chuales* (dried corn), *chile pasado* (dried chilli), or *torrejas* (dried pumpkin): all were the food of the travellers in the first instance and are now the central food on the Lenten journey.[53]

Notes

1. Beatriz Braniff Cornejo, 'Comerico e intrrelaciones entre Mesoamérica y la Gran Chichimeca', *Caminos y Mercados de México*, ed. by Janet Long Towell and Amalia Attolini Lecón (Mexico: UNAM-IIH, 2010), pp. 27–50.
2. Sergio Ortiz Hernán, 'Caminos y transportes mexicanos al comenzar el siglo XIX', *Comercio Exterior* (1973), 1247–53. All translations by the author.
3. María Luisa Pérez González, 'Los caminos reales de América en la legislación y en la historia', *Anuario de Estudios Americanos*, 58.1 (2001), 33–60.
4. Tomás Martínez, Enrique Lamadrid, and Jack Loeffler, *El camino real de tierra adentro* (Mexico: Colegio de postgraduados y Mundi-prensa México, 2009).
5. Elsa Rodríguez García, 'El Camino Real de Tierra Adentro: un sendero recorrido', *Diario De Campo*, 11 (2013), 55–59.
6. Miguel León-Portilla, 'La institucion cultural del comercio prehispánico', *Estudios de Cultura Náhuatl*, 3 (1963), 23–54.
7. 'HARRIERO. s. m. El que conduce béstias de carga, y tragína con ellas de una parte a otra' (*Diccionario de Autoridades* (Madrid: Real Academia Española, 1734).
8. Bernd Hausberger, 'En el camino. En busca de los arrieros novohispanos', *HMex*, 64.1 (2014), 65–104.
9. Hemeroteca Nacional de México (HNDM), *El Sol*, 26 November 1824. Jacobo de Villaurratia proposes the improvement of the road to Veracruz, mentioning the importance of the comfort of the passengers having established points for their rest and food.
10. George Frederick Ruxton, *Adventures in Mexico and the Rocky Mountains* (New York: Harper and Brother, 1848), p. 119.
11. Gobierno del Estado de Durango, *El Camino Real de Tierra Adentro, Travesía histórica y cultural al Septentrión Novohispano* (Mexico City: Gobierno del Estado de Durango, 2012), p. 47.
12. Ramón González, *Cuatro siglos de sabor. La cocina queretana* (Mexico City: Gobierno del Estado de Querétaro, 2005), p. 9; 'Mesón', *Academia Usual* (Madrid: Real Academia Española, 1817).
13. HNDM, 'Artículos que han de observarse', *Gazeta de México*, 7 April 1794.
14. Gobierno Durango, p. 48.; Chantal Cramoussel, 'Viajar por los caminos del norte de la Nueva España', Transición, 22 (1999), 40; Robert William Hale Hardy, *Viajes por el interior de México en 1825, 1826, 1827 y 1828*, trans. by Ernesto de la Torre Villar (Mexico: Trillas, 1997), pp. 338–40.
15. HNDM, 'Línea de postas del interior', *Diario de Avisos* (1857), p. 4.
16. Berthold Seemann, *The Botany of the Voyage of H.M.S. Herald, Under the Command of Captain Henry Kellett, R.N., C.B., During the Years 1845–51* (London: Reeve and Co., 1852), pp. 259–60.
17. Brantz Mayer, *Mexico as It Was and as It Is* (New York: J. Winchester, New World Press, 1844), p. 168.
18. Maguelonne Toussaint-Sammat, *A History of Food* (Paris: Wiley-Blackwell, 2009), p. 662.
19. Carmen Castañeda, 'Circulación de libros por el Camino Real de Tierra Adentro', Transición, 22 (1999), 21.
20. Josiah Gregg, *Commerce of the Prairies, Or, The Journal of a Santa Fé* (New York: Henry G. Langley, 1841), II, p. 123.
21. Jaime Iram Vargas Barrientos, 'Alimentación y sociedad, la cocina de Durango a través de sus recetarios a mediados del siglo XIX' (unpublished masters thesis, Universidad Juárez del Estado de Durango, 2019), p. 151; Gustavus Ferdinand von Tempsky, *Mitla: A Narrative of Incidents and Personal Adventures on a Journey in Mexico* (London: Longman, Green, 1858), p. 31.
22. Jaime Iram Vargas Barrientos, 'De la tienda a la cazuela. Tiendas y cocinas en Durango a mediados del siglo XIX', in *Ciencias Sociales y Humanidades en Durango*, ed. by Jonatan García (Mexico City: Universidad Juárez del Estado de Durango, 2021), pp. 237–74.
23. Sarah Bak-Geller Corona, 'Los recetarios afrancesados de la nación', *Modelos alimentarios y recomposiciones*

sociales en América Latina (December 2009) <https://doi.org/10.4000/aof.6464>.
24 Juan Nepomuceno, *El Tesoro de la cocina* (Mexico City: Imprenta de Juan Nepomuceno, 1866), p. 28. '*Anande para el camino*' is a species of bird endemic to Europe and Africa; the recipe was probably taken from a foreign book.
25 Clementina Días y de Ovando and Luis Mario Shneider, *Arte Culinario Mexicano. Siglo XIX* (Mexico City: Fundación de Investigaciones Sociales, 1994), p. 274. I thank Alberto Peralta for suggesting this fact.
26 Nepomuceno, p. 106, '*carne seca*'.
27 Nepomuceno, p. 260, '*jarabe de frambuesas*'.
28 Mariano Galván, *Cocinero mexicano en forma de diccionario* (Paris: Librería de Rosa y Bouret, 1858), p. 450, '*Jarabe de café*'.
29 [Anon.], *Arte nuevo de cocina y repostería acomodado al uso mexicano* (New York: Casa de Lanuza, Mendía y C., 1828), p. 202, '*Pastillas de limón para refrescar*'.
30 [Anon.], *El Cocinero Mexicano* (Mexico: Imprenta de Galván, 1831), I, p. 30, '*Pastillas de Caldo*'.
31 HNDM. *Semanario Económico de México*, 18 January 1810, p. 21.
32 HNDM. *Semanario Económico de México*, 18 January 1810, p. 23.
33 Mayer, p. 162.
34 Paul W. Sherman and Jennifer Billings, 'Darwinian Gastronomy: Why We Use Spices', *BioScience*, 49.6 (June 1999), 453–63 <https://doi.org/10.2307/1313553>.
35 [Anon.], *Arte nuevo*, p. 803, '*Sardina*'.
36 [Anon.], *Arte nuevo*, p. 17, '*Piernas de carnero a la benasón para camino*'; Galván, p. 145, '*Carnero (piernas de)*'.
37 Galván, p. 248, '*Cuñete de vaca*'.
38 Francisca Balmaceda de Redo, *Recetas de Cocina Experimentada* (Durango, 1853), II, p. 34, '*Jamón para caminar*', '*Pierna de carnero a la benasón para el camino*'.
39 Clara Elena Sánchez Arguello, '*La arriería Novohispana y las rutas de tierra adentro*', *Transición*, 22 (1999), 72.
40 HNDM, '*Avisos*', *El Sol*, 26 April 1830, p. 1.200.
41 [Anon.], *Nuevo y sencillo Arte de cocina, repostería y refrescos, dispuestos por una mexicana*, 2nd edn (Mexico City: Imprenta de Vicente García, 1842), pp. 249–51, '*Fiambre de ternera o vaca aprensada*'.
42 [Anon.], *Nuevo y sencillo arte de cocina, repostería y refrescos, dispuestos por una mexicana* (Mexico City: Imprenta de Santiago, 1836), p. 10, '*Advertencia*' II.
43 Galván, p. 639, '*Pescados vivos*'.
44 Gregg, p. 123. The expeditionary mentions a consignment of twenty sheep for daily consumption for himself and his party.
45 Hausberger, p. 84; Gobierno Durango, p. 62.
46 Cramoussel, pp. 39–47.
47 HNDM, '*Artículos que han de observarse*', *Gazeta de México*, 7 April 1794.
48 Tempsky, p. 137.
49 Francisca Balmaceda de Redo, *Recetas de Cocina Experimentada* (Durango, 1869), I, p. 26, '*Chocola de espuma*'.
50 Luís Alberto Vargas, '*Algunos hechos poco conocidos sobre los comales, las vasijas y los platillos aguados y secos*', *Cuadernos de nutrición*, 31.5 (2008), 193–97.
51 Juan de Dios Arias and others, '*Los mexicanos pintados por si mismos*' (Mexico City: Ed. M. Murguía, 1854), p. 155.
52 Jaime Iram Vargas Barrientos, *Comida de Cuaresma en Durango. Entre el ayuno y el banquete* (Mexico City: CreateSpace, 2022).
53 This reflection was possible through the discussion and feedback in the online conversation of the Oxford Food Symposium.

Contributors

Ken Albala is Tully Knoles Endowed Professor at the University of the Pacific in Stockton, California. He has published 27 books and won the 2023 Outstanding Faculty Award at the university.

Tamar Babuadze is a writer and editor in Tblisi, Georgia.

Janet Beizer, Professor of Romance Languages and Literatures at Harvard University, specializes in literature, cultural studies, and food studies. Her forthcoming book is *The Harlequin Eaters: From Food Scraps to Modernism in Nineteenth-Century France.*

Tanushree Bhowmik is a New Delhi based independent food researcher who writes about ancient to colonial food systems in the Indian subcontinent, the mythological and cultural significance of food, and food and memory.

Andrea Broomfield chairs the Department of English at Johnson County Community College in Overland Park, Kansas. Her most recent books are *Kansas City: A Food Biography* and *Iconic Restaurants of Kansas City.*

Anthony F. Buccini (PhD Cornell University) is an historical linguist and dialectologist who formerly taught at the University of Chicago in the Departments of Germanic Languages and Literatures and Linguistics. As a food historian, his research focuses on the Mediterranean and Atlantic World. He is a two-time winner of the Sophie Coe Prize in Food History.

Kathleen Burke is a doctoral student in History and Food Studies at the University of Toronto.

Voltaire Cang is an academic researcher based in Tokyo. He researches and writes about Japan's 'intangible' heritage, including food and other cultural practices and traditions.

Nattha Chuenwattana is a doctoral student in Anthropology at the University of Toronto.

Julia Fine is a PhD student in History at Stanford University, where she studies the food and environmental history of the British Empire.

Kashyapi Ghosh is a PhD scholar at the Department of Humanities and Social Sciences, Indian Institute of Technology Tirupati. For her doctoral dissertation, she is looking at the changing dynamics of the kitchen space.

Contributors

M. Fernanda Estrada González is a graduate student in Anthropology in the Faculty of Political and Social Sciences, National Autonomous University of Mexico (UNAM).

Binti Gurung is a food historian based in Kathmandu. She is from the Tamu indigenous community of Nepal and is currently researching the indigenous food history of Nepal. She maintains a digital food research and archival project called Natural Roots.

Peter Hertzmann is an autodidactic polymath with a strong contrarian bent who likes to provide an alternative approach about all aspects of food. His books include *Knife Skills Illustrated: A User's Manual*, *A Perfect Mouthful*, and *50 Ways to Cook a Carrot*.

Michael Johnson is an Associate Professor of French in the Department of World Languages and Cultures at Central Washington University. His work explores *gastronomie engagée* and food activism in the Franco-Belgian sphere.

Pırıl Kadırgan is currently a graduate student in the Design, Technology, and Society Program's Gastronomy and Design Track at Özyeğin University, İstanbul. Her research interests include Ottoman-Turkish food history, consumption studies, and the history of medicine.

Laura Kitchings is a member of the National Coalition of Independent Scholars, an elected member of the American Antiquarian Society, and holds a Master of Liberal Arts in Gastronomy from Boston University. She has worked for a variety of cultural heritage organizations.

Charlotte Kleyn, a food historian and food writer based in Amsterdam, is the author of a history of eating on the go and curator of the History of Food collections at Allard Pierson, University of Amsterdam.

Else Marie Knudsen (PhD, London School of Economics) is an Assistant Professor of Social Work at Trent University. She studies criminal justice policy, and, in particular, the experiences of children of prisoners in Canada.

Joshua Lovinger is a vascular neurologist based in New York. He has also completed graduate work in medieval and Jewish history, and is interested in the relationship between food, science, and Jewish law and custom.

Priya Mani, a designer and cultural researcher based in Copenhagen working to create gastronomical experiences, is particularly interested in the social interactions of making, presenting, and consuming food.

Contributors

Mark McWilliams, Professor of English at the United States Naval Academy, has served as Editor of the Oxford Symposium on Food and Cookery since 2011.

Nader Mehravari is a Research Associate at the College of Agriculture and Environmental Sciences, University of California, Davis. His work explores the history, principles, and practices of ancient and contemporary Persian cookery and associated foodways.

Shirin Mehrotra is a food writer, researcher and anthropologist from India. She writes about the intersection of food, society and culture, and cities and their foodscapes. Her work has been published in *Whetstone*, *The Juggernaut*, and *Condé Nast Traveller* among others.

Johanna Mendelson Forman is an Adjunct Professor at American University's School of International Service and a Distinguished Fellow at the Stimson Center, where she heads the Food Security Program.

Meher Mirza is a food, travel, and culture writer based in Mumbai. Her work has appeared in *BBC Travel*, *BBC Culture*, *CNN Travel*, *The Guardian*, *Saveur*, *Literary Hub*, *Condé Nast Traveller India*, *Food 52*, *Serious Eats*, and *Roads & Kingdoms*.

Tatsuya Mitsuda (PhD, Cambridge University) is an Associate Professor in the Faculty of Economics at Keio University. He researches the intertwined social and cultural histories of food and animals.

Jacqui Newling is a graduate of the Le Cordon Bleu Masters in Gastronomy at Adelaide University and is an Affiliate in History at the University of Sydney. She is the author of *Eat Your History: Stories and Recipes from Australian Kitchens*.

Diana Noyce holds a masters degree in Gastronomy from the University of Adelaide and researches and teaches food history and food culture. Her many areas of interest include gastronomy in Antarctica, picnics and picnic hampers, and Australia's foodways.

Ayari Pasquier (PhD, El Colegio de Mexico) is an Associate Researcher at the Center for Interdisciplinary Research in Sciences and Humanities at the National Autonomous University of Mexico. She studies social inequality and sustainability in urban food systems.

Aimée Plukker is a PhD candidate in Modern European History at Cornell University and Reviews Editor of the *Journal of Tourism History*. Her dissertation focuses on post-WWII U.S. Tourism to Western Europe.

Contributors

V. Vamshi Krishna Reddy is an Assistant Professor at the Centre for Comparative Literature, University of Hyderabad. He works broadly in the area of critical theory, film studies, and culture studies.

Simi Rezai-Ghassemi (PhD, University of Bath) is a cookery teacher, writer, organic gardener, and Iranian food tour guide.

Salma Serry, a graduate student in History at University of Toronto with a Food Studies degree from Boston University, is a food history researcher, filmmaker, and founding curator of @sufra_archive, a library dedicated to the modern food history of Southwest Asia and North Africa.

C. Thomas Shay is a senior scholar in the Department of Anthropology at the University of Manitoba and the author of *Under Prairie Skies: The Plants and Native Peoples of the Northern Plains*.

Chloe Shields is a PhD student partnered between the University of Strathclyde and the National Railway Museum in York. Her project is titled *'Eating on the Go': Cultures of Consumption and the Railways in Britain, 1879-1948*.

Sevgi Mutlu Sirakova is a PhD candidate at the Rachel Carson Center, Ludwig-Maximilians-Universität in Munich. Her broader research explores traditional fermentation practices in Bulgaria and Turkey through ethnographic methods.

Michelle A. Stinson is a Visiting Scholar in Old Testament/Hebrew Bible at Denver Seminary in Colorado. Her present research explores the topics of food, agriculture/land care, and the Psalms.

Carolyn Tillie obtained a masters in Fine Art and then promptly enrolled in cooking school. Carolyn works as an exhibiting artist, curator, and food historian who specializes in researching, creating, and presenting food as an art form.

Jaime Iram Vargas Barrientos is the author of the *Mexican Christmas Cookbook* and *Cocina de Cuaresma en Durango*.

Chonlatorn Wongrussame is a journalist, editor, and activist who currently works as a Climate Communications Campaigner at Greenpeace.